A Nexus Approach for Sustainable Development

Stephan Hülsmann • Mahesh Jampani
Editors

A Nexus Approach for Sustainable Development

Integrated Resources Management in Resilient Cities and Multifunctional Land-use Systems

Springer

Editors
Stephan Hülsmann
Institute for Integrated Management of Material Fluxes and of Resources
United Nations University
Dresden, Sachsen, Germany

Mahesh Jampani
Institute for Integrated Management of Material Fluxes and of Resources
United Nations University
Dresden, Sachsen, Germany

ISBN 978-3-030-57529-8   ISBN 978-3-030-57530-4 (eBook)
https://doi.org/10.1007/978-3-030-57530-4

This Springer imprint is published by the registered company Springer Nature Switzerland AG
The registered company address is: Gewerbestrasse 11, 6330 Cham, Switzerland

Contents

The Nexus Approach as a Tool for Resources Management in Resilient Cities and Multifunctional Land-Use Systems

Stephan Hülsmann and Mahesh Jampani

Abstract

The Nexus Approach to environmental resources management is increasingly recognized as an important vehicle to achieve sustainability as spelled out in the Sustainable Development Goals (SDGs). In particular, it was argued that the Nexus Approach is key for the sustainable use of environmental resources under conditions of global change and provides a tool to deal with challenges of global change including climate change, urbanization and population growth. Building on conceptual considerations with regard to monitoring and implementation outlined earlier, here, we explore how the Nexus Approach may provide solutions for managing resources in multifunctional land-use systems and resilient cities. In fact, the resources perspective is essential for holistic management of water, soil and waste along the urban–rural axis. Peri-urban areas provide perfect examples of multifunctional systems with manyfold opportunities to closing cycles, improve resource efficiency and mitigate trade-offs. Cases described in this book provide both positive as well as negative examples of what can be achieved by applying nexus thinking and what goes wrong if you don't. Key messages emerging include: (i) participatory approaches are a central element for successful implementation of a nexus approach, (ii) effective mechanisms of knowledge transfer are a prerequisite of adoption and upscaling of nexus approaches and (iii) the lack of economic incentives and lack of data represent major challenges for the implementation of a nexus approach. Overall, the importance of a nexus mindset of all stakeholders involved in nexus cases and of providing an enabling environment by nexus-oriented governance, including appropriate economic instruments, was confirmed.

Keywords

Rural–urban nexus · Participatory approaches · Knowledge transfer · Economic incentives

S. Hülsmann (✉) · M. Jampani
United Nations University, Institute for Integrated Management of Material Fluxes and of Resources (UNU-FLORES), Ammonstraße 74, 01067 Dresden, Germany
e-mail: stephan.huelsmann.dd@gmail.com

S. Hülsmann
Global Change Research Institute, The Czech Academy of Sciences, Bělidla 4a, 603 00 Brno, Czech Republic

M. Jampani
Technische Universität Dresden, Institute of Groundwater Management, 01069 Dresden, Germany

M. Jampani
International Water Management Institute (IWMI), 127 Sunil Mawatha, Pelawatte, Battaramulla, Colombo, Sri Lanka

S. Hülsmann and M. Jampani (eds.), *A Nexus Approach for Sustainable Development*,
https://doi.org/10.1007/978-3-030-57530-4_1

1 Introduction

The increasing human pressures on limited resources available on planet earth is pushing governments to look for sustainable natural resource management approaches and to deal with the interdependencies, synergies and trade-offs in resource management. Therefore, "the nexus is more relevant than ever" as stated in the second chapter of this volume (Bouma 2021). Giving proof of this statement is one of the major aims of this book. We do so paying special attention to a resources-oriented view of the nexus concept and to practical applications both in a rural and urban context.

Within the last decade, the nexus concept has evolved as an integrative approach that bridges the gaps between sectors and considers inter-related resources in an unbiased way to achieve sustainable resources management. It builds upon earlier integrative concepts, e.g. Integrated Water Resources Management, IWRM (Global Water Partnership (GWP) 2009), which still had a single-resource perspective, and various simultaneously evolving research initiatives and management concepts, which eventually converged into the nexus approach/concept. Mentioned for the first time in relation to resources management during the 1980s, the term nexus gained prominence in this context particularly since the year 2000 (Scott et al. 2015). Many authors, building on the seminal papers of Hoff (2011), World Economic Forum (2011) and Bazilian et al. (2011) emphasized the water-energy-food (WEF) nexus, which represents a comprehensive concept and captures the inter-linkages of water, energy and food at multiple levels. Liu et al. (2018) explored the WEF nexus to achieve various sustainable development goals (SDGs) and they also explained the complex couplings in the human-nature systems. Bleischwitz et al. (2018) discussed the interlinkages between the resources (water, energy, food, land and materials) and SDGs for developing a resource nexus perspective (Bleischwitz et al. 2017). This resources perspective has also been emphasized by the United Nations University Institute for Integrated Management of Material Fluxes and of Resources (UNU-FLORES), putting special emphasis on water, soil (basis for food and biomass production) and waste, which is interrelated with water, soil (e.g. composting), but also to energy (Kurian and Ardakanian 2015; Hettiarachchi and Ardakanian 2016a; Manschatz et al. 2016).

Overall, "the nexus is fundamentally about resource recovery, closing the loop and capturing the true efficiency gains" (Scott et al. 2015) and it is also about mitigating trade-offs and promoting synergies. As such, the Nexus Approach thus represents a path towards sustainability and indeed its relevance for sustainable development and a transition to green economy was a major focus of the Bonn 2011 nexus conference (Hoff 2011), which marked an important milestone for the Nexus Approach to become an internationally recognized concept. Since 2011, the Nexus Approach has consolidated and diversified while acknowledging that the basic concept is far from new. The Nexus Approach to the sustainable management of water, soil and waste as promoted by UNU-FLORES emphasizes the inter-relatedness of these three environmental resources along with the cycle of research to implementation. This "Dresden Nexus" concept promoted in particular in the Dresden Nexus Conference (DNC) series, which so far had issues in 2015 and 2017 (Dresden Nexus Conference 2017) and again in 2020, is strongly related to the WEF Nexus (Hoff 2011) but emphasizes on the resources perspective (Hettiarachchi and Ardakanian 2016b). Increasingly, another dimension, namely the ecosystem perspective is added to the WEF nexus, making it a WEFE nexus (e.g., Hülsmann et al. 2019).

2 The Nexus Approach as a Tool to Cope with Global Challenges

Focusing on the challenges posed by different aspects of global change (climate change, urbanization, population growth) on environmental resources management, a series of papers collected in relation to the first DNC conference explored how a Nexus Approach may help to cope with them (Hettiarachchi and Ardakanian 2016b). Obviously, the challenges of global change are interrelated, in particular urbanization and population growth (Fig. 1), while climate change is interrelated in a complex way to both of them. A strong relation can be drawn via resource use and availability. A clear conclusion by Hettiarachchi and Ardakanian (2016b) was that applying the Nexus Approach is key for the sustainable use of environmental resources under conditions of global change to achieve the Sustainable Development Goals (SDGs, United Nations 2015). The sustainable management of environmental resources is of particular relevance for goal 2 (end hunger and achieve food security), goal 6 (sustainable management of water and sanitation), goal 7 (energy security), goal 11 (resilient and sustainable cities), goal 12 (responsible consumption to reduce our ecological footprint) and goal 15 (sustainable use of terrestrial ecosystems), but strong links exist also to less resources-related SDGs (Bringezu 2018).

Clearly, sustainable resources management has to be based on effective monitoring mechanisms. Only when consistent data on resource stocks and flows are available, we can implement integrated management strategies required for achieving SDGs. Some key elements of monitoring and implementation strategies for a nexus approach were elaborated in Hülsmann and Ardakanian (2018a) as a contribution to the second DNC in 2017. In that book, with regard to the monitoring aspect, Bringezu (2018) emphasizes the need to account for material-, land- and water consumption as well as greenhouse gas emissions (the "Four Footprints") and develop respective monitoring strategies at the global level for sustainable resource management. Pikaar et al. (2018) show how these footprints are interlinked in the nitrogen cycle and argue that there is an urgent need to develop a circular nitrogen economy and bring the nitrogen cycle into balance with water and carbon cycles. Implementing a nexus approach not only requires to monitor physical entities (environmental resources), but also the involved stakeholders and their interactions, which could possibly be done via social network analysis (Kurian et al. 2018). Since implementation hinges on stakeholders, it can effectively be facilitated by participatory approaches, following established frameworks and using appropriate tools (Smajgl 2018). Creating the respective nexus mindset and awareness can be supported by serious games addressing nexus problems (Mochizuki et al. 2018). Another key aspect for nexus implementation is integration across levels, scales and regions. Such a "vertical nexus" can be achieved through mainstreaming, e.g., aligning planetary boundaries of nitrogen consumption with respective regional, national and subnational thresholds and policies (Hoff 2018).

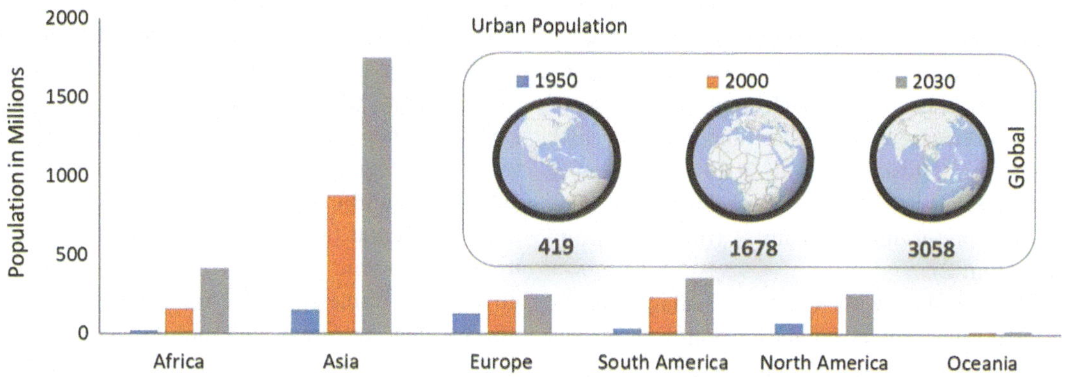

Fig. 1 Recorded and projected change in global urban population (population in millions): The bars give data separated per continent for the years 1950, 2000 and 2030, the inner box gives the total urban population for these years; *data source* UN population data (https://population.un.org/wpp/)

3 The Nexus Approach for Managing Resources in Multifunctional Land-Use Systems and Resilient Cities

Aiming at concrete cases of nexus implementation, while considering the conceptual outlines explored in Hülsmann and Ardakanian (2018a), in this volume we specifically look at examples for multifunctional land -use systems and resources management in resilient cities. This focus should allow demonstrating the close link between the Nexus Approach to the sustainable management of environmental resources and the related SDGs, evident in particular along the urban–peri-urban–rural landscape continuum (see Fig. 2).

With multifunctional land-use systems, we refer mainly to resources management in rural areas and respective ecosystem services, acknowledging that multifunctionality can also be found and achieved in urban and particularly in peri-urban systems. In fact, peri-urban areas provide one of the best examples of multifunctional land-use systems, combining various uses and sectors such as housing, agriculture, forestry (Rodenburg and Nijkamp 2004). While putting emphasis on functions and services related to the mentioned SDGs (mainly provisioning services for food, feed, energy, water), the full range of ecosystem services, including also regulatory and maintenance as well as cultural services (classification of ecosystem services according to CICES (nd)) should be considered when discussing multifunctional land use systems (Zhang and Schwärzel 2017). It is increasingly recognized that a nexus approach neglecting ecosystem services would be incomplete and misleading (Hülsmann et al. 2019). Adopting a nexus approach will lead to sustainable resource use and preserving the natural (rural) environment (Fig. 2).

When dealing with resources management in resilient cities, more emphasis should be placed on the provision, recovery and reuse of resources and the respective infrastructures. Cities are the main consumers of many primary resources, to a large extent stemming from rural areas. However, at the same time they are primary producers of many secondary resources, which are often considered as 'waste' and dissipated into the environment as urban dwellers and local decision makers are unable to recognize the value in it and/or lack the capacity to treat them appropriately (Agudelo-Vera et al. 2012). With the increasing globalization, urban pressures and limited resources availability, cities need to consider sustainable recovery and reuse options of the resources generated from human and industrial activities. For any city to become resilient, it needs to take care of its environmental resources in the urban and peri-urban landscapes in an efficient manner. Cities often neglect their peri-urban landscapes, which are multifunctional in nature and supply multiple resources to the city. The adoption of a nexus approach in cities will increase their resilience and support sustainable development by enhancing opportunities for human well-being, livelihoods, etc. (Fig. 2).

In general, it seems clear that a focus on resources and the respective SDGs is a strong integrator along the rural–urban conundrum. Discussing ways how the application of a Nexus Approach may help to achieve these SDGs implies focusing on monitoring and implementation strategies as outlined above. Specific nexus strategies have to be defined for resource problems such as inefficiency in wastewater treatment, air pollution, unstructured urban infrastructure, etc. in the urban centers to increase resilience. In rural and peri-urban areas, multifunctional land use systems are good examples for implementing integrated resources management (Zhang and Schwärzel 2017). Local planners and decision makers working towards integrated resources management in cities and multifunctional land use systems will ultimately contribute to environmental prosperity and economic growth. Moreover, developing governance frameworks for integrated resources management in cities and multifunctional land use systems can create incentives for resource recovery and efficiency. This requires considering the economic and social dimensions for the

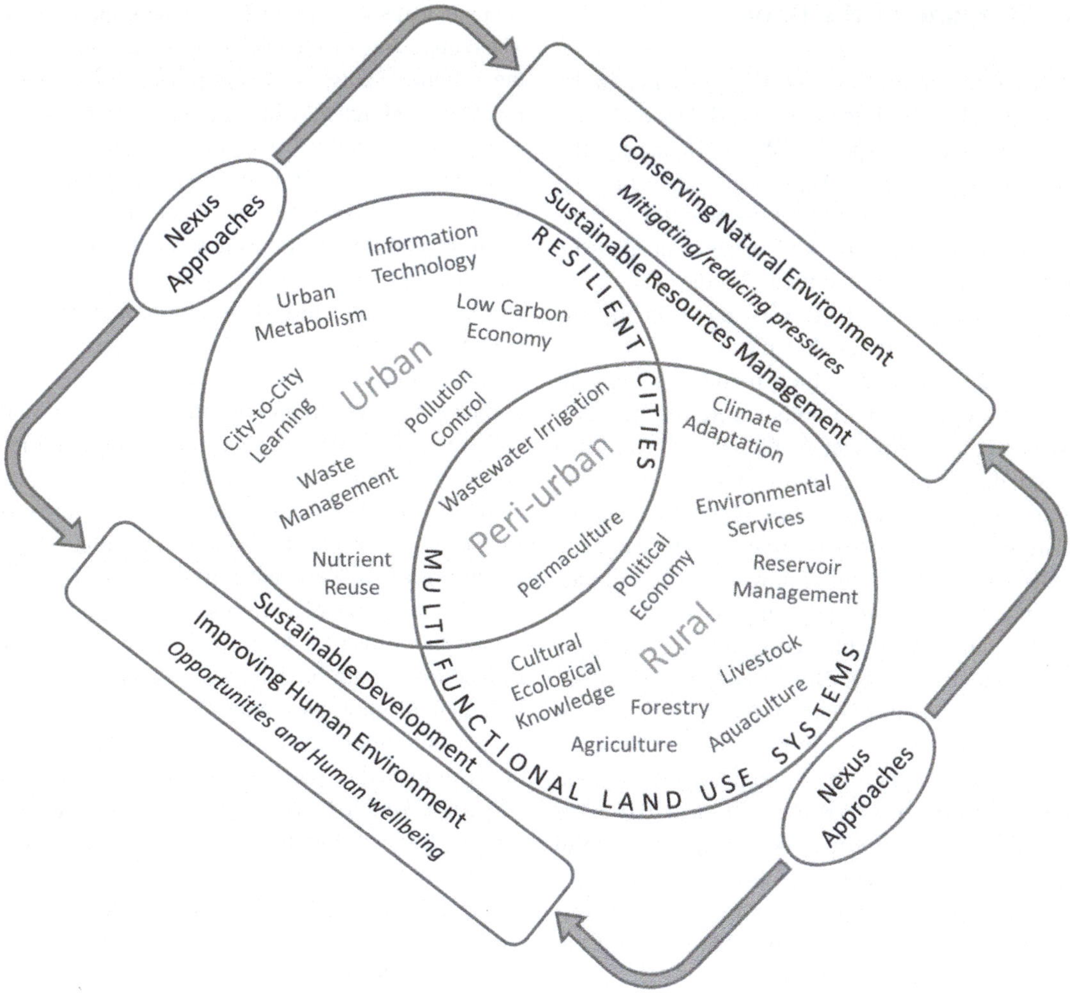

Fig. 2 Implementation of a resource nexus approach (interdependent with and complementary to a sectoral nexus) leads to sustainable development and sustainable resources management within resilient cities and multifunctional land-use systems at the urban–peri-urban–rural interface; keywords within circles refer to issues addressed in the chapters of this book

implementation of sustainable environmental resources management strategies, which need to be established and/or strengthened. This book promotes the Nexus Approach as a tool for achieving SDGs, both in resilient cities and multifunctional land use systems, paying special attention to practical applications, the socio-economic dimension and governance frameworks supporting or hindering the implementation of a Nexus Approach. While the former book (Hülsmann and Ardakanian 2018a), focused on implementation and monitoring strategies on a conceptual level, here we present practical examples on how this can be achieved in both rural and urban settings. We deliberately refrained from dividing the contributions into subsections dealing with resilient cities and multifunctional land-use systems, respectively. Instead, we aim to highlight the interconnections and point out the respective focus in our brief outline to the contributions of this book below.

4 Storyline of the Book

As a starting point, Bouma (2021) explains in the chapter "How to Integrate and Balance Water, Soil and Waste Expertise When Realizing the Corresponding Nexus Approach" that achieving the balance between the environmental resources, such as water, soil and waste, but also the respective expertise is critical for the water-soil-waste nexus approach. Reviewing recent nexus case studies he argues that the water-soil-waste nexus, currently exemplified mainly by applications of wastewater to soils, needs to be complemented by a stronger focus on solid (organic) waste, thus compost. In line with this conclusion, Hettiarachchi et al. (2020b) elaborate on this issue from various perspectives. Addressing the water-soil-waste nexus requires interdisciplinarity and Bouma particularly emphasizes the much needed, but limited interaction between soil sciences and hydrology. Both disciplines produce enormous amounts of data, making it challenging to transform them into information and knowledge. Common platforms, which could be termed nexus observatories (Kurian et al. 2016) are considered as useful, but require careful selection/development/adaptation of tools for data analysis—which may require another (sub-) platform for model selection (Mannschatz et al. 2016). Bouma argues that a focus on SDGs may be helpful to shape a nexus-oriented data analysis. However, interdisciplinarity is not enough: transdisciplinary research approaches, engaging stakeholders from the outset are required—an issue taken up in several of the contributions to this book. Eventually, they may result in case studies that should be most effective to inform and inspire governance of the nexus approach, a major aim of this book.

The chapter "Nutrient Recovery for Use in Agriculture: Economic assessment of Decentralized Compost Business Model in Nairobi" by Gebrezgabher et al. (2021) provides a successful example of implementing a water-soil-waste nexus, focusing on solid waste as promoted in the previous chapter. The study describes the development of a business model for composting urban solid waste, which can be a viable resource for urban dwellers and farmers. The composting mechanism is developed in view of the water-soil-waste nexus approach for resource recovery and reuse while taking care of socio-economic feasibility—similarly emphasized by Hettiarachchi et al. (2020a). Considering the waste streams as environmental resources, resource recovery and the reuse business model of Nairobi may provide a solution to the mismanagement of the city's waste, soil nutrient depletion, and environmental pollution.

While making use of organic solid waste for composting is required for a comprehensive water-soil-waste nexus approach, the more common approach of wastewater reuse has to be considered, too. In the chapter "Sustainable and Safe Reuse of Wastewater for Food Production in Peri-urban Areas of Karnataka, India", Ramakrishna and Hanisch (2021) discuss safe options for wastewater reuse for food production in the state of Karnataka in India using cost-effective on-farm treatment technologies. Nutrient availability in the wastewater motivated smallholder farmers to apply untreated or partially treated wastewater for irrigation, which, however, led to health risks for farmers and crop handlers, but also may cause soil and groundwater salinization and contamination (Jampani et al. 2018). Capacity development on safe wastewater reuse practices is needed to farmers and local peri-urban dwellers, and appropriate primary on-farm treatment technologies should be introduced to capture nutrients, mitigate health risks and increase crop productivity. The study also shows that the implementation and further development of nexus-oriented resources management is facilitated by a participatory approach.

Where wastewater is not available, but water is a limiting factor for agriculture, water harvesting techniques can be an option for sustainable food and biomass production. In the chapter "Usefulness of Surface Water Retention Reservoirs Inspired by 'Permaculture Design': A Case Study in Southern Spain Using Bucket Modelling", Fiebrig and van de Wiel (2021) developed a simple hydrological model to assess the efficacy of permaculture inspired surface water retention reservoirs in a water-scarce area in Southern Spain. This study gives evidence that water harvesting reservoirs can contribute to reversing desertification and to climate change mitigation. While the specific system under study turned out to suffer from ill-informed planning, the simple modelling approach indicated that permaculture inspired design and planning processes can facilitate sustainable solutions for integrated soil–water management. Water harvesting techniques are a suitable tool for integrated water-soil management and can be attributed to increasing groundwater levels, soil fertility, and biodiversity. In view of water-soil-food nexus, this type of nature-based solutions can be an economically viable land restoration approach.

Another aspect of the soil–water nexus related to reservoirs is addressed in the chapter "Multifunctional Historical data for Improved Management of Reservoirs" by Sang and Maina (2021). Sedimentation is a common threat to reservoirs, in particular in agriculture-dominated watersheds, but coherent monitoring of sedimentation, let alone sedimentation management strategies (Hülsmann et al. 2020) are often lacking, e.g., in Sub-Saharan Africa. Under such conditions, sedimentation surveys are critical to assess reservoir volumes and sediment thickness. The authors review available methodologies for sediment assessments. Using a bathymetric survey system, which can determine both the reservoir volume as well as sediment thickness, the authors investigated two water bodies in Kenya. In Ruiru Reservoir, which is critical for the water provision of Nairobi and influenced by food production in the watershed, a loss of 14% of its original volume was found. This information can inform the integrated reservoir and watershed management to secure soils for agriculture as well as maintain water storage capacities and water quality in reservoirs.

The same issue addressed in the chapter by Sang and Maina (2021), reservoir sedimentation, is subject to economic evaluation in the chapter "Economic Valuation of Environmental Services Associated with Agriculture in the Watershed of Lake Lagdo, Cameroon". Yong Ngondjeb and Ayuk (2021) evaluated the economic impact of soil erosion, considering current land use management practices and sedimentation of lake Lagdo in Cameroon, a multi-purpose water body used also for hydropower. The change in food production and associated land use practices increased soil erosion, which influenced lake sedimentation and led to increasing costs for electricity generation. This economic impact is arguably the most tangible one, but clearly not the only one. Directly or indirectly soil erosion impacts many other systems, e.g., agricultural production, which can be prevented by sustainable soil conservation projects. This study reveals interdependencies between complementary nexus approaches: resource dynamics (water, soil and waste) and sectoral dynamics (water, food and energy) (Fig. 2).

The next two papers also focus on the water-energy nexus, both looking in particular from an economic perspective considering governance. In the chapter "The Energy-Water Nexus in Iran: The Political Economy of Energy Subsidies for Groundwater Pumping", Jaghdani and Kvartiuk (2021) analyzed the interlinks between groundwater consumption and government imposed energy subsidies for the agricultural sector in Iran. Energy subsidy policies in Iran are threatening the availability of groundwater resources for irrigation and other uses since they promote overexploitation. Ultimately, they put agricultural production in the area at risk. Using the example of a failed political reform in 2010, the authors argue that the overrepresentation of rural interests in Iranian politics impeded a successful reform implementation. The political economy of energy subsidies can be considered one of the core topics governing the water-food-energy nexus in Iran at micro and macro scales.

In the second study on the economics of the water-energy nexus, Riera and Brümmer (2021) analyzed the political economy of energy subsidies that influence the groundwater aquifer system in Argentina. In their study "Political Economy of Energy Subsidies for Groundwater Irrigation in Mendoza, Argentina", the policy tools and scenarios analyzed suggest that there is a strong link between declining groundwater resources, grapevine prices, and energy subsidies. Similar to the study by Jaghdani and Kvartiuk (2021), the authors found that current economic tools (energy subsidies) promote groundwater depletion. They conclude that shifting from energy subsidy to groundwater resources optimization for agricultural irrigation could be a viable policy option for sustainable groundwater management.

Also the following chapters focus on rural areas, seizing opportunities and challenges for the implementation of integrated resources management. In the chapter "Rural Resources (Including Forestry) in the Local Development of Low Carbon Economy: A Case Study of Poland ", Wiśniewski (2021) evaluated the rural resources in an integrated manner to achieve the development of a low carbon economy in Poland. To achieve this, in rural areas carbon sequestration should be enhanced, while GHG emissions need to be reduced. The author systematically reviews the strengths and weaknesses as well as opportunities and threats (SWOT analysis) towards this end. He concludes that a number of measures would need to be implemented for a transition towards a low-carbon economy. These include, amongst others, increasing the use of manure and organic waste both for fertilization (see chapter by Gebrezgabher et al. 2021) and energy (biogas) production, preventing soil erosion (addressed also in Sang and Maina 2021; Ngondjeb and Ayuk 2021; Tesfahunegn and Ayuk 2021), increased use of intercropping, growing energy crops on marginal lands, liming of acid soils, afforestation and increased implementation of agroforestry. In addition, policies targeting the development of rural areas need to be reviewed for their potential to facilitate a low carbon economy and good practice examples in the agroforestry sector need to be promoted, in line with the recommendations by Bouma (2021).

Tesfahunegn and Ayuk (2021) analyzed "Opportunities and Challenges to Adopting Sustainable Watershed Management Interventions…" based on experiences from Ethiopia. The authors review problems in resources management and provide a list of performance indicators to evaluate nexus-oriented watershed management interventions. In some watersheds, these interventions were highly successful, resulting in e. g., increased food security, income of farmers and water resources, restoration of soil fertility and reduced soil loss. A number of opportunities were identified, which facilitate the implementation of a nexus approach. However, there has been limited uptake at a larger scale due to challenges that could be categorized as being related to traditional practices and climate related challenges; the large area of degraded land; and socio-economic and institutional challenges. In order to address these challenges and support the implementation of nexus-oriented watershed management, a general research strategy is proposed. Similar to other nexus-oriented research strategies (Mannschatz et al. 2015; Benavides et al. 2019), it highlights the importance of baseline data, decision making, implementation and monitoring and evaluation. These steps need to be done in a participatory approach. Governance should provide an enabling environment to support opportunities and decrease the challenges of nexus implementation for sustainable development.

In the chapter "The Potential Contribution of Cultural Ecological Knowledge to Resources Management in a Volcanic River Basin", Ariyanti et al. (2021) point out a rather unexplored aspect of the nexus: its relation to cultural ecological knowledge (CEK). Particularly in ancient societies, it is the embeddedness of resources management practices in the overall culture, including myths, rituals, symbols and "five senses wisdom", which is important for the implementation of natural resources management. CEK encompasses written and orally transmitted knowledge based on observations of the human-nature relationship, in this particular case from a volcanic river basin in Indonesia, including the volcano, the debris flow stemming from it, the water cycle and land use. The authors emphasize the importance of a legitimation process via institutionalized structures and dynamics to transform historical experiences into CEK. In the examined case, CEK was found to be of great importance for conservation practice, cropping patterns, fishing, etc. It can be argued that cultural aspects cannot be neglected even in "modern" societies when it comes to natural resources management. To do so, a participatory approach (Smajgl 2018; Eichhorn et al. 2021) seems to be a natural choice to ensure capturing aspects of CEK in the development of integrated, sustainable resources management.

Turning from a rural and traditional system to an urban and mainly artificial system, in the chapter "Nexus-Oriented Approach for Sharing Water Resources: Development of Eco-industrial Parks in the Catchment of Zayandeh Rud River, Iran", von Koerber et al. (2021) investigated an integrated nexus-oriented approach for addressing water efficacy, water reuse and water sharing options in eco-industrial parks (EIP) of Iran. This study explored the collaboration potential between the industries and their willingness to share the water resources of different qualities depending on their industrial water demands. Industrial development with the water resource sensitive environment and good water sharing mechanisms in the eco-industrial park can lead to considerable freshwater savings in the watershed. In view of water scarcity, industrial symbiosis with local stakeholders involvement can improve water resource efficiency. However, various obstacles to the realization of the industrial symbiosis concepts were identified jointly with involved stakeholders. One major obstacle is related to the fact that both energy and water prices were very low, making any investment in resource savings economically costly. Moreover, the actual water consumption was unknown in the studied industrial park due to the lack of meters, meaning the study had to rely on rough estimates. Overcoming the obstacles requires economic incentives, improving the data basis for better planning, but above all developing a nexus mindset of involved stakeholders. This specific case thus confirms the importance of these aspects highlighted earlier (Hülsmann and Ardakanian 2018a, b). If planned from the outset, backed by comprehensive capacity development and with appropriate economic incentives, the concept of industrial symbiosis might most easily be implemented in newly developed EIPs.

In the chapter "City-to-City Learning Within City Networks to Cater City Needs to Climate Adaptation—Results of a Preliminary Study", Haupt et al. (2021) provide an overview of existing city networks, which aim to provide platforms for peer-to-peer learning in response to common challenges, in particular climate change. Based on in-depth interviews with key stakeholders it was asked how effective these networks are in knowledge sharing and how this process can be facilitated and improved. It emerged from the responses that city-to-city networks were conceived helpful and effective, the importance of best-practice examples and of "frontrunners" was emphasized. A variety of tools and methods were considered helpful, including conferences, workshops on specific topics and webinars. In line with the nexus concept, it was argued that the inter-relations between different networks with diverging foci (e.g., region, theme) should be strengthened and involved stakeholders should diversify, e.g., with regard to the private sector.

In the chapter "A Participatory Multi-stakeholder Approach to Implementing the Agenda 2030 for Sustainable Development: Theoretical Basis and Empirical Findings", Eichhorn et al. (2021) address the question how integrated management and a participatory multi-stakeholder approach can be systematically realized at the local level (municipalities) to develop sustainability strategies. Results are based on a project in North Rhine-Westphalia, Germany, carried out by a sustainability network. They strongly confirm the importance of participatory processes for the implementation of integrated management strategies (Kirschke et al. 2016; Smajgl 2018). In particular, they found that besides the representation of critical stakeholders, the political support from administration heads and the structuring of the participatory approach is important. The latter has to ensure that working groups, steering committees, etc. work efficiently and independently.

5 Conclusions and Outlook

Coming back to the introductory statement, we indeed conclude that "the nexus is more relevant than ever" as stated by Bouma (2021) and this can be deduced both from positive examples of nexus implementations (e.g., Eichhorn et al. 2021; Gebrezgabher et al. 2021; Tesfahunegn and Ayuk 2021) as well as from those studies, which show that the lack of nexus-oriented policies results in unsustainable management practices (e.g., Jaghdani and Kvartiuk 2021; Riera and Brümmer 2021; Sang and Maina 2021) or from studies which demonstrate the feasibility of positive outcomes if a nexus approach would be adopted (e.g., von Koerber et al. 2021; Wiśniewski 2021; Yong Ngondjeb and Ayuk 2021).

The positive examples include the case of a compost business model for Nairobi (Gebrezgabher et al. 2021), the successful adoption of integrated resources management practices in Northern Ethiopia (Tesfahunegn and Ayuk 2021) and the implementation of a comprehensive participatory planning process towards SDGs (Eichhorn et al. 2021). The chapter by Ariyanti et al. (2021) stands out since they report on the case of resources management based on cultural ecological knowledge, which clearly contains nexus elements that are still in use. So here, it is rather about preserving nexus thinking than about implementing it as a new concept. Negative examples in this book include cases of purely agriculture oriented policies providing economic incentives that lead to groundwater overexploitation (Jaghdani and Kvartiuk 2021; Riera and Brümmer 2021). Other cases show that unsustainable land-use practices in watersheds of reservoirs result in high sediment loads and loss of water storage capacity (Sang and Maina 2021) and of hydropower potential (Yong Ngondjeb and Ayuk 2021). These studies, conversely, also show the potentials of switching to more holistic and sustainable policies and practices. Similarly, the studies of Ramakrishna and Hanisch (2021) and Fiebrig and van de Wiel (2021) mainly demonstrate the potentials of adopting a nexus approach. The former describes a case of wastewater use, a "golden" example of a nexus approach—if practiced in a safe manner (Hettiarachchi et al. 2018). However, this was not yet the case in the study by Ramakrishna and Hanisch (2021), but they present solutions on how to mitigate the risks of using untreated wastewater. Also, the study by Fiebrig and van de Wiel (2021) mainly shows potentials, since in the studied case the reservoir built for rainwater harvesting was badly designed, based on insufficient data. But if planned well, methods of rainwater harvesting may provide solutions for sustainable food production under water scarcity. The studies by von Koerber et al. (2021) and Wiśniewski (2021) were conceptualized as feasibility studies from the beginning. Both show convincingly the potential of nexus thinking to result in increased resource efficiency (von Koerber et al. 2021) and to foster the transition to a low carbon economy (Wiśniewski 2021).

Several studies point to the importance of knowledge transfer, education and capacity development in a broad sense. This aspect is explicitly addressed in Haupt et al. (2021) for the case of city-to-city learning with regard to climate adaptation. Knowledge transfer is also addressed by Bouma (2021), who talks about scientists as "knowledge brokers". Ramakrishna and Hanisch (2021) stress the importance of capacity development to foster safe use of wastewater, Sang and Maina (2021) point to the limited capacities for sediment monitoring both in terms of equipment and trained staff, Tesfahunegn and Ayuk (2021) report about various means of knowledge transfer such as farmer's field days.

Several studies presented here had started quite a while ago and had been originally conceptualized as IWRM projects. This holds, e.g., for Sang and Maina (2021), Tesfahunegn and Ayuk (2021), von Koerber et al. (2021) and Ariyanti et al. (2021). However, since taking a balanced approach towards at least two resources, they easily qualified convincingly as nexus-oriented projects. This of course does not mean that generally all IWRM projects can also be considered as nexus case studies.

Overall, while every single chapter of this book has its own set of lessons to learn, the following aspects stand out as being generally important, since emerging from several studies and confirming earlier results from the growing body of nexus literature:

- Participatory approaches are a central element for successful implementation;
- Effective mechanisms of knowledge transfer are a prerequisite of adoption and upscaling of nexus approaches;
- Lack of economic incentives and lack of data represent major challenges for the implementation of a nexus approach. Wrong incentives may worsen the situation. Conversely, nexus-oriented governance, including appropriate economic incentives, is key for the successful implementation of nexus approaches.

The cases presented here thus largely confirm earlier conclusions with regard to using the nexus approach as a tool to achieve SDGs (Hülsmann and Ardakanian 2018b). This refers to the importance of monitoring mechanisms (e.g., Sang and Maina 2021), closing nutrient cycles (e.g., Bouma 2021; Gebrezgabher et al. 2021), applying participatory approaches (e.g. Eichhorn et al. 2021) as well as effective and innovative means of knowledge transfer and awareness raising (e.g., Haupt et al. 2021). The presented cases are also addressing the nexus on a "vertical" axis, considering its implementation across various spatial scales (e.g., Tesfahunegn and Ayuk 2021) and hierarchical levels (e.g., Jaghdani and Kvartiuk 2021; Eichhorn et al. 2021). The importance of economical aspects, which was rather indirectly addressed in Hülsmann and Ardakanian (2018a), is clearly emerging as a key factor in a number of contributions here (Gebrezgabher et al. 2021; Yong Ngondjeb and Ayuk 2021; Jaghdani and Kviartiuk 2021; Riera and Brümmer 2021). Ultimately, nexus cases, e.g. concerning the use of waste(water) or organic waste, have to be based on a solid business model. Vice versa, without economic incentives or with economic incentives promoting unsustainable resource use, nexus approaches are bound to fail.

We hope that the cases presented here will be taken up as giving proof of the relevance of the nexus approach for achieving sustainability. Even more, we hope they may serve as inspiration for further research, but in particular for the implementation of holistic resources management strategies, providing practical examples of dos and don'ts with regard to integrated resources management.

Acknowledgements The chapters of this book are based on contributions to the Dresden Nexus Conference 2017. We are thus grateful to all colleagues who contributed to the success of the conference, be it as an organizer or participant. Big thanks go to the former UNU-FLORES team and internal and external reviewers. We also acknowledge the support of the former director of UNU-FLORES, Reza Ardakanian, and his successor, Edeltraud Guenther.

References

Agudelo-Vera CM, Leduc WRWA, Mels AR, Rijnaarts HHM (2012) Harvesting urban resources towards more resilient cities. Resour Conserv Recycl Clim Proofing Cities 64:3–12. https://doi.org/10.1016/j.resconrec.2012.01.014

Ariyanti V, Scholten PHT, Edelenbos J (2021) The potential contribution of cultural ecological knowledge to resources management in a volcanic river basin. In: Hülsmann S, Jampani M (eds) A nexus approach for sustainable development. Springer Nature, Cham, Switzerland

Bazilian M, Rogner H, Howells M, Hermann S, Arent D, Gielen D, Steduto P, Mueller A, Komor P, Tol RSJ, Yumkella KK (2011) Considering the energy, water and food nexus: towards an integrated modelling approach. Energy Policy 39:7896–7906. https://doi.org/10.1016/j.enpol.2011.09.039

Benavides L, Avellán T, Caucci S, Hahn A, Kirschke S, Müller A (2019) Assessing sustainability of wastewater management systems in a multi-scalar transdisciplinary manner in Latin America. Water 11:249. https://doi.org/10.3390/w11020249

Bleischwitz R, Hoff H, Spataru C, Voet E van der, VanDeveer SD (2017) The resource nexus: preface and introduction to the Routledge handbook. In: Routledge handbook of the resource nexus. https://www.taylorfrancis.com/. Accessed 16 Aug 2018

Bleischwitz R, Spataru C, VanDeveer SD, Obersteiner M, van der Voet E, Johnson C, Andrews-Speed P,

Boersma T, Hoff H, van Vuuren DP (2018) Resource nexus perspectives towards the United Nations sustainable development goals. Nat Sustain 1:737. https://doi.org/10.1038/s41893-018-0173-2
Bouma J (2021) How to integrate and balance water, soil and waste expertise when realizing the corresponding nexus approach. In: Hülsmann S, Jampani M (eds) A nexus approach for sustainable development. Springer Nature, Cham, Switzerland. https://doi.org/10.1007/978-3-030-57530-4_2
Bringezu S (2018) Key strategies to achieve the SDGs and consequences for monitoring resource use. In: Hülsmann S, Ardakanian R (eds) Managing water, soil and waste resources to achieve sustainable development goals, pp 11–34. https://doi.org/10.1007/978-3-319-75163-4_2
CICES (nd) Towards a Common International Classification of Ecosystem Services (CICES) for integrated environmental and economic accounting. https://cices.eu. Accessed 28 May 2019
Dresden Nexus Conference (2017) State of the nexus approach 2017: multifunctional land-use systems and resource management in resilient cities. UNU-FLORES, Dresden, Germany. https://collections.unu.edu/eserv/UNU:6340/DNC17_Report.pdf
Eichhorn S, Hans M, Schön-Chanishvili M (2021) A participatory multi-stakeholder approach to implementing the Agenda 2030 for sustainable development: theoretical basis and empirical findings. In: Hülsmann S, Jampani M (eds) A nexus approach for sustainable development. Springer Nature, Cham, Switzerland
Fiebrig I, van de Wiel M (2021) Usefulness of surface water retention reservoirs inspired by "permaculture design": a case study in southern Spain using bucket modelling. In: Hülsmann S, Jampani M (eds) A nexus approach for sustainable development. Springer Nature, Cham, Switzerland
Gebrezgabher S, Taron A, Amewu S (2021) Nutrient recovery for use in agriculture: economic assessment of decentralized compost business model in Nairobi. In: Hülsmann S, Jampani M (eds) A nexus approach for sustainable development. Springer Nature, Cham, Switzerland
Global Water Partnership (GWP) (2009) Integrated water resources management in practice: better water management for development. Earthscan, London; Sterling, VA
Haupt W, Zevenbergen C, van Herk S (2021) City-to-city learning within city networks to cater city needs to climate adaptation—results of a preliminary study. In: Hülsmann S, Jampani M (eds) A nexus approach for sustainable development. Springer Nature, Cham, Switzerland. https://doi.org/10.1007/978-3-030-57530-4_14
Hettiarachchi H, Ardakanian R (2016a) Managing water, soil, and waste in the context of global change. In: Hettiarachchi H, Ardakanian R (eds) Environmental resource management and the nexus approach: managing water, soil, and waste in the context of global change. Springer International Publishing, Cham, pp 1–7
Hettiarachchi H, Ardakanian R (eds) (2016b) Environmental resource management and the nexus approach. Springer International Publishing, Cham
Hettiarachchi H, Caucci S, Ardakanian R (2018) Safe use of wastewater in agriculture: the golden example of nexus approach. Safe use of wastewater in agriculture. Springer, Cham, pp 1–11
Hettiarachchi H, Bouma J, Caucci S, Zhang L (2020a) Organic waste composting through nexus thinking: linking soil and waste as a substantial contribution to sustainable development. In: Hettiarachchi H, Caucci S, Schwärzel K (eds) Organic waste composting through nexus thinking: practices, policies, and trends. Springer International Publishing
Hettiarachchi H, Caucci S, Schwärzel K (eds) (2020b) Organic waste composting through nexus thinking: practices, policies, and trends. Springer International Publishing
Hoff H (2011) Understanding the NEXUS, background paper for the Bonn 2011 conference: the water, energy and food security nexus. Stockholm Environment Institute, Stockholm
Hoff H (2018) Integrated SDG implementation—how a cross-scale (vertical) and cross-regional nexus approach can complement cross-sectoral (horizontal) integration. In: Hülsmann S, Ardakanian R (eds) Managing water, soil and waste resources to achieve sustainable development goals, pp 149–163. https://doi.org/10.1007/978-3-319-75163-4_7
Hülsmann S, Ardakanian R (eds) (2018a) Managing water, soil and waste resources to achieve sustainable development goals: monitoring and implementation of integrated resources management. Springer
Hülsmann S, Ardakanian R (2018b) The nexus approach as tool for achieving SDGs: trends and needs. In: Hülsmann S, Ardakanian R (eds) managing water, soil and waste resources to achieve sustainable development goals, pp 1–9. https://doi.org/10.1007/978-3-319-75163-4_1
Hülsmann S, Rinke K, Paul L, Diez Santos C (2020) Storage reservoir management and operation including complex multiunit and multipurpose systems. In: Bogardi (ed) Springer handbook of water resources management (in press)
Hülsmann S, Sušnik J, Rinke K, Langan S, van Wijk D, Janssen AB, Mooij WM (2019) Integrated modelling and management of water resources: the ecosystem perspective on the nexus approach. Curr Opin Environ Sustain 40:14–20. https://doi.org/10.1016/j.cosust.2019.07.003
Jaghdani T, Kvartiuk V (2021) The energy-water nexus in Iran: the political economy of energy subsidies for groundwater pumping. In: Hülsmann S, Jampani M (eds) A nexus approach for sustainable development. Springer Nature, Cham, Switzerland. . https://doi.org/10.1007/978-3-030-57530-4_8
Jampani M, Hülsmann S, Liedl R, Sonkamble S, Ahmed S, Amerasinghe P (2018) Spatio-temporal distribution and chemical characterization of groundwater quality of a wastewater irrigated system: a case

study. Sci Total Environ 636:1089–1098. https://doi.org/10.1016/j.scitotenv.2018.04.347

Kirschke S, Horlemann L, Brenda M, Deffner J, Jokisch A, Mohajeri S, Onigkeit J (2016) Benefits and barriers of participation: experiences of applied research projects in integrated water resources management. In: Borchardt D, Bogardi JJ, Ibisch RB (eds) Integrated water resources management: concept, research and implementation. Springer International Publishing, Cham, pp 303–331

Kurian M, Ardakanian R (eds) (2015) Governing the nexus—water, soil and waste resources under conditions of global change. Springer, Berlin, Heidelberg

Kurian M, Ardakanian R, Gonçalves Veiga L, Meyer K (2016) Resources, services and risks: how can data observatories bridge the science-policy divide in environmental governance? Springer, Switzerland

Kurian M, Portney KE, Rappold G, Hannibal B, Gebrechorkos SH (2018) Governance of water-energy-food nexus: a social network analysis approach to understanding agency behaviour. In: Hülsmann S, Ardakanian R (eds) Managing water, soil and waste resources to achieve sustainable development goals, pp 125–147. https://doi.org/10.1007/978-3-319-75163-4_6

Liu J, Mao G, Hoekstra AY, Wang H, Wang J, Zheng C, van Vliet MTH, Wu M, Ruddell B, Yan J (2018) Managing the energy-water-food nexus for sustainable development. Appl Energy 210:377–381. https://doi.org/10.1016/j.apenergy.2017.10.064

Mannschatz T, Buchroithner MF, Hülsmann S (2015) Visualization of water services in Africa: data applications for nexus governance. In: Kurian M, Ardakanian R (eds) Governing the nexus. Springer International Publishing, pp 189–217

Mannschatz T, Wolf T, Hülsmann S (2016) Nexus tools platform: web-based comparison of modelling tools for analysis of water-soil-waste nexus. Environ Model Softw 76:137–153. https://doi.org/10.1016/j.envsoft.2015.10.031

Mochizuki J, Magnuszewski P, Linnerooth-Bayer J (2018) Games for aiding stakeholder deliberation on nexus policy issues. In: Hülsmann S, Ardakanian R (eds) Managing water, soil and waste resources to achieve sustainable development goals, pp 93–124. https://doi.org/10.1007/978-3-319-75163-4_5

Pikaar I, Matassa S, Rabaey K, Laycock B, Boon N, Verstraete W (2018) The urgent need to re-engineer nitrogen-efficient food production for the planet. In: Hülsmann S, Ardakanian R (eds) Managing water, soil and waste resources to achieve sustainable development goals, pp 35–69. https://doi.org/10.1007/978-3-319-75163-4_3

Ramakrishna G, Hanisch M (2021) Sustainable and safe reuse of wastewater for food production in peri-urban areas of Karnataka, India. In: Hülsmann S, Jampani M (eds) A nexus approach for sustainable development. Springer Nature, Cham, Switzerland

Riera FS, Brümmer B (2021) Political economy of energy subsidies for groundwater irrigation in Mendoza, Argentina. In: Hülsmann S, Jampani M (eds) A nexus approach for sustainable development. Springer Nature, Cham, Switzerland. https://doi.org/10.1007/978-3-030-57530-4_9

Rodenburg CA, Nijkamp P (2004) Multifunctional land use in the city: a typological overview. Built Environ (1978-) 30:274–288

Sang JK, Maina CW (2021) Multifunctional historical data for improved management of reservoirs. In: Hülsmann S, Jampani M (eds) A nexus approach for sustainable development. Springer Nature, Cham, Switzerland. https://doi.org/10.1007/978-3-030-57530-4_6

Scott CA, Kurian M, Wescoat JL Jr (2015) The water-energy-food nexus: enhancing adaptive capacity to complex global challenges. In: Kurian M, Ardakanian R (eds) Governing the nexus. Springer International Publishing, pp 15–38

Smajgl A (2018) Participatory processes and integrated modelling supporting nexus implementations. In: Hülsmann S, Ardakanian R (eds) Managing water, soil and waste resources to achieve sustainable development goals, pp 71–92. https://doi.org/10.1007/978-3-319-75163-4_4

Tesfahunegn GB, Ayuk ET (2021) Opportunities and challenges to adopting sustainable watershed management interventions: an overview of experiences from Ethiopia. In: Hülsmann S, Jampani M (eds) A nexus approach for sustainable development. Springer Nature, Cham, Switzerland. https://doi.org/10.1007/978-3-030-57530-4_11

United Nations (2015) Transforming our world: the 2030 Agenda for sustainable development. UN, New York

von Koerber J, Raber W, Schneider P (2021) Nexus-oriented approach for sharing water resources: development of eco-industrial parks in the catchment of Zayandeh Rud river, Iran. In: Hülsmann S, Jampani M (eds) A nexus approach for sustainable development. Springer Nature, Cham, Switzerland. https://doi.org/10.1007/978-3-030-57530-4_13

Wiśniewski P (2021) Rural resources (including forestry) in the local development of low carbon economy: a case study of Poland. In: Hülsmann S, Jampani M (eds) A nexus approach for sustainable development. Springer Nature, Cham, Switzerland. https://doi.org/10.1007/978-3-030-57530-4_7

World Economic Forum (2011) Water security: the water-energy-food-climate nexus | World economic forum water initiative. Island Press, Washington D.C.

Yong Ngondjeb D, Ayuk E (2021) Economic valuation of environmental services associated with agriculture in the watershed of lake Lagdo, Cameroon. In: Hülsmann S, Jampani M (eds) A nexus approach for sustainable development. Springer Nature, Cham, Switzerland. https://doi.org/10.1007/978-3-030-57530-4_7

Zhang L, Schwärzel K (2017) Multifunctional land-use systems for managing the nexus of environmental resources. Springer, Cham

How to Integrate and Balance Water, Soil and Waste Expertise When Realizing the Corresponding Nexus Approach

Johan Bouma

Abstract

The water-soil-waste nexus is more relevant than ever. UN Sustainable Development Goals (SDGs) covering food, water, climate and biodiversity can all significantly be served by applying wastewater and compost to soils, thereby potentially increasing food production, combating water scarcity while higher contents of soil organic matter are effective for climate mitigation and preserving biodiversity. The hydrology and soil science disciplines produce enormous amounts of methods and data, but the interaction between both disciplines is, unfortunately, rather limited. Interdisciplinarity, let alone transdisciplinarity, tends to suffer. UNU-FLORES initiated reports on 17 case studies from all over the world dealing with wastewater application to agricultural soils. These highly informative studies showed that the water-soil-waste nexus is still quite skewed with major emphasis on waste composition and quality focused on agricultural production, including health and safety, but hardly any information on hydrology and soils. The cases also indicated that policy studies focusing on rules and regulations are still in an infant stage, the more so since waste application to soils not only involves health risks but also faces unique emotional and psychological barriers. Successful waste application systems to the soil can only be developed with true and genuine engagement of stakeholders to the research process as part of transdisciplinary case studies. Presenting successful results of such case studies to the policy arena, based on a thorough analysis of both technical and socio-economic aspects, are potentially quite effective and can also be the source of innovative research ideas.

Keywords

Hydrology · Soil sciences · Case studies · Compost · Organic waste

1 Introduction

The Water-Soil-Waste nexus requires effective interaction between hydrologists, soil scientists and sanitary engineers, while implementation of system designs will only be feasible when stakeholders and policy makers are seriously involved in an interactive design process (e.g. UNU-FLORES 2015; Hettiarachchi and Ardakanian 2016a). The disciplines involved produce large volumes of data and the

J. Bouma (✉)
Formerly Soils Dept., Wageningen University, Wageningen, The Netherlands
e-mail: johan.bouma@planet.nl

S. Hülsmann and M. Jampani (eds.), *A Nexus Approach for Sustainable Development*,
https://doi.org/10.1007/978-3-030-57530-4_2

establishment of "Data Observatories" is attractive, in principle, to allow effective interdisciplinarity, forming an essential bridge over the deep science-policy divide when considering environmental governance (Kurian et al. 2016). In fact, more scientific disciplines are involved than the three core disciplines mentioned above because of the UN Sustainable Development Goals (SDGs) (https://sustainabledevelopment.un.org) can be seen as the ultimate goals to be reached and this would also involve agronomists, climatologists, ecologists, economists and sociologists, to just mention a few (e.g., Bouma 2014; Blum 2016; Lal 2014; Keesstra et al. 2016). In 2015, 193 governments have signed a commitment to reach these goals by 2030.

Data gathering and the associated information technology involves a number of aspects: (i) legacy data obtained from published reports and other publications can, if necessary and relevant, be digitized and stored; (ii) data obtained by modern sensing and monitoring techniques are transmitted to central data storage facilities; (iii) data are next systematically stored and made accessible; (iv) algorithms can be developed that allow combinations of data that are effective for any design process, often requiring computer modeling.

The objective of this paper is to present: (i) a broad-brush analysis of the four data aspects, mentioned above, for the three core scientific disciplines: hydrology, soil science and engineering as it relates to waste management; (ii) a discussion as to whether developments in these disciplines serve the purpose to develop effective inter- and transdisciplinary procedures characterizing the water-soil-waste nexus, and (iii) an exploratory discussion on attractive future developments to ensure that both the stakeholder and policy arena are not only engaged but that research also contributes to effective results in the real world. To avoid a discourse that could easily become too theoretical, conceptual and abstract, a number of case studies will be considered, as mentioned by Kurian et al. (2016) and Hettiarachchi and Ardakanian (2016b).

1.1 Developments in Hydrology

Soil and water regimes in the regions around the world have been studied by many authors for at least a hundred years (e.g., Hoekstra and Mekkonen 2012; World Water Assessment Programme 2009). Measurements of soil water contents and fluxes have been refined over the years and now automated monitoring equipment is widely available and applied. Automated tensiometers with transducers allow measurement of soil water potentials in unsaturated soil; Time Domain Reflectometry, Neutron probes and new proximal and remote sensors based on the reflection of specific wavelengths are widely used to measure soil water contents (e.g., SSSA 2002; Viscarra Rossel et al. and Minashy 2010). In addition, the advance of computers has enabled the development of simulation models for water regimes in soils and landscapes for actual but also for future conditions, using projected future climatic data as input (e.g. Bonfante and Bouma 2015; Bonfante et al. 2019, 2020). This work is particularly relevant because the future effects of climate change need to be explored to propose appropriate action. The implicit assumption in these models that soils are isotropic and homogeneous needs modification and soil survey data can be used to account for the occurrence of heterogeneous flows due to the occurrence of macropores or slowly permeable soil horizons (e.g. Bouma 2016a). So far, such soil survey data are hardly used in hydrological modeling, illustrating the disciplinary gap between soil science and hydrology. The following quote from Droogers and Bouma (2014), is relevant in the context of considering modeling: "*A huge number of hydrological models exist, and applications are growing rapidly. The number of pages on the Internet including "hydrological model" is over 5.8 million, using Google in January 2014. Using the same search engine with "water resources model" returns 150 million pages. The number of existing hydrological simulation models is probably in the tens of thousands. Even if we exclude the one-*

off models developed for a specific study and count only the more generic models, it must exceed a thousand. Some existing model overviews cover numerous models. Amongst others (with the number of models mentioned) are: IRRISOFT (2014), 114; USGS (2014), 110; EPA (2014), 211; USACE (HEC, 2014), 18; and REM (2014), 681" (for details on these references the reader is referred to Droogers and Bouma 2014*).*

The conclusion can be that there is an enormous and overwhelming amount of methods and models available for measuring and simulating water movement in soils and landscapes. Which ones are particularly suitable to assess the water-soil-waste nexus? Mannschatz et al. (2016) have provided a first attempt to present a web-based comparison tool for analysing models characterizing the water-soil-waste nexus.

1.2 Developments in Soil Science

The various sub-disciplines in soil science (soil survey/pedology, soil chemistry, -physics and – biology) have operated rather independently in the past and still do. Currently, much effort is spent on obtaining databases with a global extent, derived from pedological data. Some are based on existing soil maps, such as the Harmonized World Soil Database (FAO et al. 2012; Batjes 2016) or S-World (Stoorvogel et al. 2017). Others are based on digital soil mappings, such as Soil Grids (Hengl et al. 2014) and the Global Soil Map (Arrouays et al. 2014). All include a reference to worldwide soil classification systems (IUSS-WRB 2015; Soil Survey Staff 2010). The following functional soil characteristics are distinguished: *Depth to rock, plant exploitable depth, organic carbon, pH, Clay, Silt, Sand, coarse fragments, ECEC, Bulk density, Bulk Density of the fine earth, available water capacity (AWC), Electrical Conductivity (EC).* These data are relevant for the nexus as they allow direct or indirect assessments of soil moisture regimes and the capacity of soils to absorb and filter waste components.

In recent years many new proximal sensing techniques have been developed that allow rapid measurement of soil properties in the field. This implies a major improvement as compared with cumbersome, costly and time consuming laboratory measurements as made in the past (e.g., Viscarra Rossel et al. 2010; Vicarra-Rossel and Bouma 2016). The link of soil data with functionality was initially focused on empirical land evaluation, later quantified by simulation of soil–water-plant processes (Bouma et al. 2011, 2012, 2016a; Stoorvogel et al. 2015). Functionality is now also emphasized by the Soil Security concept that distinguishes soil condition, -capability, -capital, -connectivity (links with stakeholders and policy makers) and -codification (links with legislation) (Field et al. (2017) with many examples; Bouma et al. (2017)). One major activity is to link elementary soil data (often texture, bulk density and %C) by regression analysis to parameters needed for simulating water and nutrient regimes in soils (for example, hydraulic conductivity and moisture retention) developing so-called pedotransfer functions (ptf's) (e.g., Bouma 1989; van Looy et al. 2017). These ptf's provide a major link between soil science and hydrology.

1.3 Engineering and Waste Generation

Installations to purify wastewater and treat solid waste have been built all over the world and can by now be based on well established and proven technologies as well as quality indicators. Composting facilities are frequently attached to solid waste plants, utilizing only decomposable parts of solid waste. Bouma (2016b) mentioned the Edmonton Composting Facility in Canada, the largest in North America, and the Qatar Domestic Solid Waste Management Centre, the largest in the Middle East and the large Lahore Composting Facility. The quality of purified wastewater is a function of the sources of sewage. Urban inputs, also by industry, may result in relatively high contents of heavy metals, drug remnants or hormones that are difficult to remove. Even though on-site liquid waste disposal in septic tank systems is common in rural areas all over the world (Bouma 2016b), central sewage systems in rural areas may contain excess nutrients or pesticides that are also difficult to remove. Quality control of both purified wastewater and compost are therefore very important before application to land is considered and threshold data are crucial to define quality. As illustrated in the following case studies, such threshold and quality data related to health and safety aspects of the waste application are not only widely available but also widely applied. Also, the engineering of wastewater treatment facilities follows standardized practices. Unanswered questions arise when waste or wastewater is applied to a wide variety of soils in different climate zones with different soil moisture regimes.

2 A Brief Review of Case Studies on the Water-Soil-Waste Nexus

Hettiarachchi and Ardakanian (2016b) edited an interesting UNU-FLORES sponsored book with 17 worldwide case studies focusing on good practice examples of safe use of wastewater in agriculture. The case studies allow an evaluation of the state-of-the-art in terms of the input of hydrology and soil science into systems that operate in the real world, "warts and all". Examples describe the application of untreated liquid wastewater for irrigation purposes, intended to overcome water shortages while also nutrient contents are beneficial for plant growth. Clearly, health risks are substantial when applying untreated wastewater and, increasingly, treatment by Wastewater Treatment Plants (WWTPs) occurs before applying to the soils is considered. But this results in many financial and operational problems. For example, in Bolivia (case 12) 31 of 84 WWTPs don't work, and the remaining plants have efficiencies below 50%. Different systems are described: (i) wastewater, either untreated or treated, is directly applied to the soil; (ii) an enclosed "sandfilter" with growing plants on top is used to purify the effluent before it is discharged and collected; and (iii) constructed wetlands guide wastewater through an open area with plants and treated water is collected at the point of outflow.

Regarding the water-soil-waste nexus, a number of general observations can be made:

(i) Excellent analyses are reported of wastewater in many of the cases in terms of its chemical composition, including heavy metals and various organic compounds, contents of pathogenic viruses and bacteria. Measurements of groundwater and surface water quality are widely reported as well. Obviously, standard analytical procedures recommended by the WHO are not only widely available but also applied in practice.

(ii) A strong emphasis is on the importance of wastewater irrigation for agriculture with major concerns for health effects when consuming irrigated crops or vegetables, which is, of course, justified (see, for example, case 4, Lima, Peru). The case studies show that significant economic advantages obtained by farmers applying wastewater for irrigation are a prime driver leading to acceptance of these practices.

(iii) Little data on soil water regimes are provided that are crucial for wastewater purification. Case 1 reports infiltration rates of sands where wastewater is infiltrated but purification is associated with the unsaturated flow and travel times as wastewater moves down to the groundwater (Bouma 1979, 2016b). No data are provided on these processes. Virus removal rates are reported but these values are difficult to interpret because data on the flow system are lacking. None of the other studies report data on the movement of wastewater into and through the soil. Water use efficiency by applying drip irrigation is emphasized in several cases (e.g., case 9 on irrigation of sugarcane in Colombia applying rates on the basis of plant demands and activities in Mexico, reported in case 16). But lack of soil physical data does not allow a judgement as to the efficiency reached. Lack of information on flow regimes implies that measurements of water quality at different locations differ without any possible explanation (e.g., case 7 in Mexico City).

(iv) Only a few studies report soil data (e.g. case 1 very broadly in terms of "sands and gravel", case 6 in terms of various "clayey" textures, case 7 in Mexico City as soil classifications: Leptosols, Phaeozems and Vertisols and case 9 in Colombia as Inceptisols with vertic properties. But the link between Taxonomic soil classes and their hydraulic and purifying behaviour remains obscure which not only limits the interpretation of local processes but also the possibility to extrapolate results to other areas with comparable soils.

(v) Major problems are encountered when judging rules and legislation related to applying wastewater to soils. Aside from malfunctioning WWTPs, which appears to be common, top-down rules and regulations clearly don't work (see descriptions in cases 2 (Lima), 3 (Egypt), 5 (Brasilia), 11 (South Africa), 12 (Bolivia), 13 (Nepal), 14 (Argentina) and 15 (Iran)). Carlos Antonio Pailles Bouchez of the "Council for certification of irrigation with treated water" in Mexico, presents a crucial remark: "*the culture of water in our countries does not include the importance of the treatment of wastewater*". Only education and demonstration of successful examples of wastewater application to soils, that consider the complete flow system, can turn this around. This not only applies to stakeholders but to the policy arena as well. In fact, the policy arena can best be approached by presenting specific case studies demonstrating that well guided waste applications to soil represent significant contributions towards reaching SDGs. Such case studies should go beyond anecdotal data, applying modern measurement, monitoring and modeling techniques for soil and water. This was lacking in the case studies presented by Hettiarachchi and Ardakanian (2016b) but can show the way for effective future research. Usually, politicians are more than willing to share the success stories, rather than failures. Top-down rules and regulations based on theoretical and conceptual considerations can hardly be effective as they bypass stakeholder opinions and experiences in the real world.

Clearly, data on liquid waste application to soil is strongly restricted to the first part of the chain: waste generation and treatment, chemical characteristics of the waste in different phases of the particular treatment being followed and, less so, to application to the soil by flooding or drip irrigation. Also, groundwater quality is being measured, but not everywhere. Quality of crops and vegetables that are grown receive much emphasis. How water infiltrates into the soil and the travel path to the groundwater, which determines the purification process that is quite variable in different soils, does not receive any attention at all. If infiltration rates are too low, surface runoff into surface water may occur as well as erosion both leading to significant pollution of surface waters. Nor is there any attention for the water regime in a more regional context: where does the irrigation water move once it has reached the groundwater aquifer or surface waters? The conclusion can only be that the water-soil-waste nexus is as yet quite incomplete and fragmented.

3 How to Complete the Water-Soil-Waste NEXUS Chain?

The cases reported by Hettiarachchi and Ardakanian (2016b) are most valuable as they allow a realistic appraisal of the state-of-the-art dealing with the application of wastewater to soils. A similar publication on applying solid waste derived compost to soils would be very welcome because of the similar problems likely to be found here. Compost may increase the organic matter content of soils, increasing its productivity and contributing to climate mitigation (the "4per1000" proposal of Paris climate conference in 2015).

Before continuing, the basic question has to be raised as to the significance of soil water regimes when dealing with wastewater use. Some may argue that hydrologists and soil scientists only pursue and push their own disciplinary activities that may be relatively unimportant in the broader context. Such comments should be taken seriously by showing, for example, that only measuring the quality of groundwater at certain locations does answer the question as to "what". But also the "why" is important because only this will increase understanding of processes involved and allow extrapolation of results. Some of the case studies report that chemical concentrations in some wells in a given area were below environmental thresholds while others were above. So what's the conclusion? Only when differences can be explained because of different application rates, different hydraulic properties of soils or different travel times, results can be interpreted and extrapolated and can be used to improve designs.

This is important because the overall significance of the water-soil-waste nexus is very high: SDGs on food (2), water (6), climate (13) and biodiversity (15) are directly related to the wastewater issue as water and food shortages become increasingly urgent in future, as climate change threatens the existence of mankind. But an additional emphasis on technical aspects of soil water regimes in different soils only covers part of the challenges ahead. The 17 case studies dramatically illustrate that the issue of acceptance by citizens and introduction of effective rules and regulations urgently needs attention as well.

4 The Challenge to Reach Effective Governance

Kurian et al. (2016) provide a well-documented study on relations between science and policy in the environmental arena. They correctly emphasize the importance of transdisciplinarity, where scientists of different disciplines work together with stakeholders and policy makers. Kurian (2017) also emphasizes the importance of trade-offs and thresholds when dealing with the water-energy-food nexus and defined a wastewater reuse effectiveness index (WREI) that considers physical aspects in an institutional and socio-economic context, this deserves further development. In line with this, Bouma et al. (2015) analyzed six already published case studies related to land use in the Netherlands and Italy, not only showing which SDGs were being addressed but also how problems were defined and worked out in interaction with stakeholders, operating in a policy-defined context. Linking soil and water data directly with the policy arena, defacto bypassing the involved stakeholders or just paying lip service to their contributions, was shown to be undesirable and this would certainly apply when addressing a sensitive issue such as waste application to soils. A primary focus on stakeholders is advisable but only when based on a "joint learning" approach. Results can then next be communicated to the policy arena, preferably by stakeholders and not by researchers (e.g., Bouma 2015, 2019; De Vries et al. 2015). The concept of Data Observatories, as proposed by Kurian et al. (2016), is attractive as it emphasizes a bottom-up and transdisciplinary approach, but the question remains which data should be deposited, how data should be combined and balanced as they come from different sources and how they will be integrated aiming at information (seen as data with meaning) and knowledge (seen as internalized information). Next to this approach, presentation and a thorough analysis of completed and successful case studies, as reported by Hettiarachichi and Ardakanian (2016b), would appear to be also suitable to be used as examples for others to follow, while an analysis of the many opportunities and pitfalls during the realization of the case study would present inspiration for both practitioners and future research.

Future water-soil-waste research could, next to emphasizing data observatories, be focused on such case studies covering technical aspects in a socio-economic context, including increasingly important modes of communication (e.g., Bouma

2018). Kurian et al. (2016) present important studies on governance focusing on soil conservation (Laos), watershed management (India), water supply (Manilla) and social investment (Honduras). But only the Indian study covers wastewater reuse in a peri-urban agricultural setting and does so in a rather abstract manner with no attention for the complete sequence starting with waste generation and ending with application to soil and its effects. Considering all the implications of waste disposal on soils already offers many, often rather unique problems. Broadening the scope to other land use problems is intellectually attractive but may divert attention from the specific issues associated with waste treatment and the water-soil-waste nexus.

A final remark about transdisciplinarity is wastewater application to soils results in "wicked" problems that have no single solutions. Only compromises between often quite different points of view of various groups of stakeholders have to be found, to be expressed by a series of alternative scenarios. Scientists, acting as knowledge brokers, can play a crucial role here. A major study in the Netherlands on sustainable land use showed that interaction with stakeholder groups and policy makers was very time consuming, but developing joint research proposals and working together with researchers in a "joint-learning" mode to realize those proposals turned out to be a recipe for success (e.g., Bouma et al. 2011). "Joint learning" particularly applied to the stakeholders, but interaction with the policy arena was maintained during the process, avoiding unexpected confrontations at the end of the process.

5 Conclusions

1. The water-soil-waste nexus is of high strategic importance for achieving several SDG's because solid and liquid waste can provide important contributions to achieve biomass production, water quality, climate mitigation and biodiversity preservation as affected by soil management that needs to be defined for different soils, water regimes and climate zones.
2. Case studies on waste applications to soils show a structural imbalance within the nexus, with so far little emphasis on hydrology and soils and their interaction. A meaningful nexus can only be achieved when the constituting parts of the nexus are in balance.
3. Current databases for hydrology and soils contain many data and monitoring methods that are widely available. Operational ICT makes all this well accessible. Scientists can hardly cope with this flood of data and the transformation of data into meaningful information and knowledge, that is relevant for waste management, presents a major challenge. A focus on Goals (the SDGs) may point the way to a structured approach to the use of data for studying the water-soil-waste nexus.
4. The policy arena, responsible for environmental rules and regulations, can most effectively be inspired by successful case studies of waste application to soils, developed in close interaction with diverse stakeholder groups. Such case studies should, however, be documented applying a thorough technical and socio-economic analysis using modern data, methods and models with a focus on the entire water-soil-waste nexus.

References

Arrouays D, Grundy MG, Hartemink AE, Hempel JW, Heuvelink GBM, Young Hong S, Lagacherie P, Lelyk G, Mc Bratney AB, Mc Kenzie NJ, Mendonca-Santos M, Minashy B, Montanarella L, Odeh IOA, Sanchez PA, Thompson JA, Zhang G (2014) Global soil map: toward a fine-resolution grid of soil properties. Adv Agron 125:93–134

Batjes NH (2016) Harmonised soil property values for broad-scale modelling (WISE30sec) with estimates of global soil carbon stocks. Geoderma 269:61–68

Blum WEH (2016) Role of soils for satisfying global demands for food, water and bioenergy. In: Hettiarachchi H, Ardakanian R (eds) Environmental resource management and the nexus approach. Springer International Publishing, Switzerland, pp 143–179

Bouma J (1979) Subsurface applications of sewage effluent. In: Beatty MT, Petersen GW, Swindale LD (eds) Planning the uses and management of land. Agronomy 21, ASA-CSSA-SSSA, Madison, WI, USA, pp 665–703

Bouma J (1989) Using soil survey data for quantitative land evaluation. In: Stewart BA (ed) Advances in soil science, vol 9. Springer, pp 177–213

Bouma J (2014) Soil science contributions towards sustainable development goals and their implementation: linking soil functions with ecosystem services. J Plant Nutr Soil Sci 177(2):111–120

Bouma J (2015) Engaging soil science in transdisciplinary research facing wicked problems in the information society. Soil Sci Soc Am J 79:454–458. https://doi.org/10.2136/sssaj2014.11.0470

Bouma J (2016a) Hydropedology and the societal challenge of realizing the 2015 United Nations sustainable development goals. Vadose Zone J 15(12):36–48. https://doi.org/10.2136/vzj2016.09.0080

Bouma J (2016b) Implications of the NEXUS approach when assessing water and soil quality as a function of solid and liquid waste management. In: Hettiarachchi H, Ardakanian R (eds) Environmental resource management and the nexus approach. Springer International Publisher, Switzerland, pp 179–209

Bouma J (2018) The challenge of soil science meeting society's demands in a "post-truth", "fact-free" world. Geoderma 310:22–28. https://doi.org/10.1016/geoderma2017.09.017

Bouma J (2019) How to communicate soil expertise more effectively in the information age when aiming at the UN sustainable development goals. Soil Use Manag 35(1):32–38. https://doi.org/10.1111/sum.12415

Bouma J, van Altvorst AC, Eweg, Smeets RPJAM, van Latesteijn HC (2011) The role of knowledge when studying innovation and the associated wicked sustainability problems in agriculture. Adv Agron 113:285–314

Bouma J, Stoorvogel JJ, Sonneveld WMP (2012) Land evaluation for landscape units. In: Huang PM, Li Y, Summer M (eds) Handbook of soil science, 2nd edn., Chapter 34. CRC Press, Boca Raton, London, New York, pp 34-1–34-22

Bouma J, Kwakernaak C, Bonfante A, Stoorvogel JJ, Dekker LW (2015) Soil science input in transdisciplinary projects in the Netherlands and Italy. Geoderma Reg 5:96–105. https://doi.org/10.1016/j.geodrs.2015.04.002

Bouma J, van Ittersum MK, Stoorvogel JJ, Batjes NH, Droogers P, Pulleman MM (2017) Soil capability: exploring the potentials of soils. In: Field DJ, Morgen CLS Mc Bratney AB (eds) Global soil security. Springer International, Switzerland, pp 27–44

Bonfante A, Bouma J (2015) The role of soil series in quantitative land evaluation when expressing effects of climate change and crop breeding on future land use. Geoderma 259–260:187–195

Bonfante A, Terribile F, Bouma J (2019) Refining physical aspects of soil quality and soil health when exploring the effects of soil degradation and climate change on biomass production: an Italian case study. SOIL 5:1–14. https://doi.org/10.5194/soil-5-1-2019

Bonfante A, Basile A, Bouma J (2020) Exploring the effect of varying soil organic matter contents on current and future moisture supply capacities of six Italian soils. Geoderma 361. https://doi.org/10.1016/j.geoderma.2019.114079

De Vries W, Kros J, Dolman MA, Vellinga TV, de Boer HC, Sonneveld MPW, Bouma J (2015) Environmental impacts of innovative dairy farming systems aiming at improved internal nutrient cycling: a multiscale assessment. Sci Total Environ 536:432–442

Droogers P, Bouma J (2014) Simulation modeling for water governance in basins. Int J Water Res Dev 30 (3):475–494

FAO, IIASA, ISRIC, ISSCAS, JRC (2012) Harmonized World Soil Database (version 1.2). Food and Agriculture Organization of the United Nations (FAO), International Institute for Applied Systems Analysis (IIASA), ISRIC - World Soil Information, Institute of Soil Science - Chinese Academy of Sciences (ISSCAS), Joint Research Centre of the European Commission (JRC), Laxenburg, Austria. https://www.iiasa.ac.at/Research/LUC/External-World-soil-database/HWSD_Documentation.pdf.

Field DJ, Morgan CES, McBratney AB (2017) Global soil security. Progress in soil science. Springer International Publisher, Switzerland

Hengl T, Mendes de Jesus J, McMillan RA, Batjes NH, Heuvelink GBM, Ribeiro E et al (2014) Soil Grids 1km: global soil information based on automated mapping. PLoS One 9(8):e105992. (12):e114788. https://doi.org/10.1371/journal.pone.0105992

Hettiarachchi H, Ardakanian R (eds) (2016a) Environmental resource management and the NEXUS approach. Managing water, soil and waste in the context of global change. Springer International Publishers, Switzerland

Hettiarachchi H, Ardakanian R (eds) (2016b) Safe use of wastewater in agriculture: good practice examples. UNU-FLORES, Dresden, Germany

Hoekstra AY, Mekonnen MM (2012) The water footprint of humanity. Proc Natl Acad Sci 109:3232–3237. https://doi.org/10.1073/pnas.1109936109

IUSS Working Group WRB (2015) World reference base for soil resources 2014: international soil classification system for naming soils and creating legends for soil maps. World Soil Resour Rep 106. Update 2015. FAO, Rome. https://www.fao.org/3/a-i3794e.pdf

Keesstra SD, Bouma J, Wallinga J, Tittonell P, Smith P, Cerda A, Montanarella L, Quinton J, Pachepsky Y, van der Putten WH, Bardgett RD, Moolenaar MG, Fresco LO (2016) The significance of soils and soil science towards realization of the United Nations sustainable development goals. SOIL 2:111–128. https://doi.org/10.5194/soil-2-111-2016

Kurian M, Ardakanian R, Veiga LGW, Meyer K (2016) Resources, services and risks. How can data observatories bridge the science-policy divide in environmental governance. Springer briefs in environmental science. UNU-FLORES & Springer International Publishers, Switzerland

Kurian M (2017) The water-energy-food nexus: trade-offs, thresholds and transdisciplinary approaches to sustainable development. Environ Sci Policy 68:97–106. https://doi.org/10.1016/j.envsci.2016.11.006

Lal R (2014) World soils and the carbon cycle in relation to climate change and food security. In: Weigelt J, Muller A, Bekh C, Töpfer K (eds) Soils in the nexus. A crucial resource for water, energy and food security. Oekom, Munchen, pp 31–67

Mannschatz T, Wolf T, Hülsmann S (2016) Nexus tools platform: web-based comparison of modeling tools for analysis of water-soil-waste nexus. Environ Model Softw 76:137–153. https://doi.org/10.1016/envsoft.2015.10.031

Soil Science Society of America (SSSA) (2002) In: Dane JH, Topp GC (eds) Methods of soil analysis. Part 4—physical methods. SSSA Book Series no. 5. Madison, WI, USA

Soil Survey Staff (2010) Keys to soil taxonomy, 11th edn. United States Government Publishing Office, Washington, DC

Stoorvogel JJ, Kooistra L, Bouma J (2015) Managing soil variability at different spatial scales as a basis for precision agriculture, Chapter 2. In: Lal R, Stewart BA (eds) Soil specific farming: precision agriculture. Advances in soil science. CRC Press, Taylor Francis Group, Boca Raton, FL, USA, pp 37–73

Stoorvogel JJ, Bakkenes M, Temme AJAM, Batjes NH, ten Brink BJE (2017) S-world: a global soil map for environmental modelling. Land Degrad Dev 28:22–33. https://doi.org/10.1002/ldr.2656

UNU-FLORES (2015) White Book. Advancing a NEXUS approach to the sustainable management of water, soil and waste. Dresden, Germany.

Van Looy K, Bouma J, Herbst M, Koestel J, Minasny B, Mishra U, Montzka C, Nemes A, Pachepsky Y, Padarian J, Schaap M, Tóth B, Verhoef A, Vanderborght J, van der Ploeg M, Weihermüller L, Zacharias S, Zhang Y, Vereecken H (2017) Pedotransfer functions in Earth system science: challenges and perspectives. Rev Geophys. https://doi.org/10.1002/2017RG000581

Viscarra Rossel RA, Mc Bratney A, Minashy B (Eds) (2010) Proximal soil sensing. Progress in soil science. Springer International Publisher, Switzerland

Viscarra Rossel RA, Mc Bratney A, Minashy B (Eds) (2010) Proximal soil sensing. Progress in soil science. Springer Verlag. Dordrecht, Heidelberg, London, New York.

Viscarra-Rossel RA, Bouma J (2016) Soil sensing: a new paradigm for agriculture. Agric Syst 148:71–74

World Water Assessment Programme (2009) The United Nations world water development report 3: water in a changing world. In: Earthscan (ed) World water, vol 3. UNESCO, Paris, p 349

Nutrient Recovery for Use in Agriculture: Economic Assessment of Decentralized Compost Business Model in Nairobi

Solomie Gebrezgabher, Avinandan Taron, and Sena Amewu

Abstract

Large cities in developing countries are facing the challenge of rapid urban population growth, which results in increasing waste generation. In Nairobi, the solid waste situation is characterized by low coverage of collection, pollution from uncontrolled dumping, inefficient public services, unregulated and uncoordinated private sector operators and lack of key solid waste management infrastructure. About 3,121 tons of municipal solid waste (MSW) is generated daily, of which about 850 tons are collected and the remaining is burnt or dumped in unauthorized sites or landfilled in the Dandora dumpsite causing health and environmental problems. The recovery of nutrients from the organic content of MSW for reuse in agriculture has the potential to address the dual challenge of waste management and soil nutrient depletion. This study assessed the economic and environmental impact of decentralized composting business model in Nairobi based on a comparison with the baseline scenario using an indicator expressed in tons CO_2 equivalent. The cost–benefit analysis was based on data collected from existing compost plants in Kenya. To assess the sensitivity of the results to variation in input variables, a simulation model was developed using the Monte Carlo method. The decentralized composting business model resulted in a net GHG emission saving of 1.21 tons CO_2-eq/ton of compost, being both financially and economically feasible with more than 70% chance of economic success. Assessing the economic and environmental impact is an important tool for decision making and to ensure that the business model results in desired benefits to society.

Keywords

Nutrient recovery · Nexus · Decentralized compost · Costs and benefits · Monte Carlo simulation

1 Introduction

The increasing quantity of urban waste in urban agglomerations of developing nations coupled with the inadequate sanitation services is a growing concern to the deteriorating urban environment (Oyoo et al. 2011). Solid waste management (SWM) is a major public health and environmental concern in urban areas of many developing countries. Nairobi's solid waste

S. Gebrezgabher (✉) · S. Amewu
International Water Management Institute,
PMB CT 112, Cantonments, Accra, Ghana
e-mail: s.gebrezgabher@cgiar.org

A. Taron
International Water Management Institute,
P.O. Box 2075, Colombo, Sri Lanka

S. Hülsmann and M. Jampani (eds.), *A Nexus Approach for Sustainable Development*,
https://doi.org/10.1007/978-3-030-57530-4_3

situation, which could be taken to generally represent Kenya's status, is largely characterized by low coverage of solid waste collection, pollution from uncontrolled dumping of waste, inefficient public services, unregulated and uncoordinated private sector and lack of key solid waste management infrastructure (Njoroge et al. 2014). These practices of open dumping of MSW have negative impacts on the environment and people's health.

Increasingly the "nexus-approach" is gaining importance in environmental resources management and it is thought to be one of the means to achieve sustainability (Hoff 2011). This approach goes beyond looking at the individual components and integrates the functioning, productivity and management of the interlinkages and interdependencies across sectors. One such nexus link is water, soil and waste, where the production of food relies on water and soil, and waste contributes an important factor to the provision of nutrients and other organic material (Hettiarachchi and Ardakanian 2016). The conversion of the organic fraction of MSW into secondary valuable products such as compost reflects another linkage of the nexus. Recovering and reuse of resources is at the core of thinking when applying the nexus approach (Schwärzel et al. 2015). This approach creates financial and environmental sustainability and presents a great opportunity to mitigate the prevailing waste management problems in developing countries while addressing soil nutrient depletion and environmental pollution.

Composting of MSW has been done for several years; however, the success of such businesses has been limited in developing countries. The potential economic, environmental and social impacts of composting MSW need to be assessed to ensure its sustainable development. While composting can be a proposed mechanism to meet the water, soil and waste nexus (see Bouma 2021), it is important to assess the socio-economic feasibility of composting options based on the scale of operation. In many developing countries, large centralized and highly mechanized composting plants have often failed to produce good quality compost. Centralized composting plants are often abandoned due to high operational and maintenance costs. In contrast, decentralized composting ensures separation of organic waste either at the household level by creating awareness of resource recovery or at community levels and thus creating employment opportunities for the urban poor. The decentralized composting approach reduces high capital investments on mechanization, operation and maintenance costs, which can make use of low-cost technologies based on manual labor to ensure waste is well sorted before it is composted. The transportation of waste from densely populated areas is an additional challenge and most of the waste remains uncollected posing negative environmental and health externalities to the population at large. This study, therefore, hypothesizes that setting up of decentralized plants in areas where population density is higher, significantly reduces environmental and health externalities and provides a business opportunity in the urban context. Therefore, the main focus of the study is to assess the financial and environmental viability of a decentralized compost business model in the densely populated areas of Nairobi.

2 Generation and Composition of Municipal Solid Waste in Nairobi

In Nairobi, 3,125 tons of household waste is generated daily[1] along with 1,467 tons of non-household waste indicating that household waste comprises about 68% of the total waste generated in the city (UNEP 2010). Table 1 shows the characteristics and composition of the MSW generated in Nairobi city. The organic fraction of the waste comprises 50.9% while the rest is paper (17.5%), plastic (16.1%), glass (2%), metal (2%) and other miscellaneous waste (11.4%). The estimated current reuse and recycling is merely 100–150 tons/day, while about 830 tons of waste

[1]Projected household waste for 2016 considering 0.69 kg/capita/day of waste generated (UNEP 2010).

Table 1 Composition of MSW at the source and community waste collection points

Waste	City-wide waste composition (%)	
	At immediate source	At communal waste collection pits
Organics	50.9	43.0
Paper	17.5	12.1
Plastics	16.1	15.1
Glass	2.0	5.6
Metals	2.0	2.7
Others	11.4	21.7

Source UNEP (2010)

is landfilled in Dandora dumpsite and the rest is open-dumped (UNEP 2010). The Nairobi City Council (NCC) has the primary responsibility for the provision and regulation of solid waste management services in the city. However, the collection efficiency is quite low and the total amount of waste that is being improperly disposed of is 2,140 tons/day (UNEP 2010). With time, other actors such as private companies and community-based organizations have come into play. These actors handle waste on a smaller scale and are irregular in their services.

Nairobi is divided into administrative districts, each having different population density. The Dandora (Nairobi East district) and Kibera (Nairobi West district) are considered as high-density zones with 36% and 22% of the population living in those districts, respectively (Khamala and Alex 2013). In this study, we conduct the economic and environmental assessment of composting the waste generated in the densely populated area of the Nairobi West district. We assume that of the 2,140 ton/day of waste generated in Nairobi, 22% of the waste is generated from the West district and is targeted for composting. The MSW is to be composted in small decentralized compost plants, each having a capacity of 10.5 ton/day of incoming waste. Therefore, to absorb the waste generated in the West district (470 ton/day), 45 small scale decentralized compost plants are needed. Figure 1 shows the location of the proposed decentralized compost plants and as shown in the figure, there are already existing compost plants in the Nairobi Western district.

3 The Nexus Approach to Economic Assessment of Nutrient Recovery Models

Human interactions with the components of the ecosystems lead to changes in the services derived from that particular ecosystem. The shift towards sustainable production and consumption practices related to waste management programmes have received wide attention in the nexus of resources conservation, recovery and reuse (Pires et al. 2011). The core idea of nexus thinking is maximizing synergies and reducing trade-offs through integrated management of more than one resource (Hettiarachchi and Ardakanian 2016). In the case of resource recovery and reuse, waste generated from different waste streams is being tapped for recovering resources such as nutrients, energy and water. Processing of organic waste for nutrient recovery for use as a soil amendment has a positive impact on soil quality indicating that there exists a strong inter-connectivity in water-soil-waste nexus (Fig. 2) (Lal 2015).

Application of nutrients recovered from organic waste in agriculture contributes to counteracting issues such as soil nutrient depletion or affordability of fertilizers and will result in improved soil quality and farm productivity thereby increasing food security and conserving ecosystem services. Through nutrient recovery models, synergies are created among different sectors such as the waste management and food

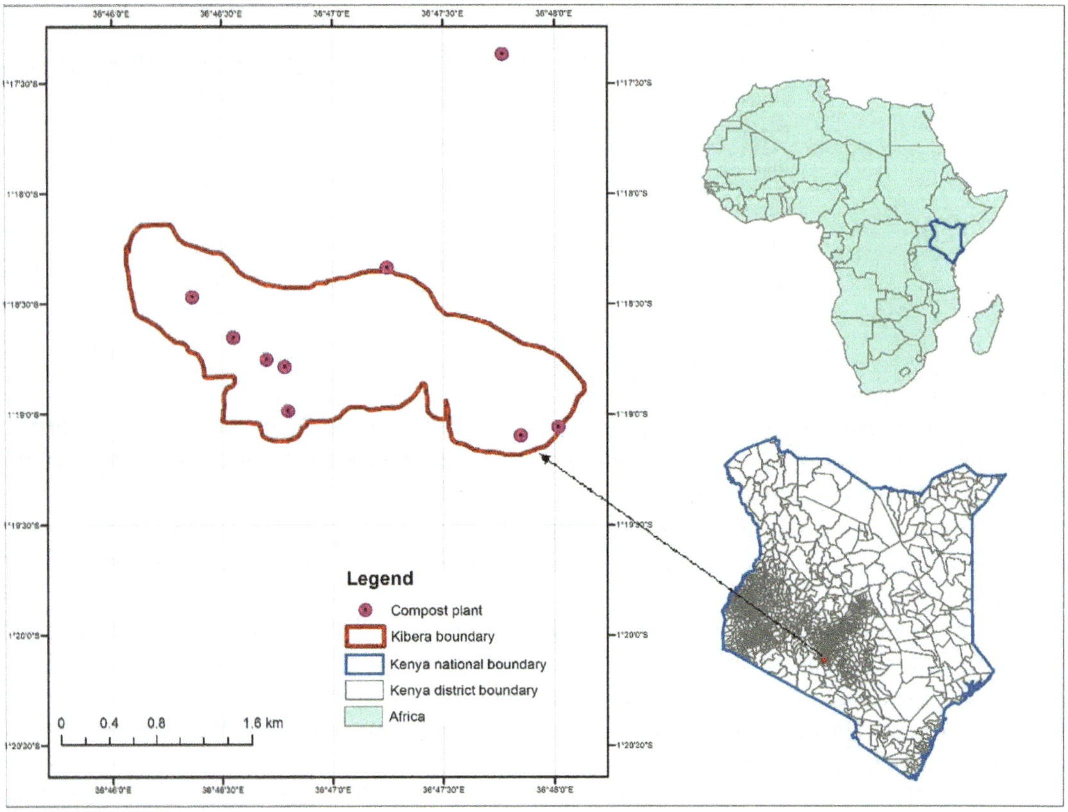

Fig. 1 Study location map and existing compost plants at Nairobi Western District

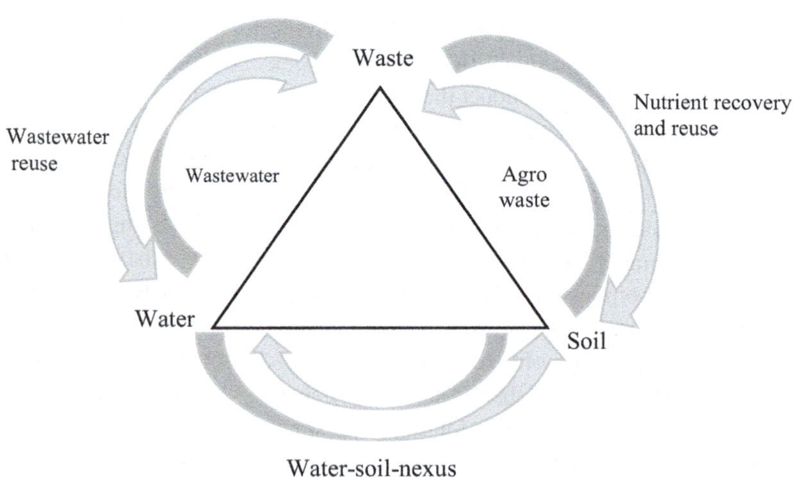

Fig. 2 Soil, water and waste nexus (Adapted from Lal 2015)

sector. These synergies with a nexus approach in mind present a great opportunity to mitigate the prevailing waste management problems while addressing soil nutrient depletion and environmental pollution (Sallwey et al. 2017).

The link between waste and water is also equally important as wastewater can be used as a source of water and plant nutrients (Hettiarachchi et al. 2018; Weckenbrock and Alabaster 2015). Recovering and reuse of waste as a source of nutrient not only adds to the restoration of ecosystem services but also reduces negative externalities such as GHG emissions associated with improper management of waste in developing countries (Kong et al. 2012).

4 Materials and Methods

4.1 Economic Assessment

The economic assessment is done based on Return on Investment (ROI), Net Present Value (NPV) and Benefit Cost Ratio (BCR) valuation criteria. ROI measures per period; the rates of return on money invested in the business and is determined by dividing net profit by total investment. The NPV is the total sum of the present value of expected future cash flows. Higher ROI and NPV values represent greater economic benefits. The BCR is the ratio of the total benefit of a project relative to its costs, both benefits and costs expressed in discounted present values.

The economic analysis of the decentralized compost business model is based on the data collected in 2015 from 22 operating compost plants located in different counties and cities of Kenya. In this section, we first provide descriptive statistics of the surveyed plants, which will be used as a basis for building the financial model for the decentralized compost business model.

4.1.1 Technical and Financial Data of Compost Plants in Kenya

Table 2 shows data from the year 2014 on selected parameters of surveyed compost plants in Kenya. Data was collected pertaining to the production capacity, capacity utilization, quantity

Table 2 Technical and economic data of compost plants in Kenya

Item	Capacity of the plant (tons of compost/year)						All	
	<100 tons (n = 6)		100–500 tons (n = 11)		>500 tons (n = 3)			
	Mean	SD	Mean	SD	Mean	SD	Mean	SD
Technical data								
Land size (acres)	0.70	0.39	1.84	1.99	1.74	1.04	1.64	1.75
Capital utilization (%)	53	16	51	21	75	35	52	23
Quantity of waste (ton/year)	139	68	466	154	1050	212	420	311
Input–output ratio (%)	0.55	0.24	0.51	0.16	0.50	0.25	0.53	0.19
Financial data								
Investment cost (KES[a]/ton)	15,433	18,092	14,365	12,891	12,583	16,381	14,523	14,170
Production cost (KES/ton)	4,833	2,020	3,908	3,222	1,784	422	3,831	2,620
O & M cost (%)	1.2	1.6	1.5	1	3	0	2	3
Depreciation (%)	3	4	2	2	7	9	3	4
Price of compost (KES/ton)	20,000	6,324	14,000	5,676	22,500	10,606	19,210	11,933

[a]1 USD = 100.76 KES

of waste used to produce the final product and the land size of the compost plants. The amount of waste processed varied between less than 100 ton/year in the smallest plants up to more than 1,000 ton/year in large plants. In this study, the plants were categorized into three groups based on their production capacity as small (<100 tons), medium (100–500 tons) and large (>500 tons) scale. The dominant feedstocks used for compost production are household and market waste with agro waste including animal dung as additional feedstocks. The majority of the compost plants have an annual production capacity ranging between 100 and 500 tons with only a few of them having an annual production capacity of either less than 100 ton or greater than 500 tons. The annual average quantity of waste processed is 139 tons in the smallest case and more than 1,000 tons in the largest case. The average input–output conversion rate for the small scale plants is 0.55 ton/ton of input processed while for the larger plants, it is 0.50 ton/ton of input. The plants operate at an average capacity utilization of 52% with the larger plants operating at an average capacity utilization of 75% indicating that the plants are not operating at full capacity. The average land size for the plants is 0.70 acres in the smallest case while in the larger cases, the average land size is 1.74 acres indicating that land size increases as the plant capacity increases. The majority of the plants use open windrow composting method in their production process, which requires more land as production capacity increases.

Investment costs are a total of the whole installation inclusive of land, building, machine, equipment and other costs. The average investment cost per unit of capacity in the smallest plants is 15,433 KES/ton (153 USD/ton) while in the largest cases, it is 12,583 KES/ton (125 USD/ton). It can be observed that as the capacity of the plant increases, the average investment cost per unit of capacity decreases. Moreover, investment cost per unit of capacity showed variation among different plants within their respective scales as manifested in the large standard deviations. Mean investment cost, looking at all the plants is 14,523 KES/ton (144 USD/ton). Production and operational cost, which comprises raw material, labor, utilities, administration and marketing cost, is 4,833 KES/ton (48 USD/ton) in the smallest plants and 3,908 (39 USD/ton) in the medium sized. Looking at all the plants, the mean production cost is 3,831 KES/ton (38 USD/ton). The depreciation rate for the small plants is 3% while for the largest plants it is at 7%. The operational and maintenance costs for small plants are 1.2% and for the larger plants, it is 3% of the total investment cost excluding land. The average price of compost varies among the plants within their respective ranges. The average price of the compost is 14,000 KES/ton (139 USD/ton) in the case of medium sized plants. The corresponding figures for the small and large plants are 20,000 (198.5 USD/ton) and 22,500 KES/ton (223 USD/ton), respectively. The average price of the compost for all the plants is 19,210 KES/ton (191 USD/ton). The variation in the price of compost among the different scales of compost plants could be due to different factors such as variation in the way the compost is distributed to the final buyer, variation in the marketing strategies of different plants and variation in the quality and cost of production of the compost product.

The cost of land accounts for 38% and 35% of the total investment cost for the smallest and largest compost plants respectively, whilst it accounts for about 51% of the investment cost for the medium sized plants (Fig. 3). Land accounts for 45% of the total investment cost for all the compost plants and it is a major cost for the plants. The majority of the plants use low technology processes, which require more space to sort and process raw materials using manual labor and low cost machinery. In addition to land, plant infrastructure is a major cost for large scale plants accounting for more than one-third of the total investment cost. Machines and equipment account for about one-third of the total investment cost of all surveyed plants.

The production and operational cost components include raw material, labor, utilities, marketing and administration costs (Fig. 4). Labor cost stands out as the major cost for all the surveyed plants, which is expected as most of the

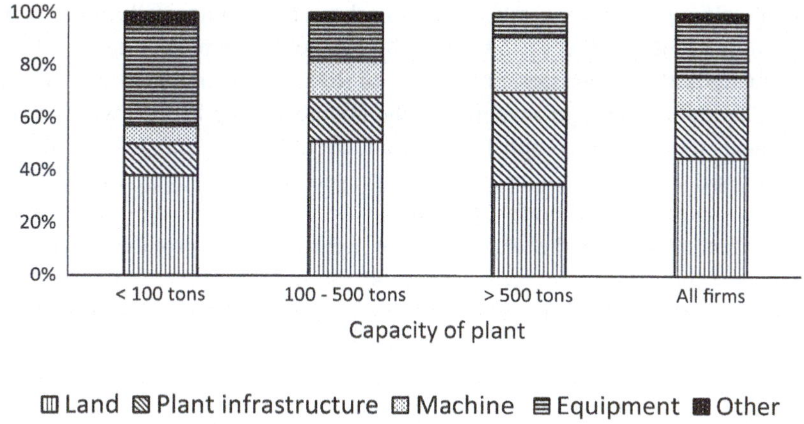

Fig. 3 Share of cost components in the total investment cost of surveyed compost plants

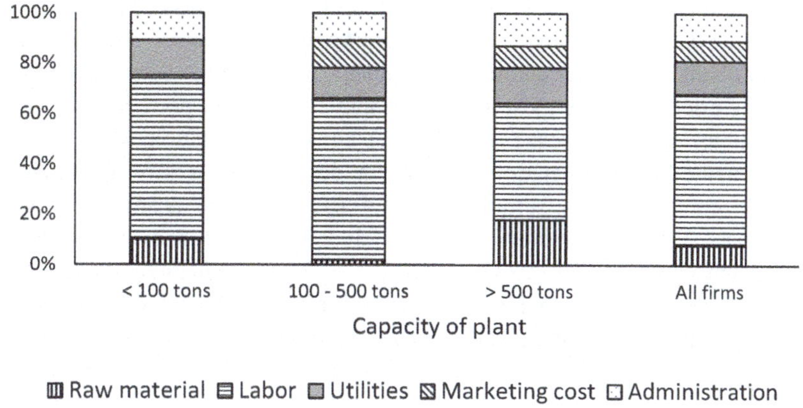

Fig. 4 Production and operational cost components of the surveyed compost plants

plants use low technology and labor intensive processes. The raw material is not a major cost and accounts for only 2% of the total production cost in the case of medium scale plants while it accounts for 18% of the total production cost in the case of large scale plants. Utilities and administration costs account for about 11% and 14% of the total production costs respectively, for all the plants. Marketing cost is a minor cost for all the plants accounting for only 11% of the total cost in case of medium scale plants and 8% in case of the large scale plants while it is nil in case of small scale plants indicating that compost plants are not investing in the marketing of their products.

4.1.2 Key Assumptions and Parameters for Economic Analysis of Decentralized Compost Business

Investment cost per unit of compost is assumed to be 213 USD/ton based on similar existing plants in Kenya. The Investment cost includes land, building, machine, equipment and other costs. The useful life of the compost plants is assumed to be 15 years. Production cost, which includes mainly labor, raw material and other costs is 55 USD/ton. The conversion of input to compost is 50%. The capacity of individual compost plants is 10.5 tons/day of incoming waste and with 50.9% of organic fraction and input–output conversion ratio is 0.5; each composting plant produces 534 tons of compost annually with annual operational days of 200.

The major source of revenue for the compost plants is the sales of compost to different clients. Based on existing compost plants surveyed in Kenya, the average price of compost is 162 USD/ton. Based on the experience of existing compost businesses in Kenya, not all of the compost produced is sold in the first few years of operation and thus it is assumed that 50% of production is sold in the first two years, 70% from third to fifth year and 85% in rest of the period. Operating and maintenance costs are assumed to be 3% of the total investment cost. A discount rate of 12% is assumed based on the Kenya Central bank rate (https://www.centralbank.go.ke/rates/central-bank-rate/). The selling price of the compost and other input costs are subjected to an escalation of 2%. A straight line method of depreciation is used for depreciable capital costs assuming a useful life of 15 years. Taxes and interest were not included in the analysis.

Another potential source of revenue for the decentralized compost business model is from trading carbon credits. Depending on their eligibility, carbon credits are traded on either the regulatory Clean Development Mechanism (CDM) market or the voluntary carbon market. Based on the World Bank (2014), carbon credit prices in the EU ETS range about USD 5–9 (€ 4–7) in 2014, while prices were USD 18 (€13) in 2011. In this study, it is assumed that carbon credits are worth on average USD 7 per ton of CO_2 equivalent.

4.2 Environmental Assessment

The environmental assessment of a decentralized compost business model is conducted based on comparison with a reference or baseline scenario. The baseline scenario is the benchmark to compare project alternatives. The current operation of open dumping is assumed to be the baseline scenario for comparison. Under the alternative decentralized compost business model, MSW is composted in 45 small scale compost plants, each having a capacity of 10.5 ton/day of incoming waste. The process used for composting is open windrow composting. Other fractions remaining from the composting business model are dumped in a nearby dumpsite. Decentralized composting helps in local collection of the waste and provides savings in terms of the transportation cost of the waste. At the same time with 50.9% of the organic fraction of the waste being diverted to the compost plant, the environmental effects of dumpsites are also restricted (UNEP 2010).

The total emissions under the baseline scenario include emissions from the open dumping of MSW. The emissions from the decentralized compost business scenario include emissions from composting of MSW, emissions from transportation of inorganic fractions and transportation of compost to end users. The study focuses on GHG emissions and other emissions such as ammonia (NH_3) from composting are not included in this study as it is assumed that the plants would use proper turning of compost piles, which would reduce NH_3 emissions (Aye and Widjaya 2006). The compost is sold to farmers about 20 km from the plants. It is assumed that the compost is to be used as complementary input to other inorganic fertilizers and thus no inorganic fertilizer is assumed to be replaced by the compost. Moreover, emissions associated with machines or equipment used in the compost business are excluded from the scope of this study.

The estimation of environmental emission was based on a number of studies (Shafie et al. 2014; Ruiz et al. 2013; Aye and Widjaya 2006). The GHG emissions avoided as a result of diverting MSW to composting are measured in terms of the avoided tons of CO_2 per ton of incoming waste (Table 3). The avoided emission from the open dumping of MSW is assumed to be 0.47 ton/ton of input while during composting, GHG emissions of 0.32 ton/ton of organic fraction of MSW is assumed. Under the alternative scenario, a truck with a maximum capacity of 15 tons is assumed to be used to transport the inorganic fraction of the MSW and the final

Table 3 GHG emissions under baseline and alternative scenario (ton CO_2-eq/ton of waste)

	Value
Avoided GHG emissions under baseline scenario	
Emissions from open dumping	0.47
Emissions from transporting MSW to dumpsite	0.0036
GHG emissions under decentralized compost business scenario	
Emissions from composting MSW (organic fraction)	0.32
Emissions from transporting inorganic fraction of MSW	0.0009
Emissions from transporting compost to end users	0.0005

compost product to end users within an average distance of 20 km each way. The use of trucks results in CO_2 emissions ranging from 2.6 to 3 kg/L of diesel fuel (Ruiz et al. 2013). In this study, CO_2 emissions of 3 kg/L of diesel fuel were used and the CO_2 emissions are calculated based on a mean distance of 20 km and diesel consumption of 0.45 L/km.

4.3 Sensitivity Analysis Using Monte Carlo Simulation

To assess the sensitivity of the results to variation in input variables such as variation in unit investment cost, production cost, price of compost and other variables, a simulation model is developed using the Monte Carlo approach. Using this simulation model, the uncertainty of variations in input cost and price of outputs was incorporated into the economic assessment. Using this simulation model, the uncertainty of variations in input cost and price of outputs was incorporated into the economic assessment.

Table 4 presents the stochastic variables and the distribution assumed for each variable. The unit investment cost, production cost and price of compost showed variations across the surveyed compost plants in Kenya (Table 2). A distribution was fitted on the data collected from the field to determine the probability distributions for the value of investment cost, production cost and price per ton of compost. The mean value of unit investment cost is 213 USD/ton of compost, mean production cost is 55 USD/ton and mean selling price of compost is 162 USD/ton based on the observed values from the field survey. Normal and lognormal distributions were assumed to model the variations in investment cost, price of compost and production cost. For

Table 4 Stochastic variables and distribution assumed for each variable in the simulation model

Variable	Unit	Distribution type	Description/value
Investment cost	USD/ton	Normal	Mean = 213 SD = 139
Production cost	USD/ton	Lognormal	Mean = 55 SD = 43
Price of compost	USD/ton	Normal	Mean = 162 SD = 73
Price of carbon credit	USD/ton CO_2-eq	Triangular	Minimum = 7 Most likely = 10 Maximum = 14
Discount factor	%	Triangular	Minimum = 10 Most likely = 12 Maximum = 14.5

the price of carbon credit and the discount factor, since there is limited data, a triangular distribution was assumed.

5 Results

5.1 Results of Environmental Assessment

The overall GHG emissions from composting and transporting of 470 ton/day of MSW in 45 decentralized compost plants are shown in Fig. 5. GHG emissions from transporting and dumping MSW are negative, representing GHG emission savings from the decentralized composting business model. The savings are mainly from avoided dumping of MSW. Under the decentralized composting scenario, the highest GHG impact is from the process of composting MSW. Other processes such as transport of inorganic fraction to dumpsite and compost to end users did not contribute significantly to the total environmental impacts of the compost business.

5.2 Results of Economic Assessment

The total quantity of compost produced from 45 small scale compost plants is 24,030 ton/year (Table 5). The total investment cost is USD 5.09 million. Assuming a discount rate of 12% and useful life of 15 years, the business model resulted in a mean NPV of USD 4.38 million and ROI of 29% and BCR of 1.96 indicating that the business model is financially viable. Moving from the financial results to including the environmental impacts, the incremental benefit from carbon credit increases NPV by 30%.

5.3 Simulation Results

Simulation results provide the expected values for each of the valuation criteria, estimates of the criteria variability and the probability of economic success. The risk analysis showed a mean NPV of USD 4.32 million with variability from USD −14.86 million to USD 22.27 million (90% confidence interval) when only direct benefits and costs are taken into account (Table 6). The mean BCR is 2.91 but could be as high as 14 under an optimistic scenario and as low as −7 under a pessimistic scenario. When environmental benefits and costs were taken into account, the business model performed better and resulted in a mean NPV of USD 6.36 million and a BCR of 3.26. The risk analysis showed that the most important variable with a significant effect on NPV values is the price of compost followed by the production cost and investment cost.

The simulation model provides a clear understanding of the variability and the probability distribution of the valuation criterion. Figure 6 shows the probability density function of NPVs when direct and indirect benefits and costs are taken into account. The probability of a negative NPV when only direct benefits and costs are taken into account is 33% and this decreases to

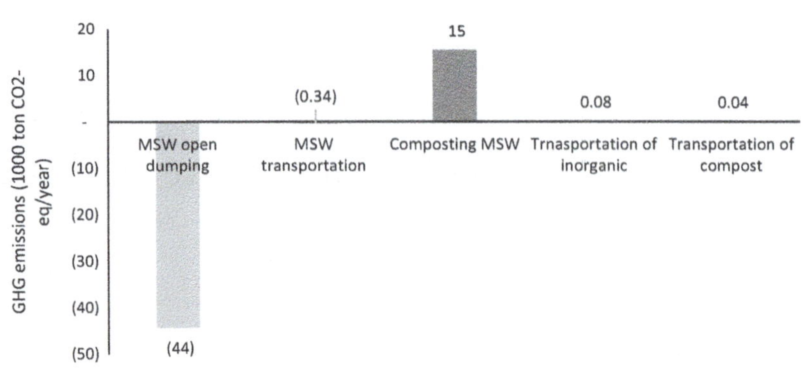

Fig. 5 GHG emissions and saving from decentralized compost business model

Table 5 Technical and economic results of decentralized composting business (USD)

	Value
Technical data	
Total quantity of MSW to be composted (ton/day)	470
Capacity of decentralized compost plant (ton/day)	10.5
Number of small-scale decentralized compost plants	45
Land area for decentralized compost plant (acres)	3.21
Total quantity of compost (ton/year)	24,030
Financial data (Million USD)	
Total investment cost	5.09
Total revenue from compost sales	3.64
Production cost	1.58
O & M cost	0.22
Financial results	
NPV (Million USD)	4.38
ROI	29%
BCR	1.96
Financial and environmental results	
Present value of carbon credit (Million USD)	1.39
NPV (Million USD)	5.77
ROI	31%
BCR	2.24

Table 6 Results of the simulation model[a]

	Financial value	Financial and environmental value
NPV-mean (Million USD)	4.32	6.36
5%	−14.86	−12.84
95%	22.27	24.37
BCR-mean	2.91	3.26
5%	−7	−7
95%	14	15

[a]5000@Risk iterations

27.5% when environmental costs and benefits are included. Thus investing in the decentralized composting business model is economically viable with the potential of attaining an NPV of USD 24 million under optimistic scenarios and has a probability of economic success of more than 70%.

There is widespread interest in composting of MSW to create economic and environmental benefits and ultimately reduce the pressure on local governments in managing MSW. However, composting is still seldom considered as a strategic element and there is very little evidence available of its economic feasibility (Pandyaswargo and Premakumara 2014). The results of this study are comparable to other studies such as the study by Aye and Widjaya (2006), which assessed the economic and environmental feasibility of different waste reuse options for market waste disposal in Indonesia. The study concluded that composting at a centralized plant is the most economically feasible option under Indonesian

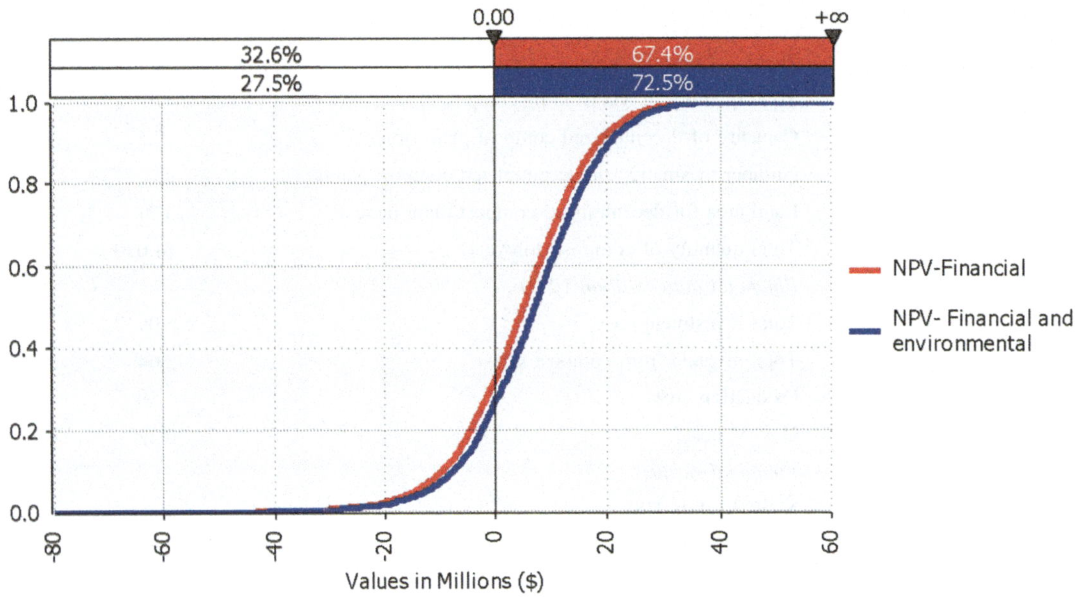

Fig. 6 Probability density function of NPVs when direct and indirect benefits and costs are taken into account

conditions. Likewise, Pandyaswargo and Premakumara (2014) conducted a cost–benefit-analysis of composting plants in Asia and concluded that medium-scale and lower large-scale composting plants are financially feasible as compared with smaller and larger capacity plants. In contrast, Galgani et al. (2016) concluded that composting, anaerobic digestion and biochar business models were found not to be economically viable without external subsidies and access to carbon markets would not change the situation significantly under Ghana conditions. This indicates that the economic and environmental sustainability of the nutrient recovery and reuse models vary across countries depending on a number of factors such as local market conditions, processing methods, scale, technologies and marketing strategies.

6 Conclusions

The potential economic, environmental and social impacts of nutrient recovery and reuse models that promote synergies in the waste management and food sectors need to be assessed to ensure sustainable development. These synergies with a nexus approach promote sustainable production and consumption patterns by mitigating the prevailing waste management problems while addressing soil nutrient depletion and environmental pollution.

This study examined the economic and environmental viability of a decentralized compost business model in Nairobi. The analysis was conducted based on the valuation of environmental and financial benefits and costs associated with the business model. The environmental impacts associated with the business model were estimated based on GHG emissions avoided from open dumping of MSW and emissions from the business model, which included emissions from composting, transportation of inorganic fraction of MSW to dumpsite and compost to end users. The highest GHG impact under the decentralized composting business model is from the process of composting MSW. Other processes such as transport of inorganic fraction to dumpsite and compost to end users did not contribute significantly to the total environmental impacts of the decentralized compost business. The GHG emissions from the business model were lower than the emissions under the baseline scenario and thus the business model resulted in net savings in

GHG emissions expressed in ton CO_2-equivalent. Compared to the baseline scenario, the decentralized composting business model resulted in a net GHG emission saving of 1.21 ton CO_2-eq/ton of compost. Looking at the economic and environmental impacts, the business model is both financially and economically feasible with more than 70% chance of economic success.

The establishment of resource recovery and reuse business models considers different waste streams as a resource. The promotion of such businesses has implications for the developing countries especially for micro, medium and small scale businesses in terms of using sustainable production and consumption patterns and behaviors. Thinking in terms of society as a whole, the development of a waste reuse industry such as the waste to nutrient business has a spillover effect since it induces both direct and indirect costs and benefits to different sectors of the economy. Assessing the economic and environmental viability of waste to nutrient business models is an important tool for decision making in order to ensure that the business models result in desired socio-economic benefits to society and thus justify their development and promotion. This study focused on the economic and environmental aspects of a nutrient business model, however, social benefits and costs associated with the business model need to be considered in assessing the overall socioeconomic assessment of waste reuse businesses.

Acknowledgements This research was carried out with funding by the European Union (EU) and technical support of the International Fund for Agricultural Development (IFAD) and the Water Land and Ecosystem (WLE) research programme of the CGIAR.

References

Aye L, Widjaya ER (2006) Environmental and economic analyses of waste disposal options for traditional markets in Indonesia. Waste Manag 26:1180–1191

Bouma J (2021) How to integrate and balance water, soil and waste expertise when realizing the corresponding nexus approach. In: Hülsmann S, Jampani M (eds) A nexus approach for sustainable development. Springer Nature, Cham, Switzerland. https://doi.org/10.1007/978-3-030-57530-4_1

Galgani P, van der Voet E, Korevaar G (2016) Composting, anaerobic digestion and biochar production in Ghana. Environmental–economic assessment in the context of voluntary carbon markets. Waste Manag 34:2454–2465

Hettiarachchi H, Caucci S, Ardakanian R (2018) Safe use of wastewater in agriculture: the golden example of nexus approach. In: Hettiarachchi H, Ardakanian R (eds) Safe use of wastewater in agriculture. Springer, Cham

Hettiarachchi H, Ardakanian R (2016) Managing water, soil, and waste in the context of global change. In: Hettiarachchi H, Ardakanian R (eds) Environmental resource management and the nexus approach. Springer, Switzerland, pp 1–7

Hoff H (2011) Understanding the nexus. In: Background paper for the Bonn2011 conference: the water, energy and food security nexus. Stockholm Environment Institute, Stockholm

Khamala EM, Alex AA (2013) Municipal solid waste composition and characteristics relevant to the waste-to-energy disposal method for Nairobi city. Glob J Eng Des Technol 2:1–6

Kong D, Shan J, Iacoboni M, Maguin SR (2012) Evaluating greenhouse gas impacts of organic waste management options using life cycle assessment. Waste Manag Res 30:800–812

Lal R (2015) The nexus approach to managing water, soil and waste under changing climate and growing demands on natural resources. In: White book: advancing a nexus approach to the sustainable management of water, soil and waste. United Nations University

Njoroge BNK, Kimani M, Ndunge D (2014) Review of municipal solid waste management: a case study of Nairobi, Kenya. Int J Eng Sci 4:16–20

Oyoo R, Leemans R, Mol APJ (2011) Future projections of urban waste flows and their impacts in African metropolises cities. Int J of Environ Resour 5:705–724

Pandyaswargo AH, Premakumara DGJ (2014) Financial sustainability of modern composting: the economically optimal scale for municipal waste composting plant in developing Asia. Int J Recycl Org Waste Agricult 3:4

Pires A, Martinho G, Chang NB (2011) Solid waste management in European countries: a review of systems analysis techniques. J Environ Manag 92:1033–1050

Ruiz JA, Juarez MC, Morales MP, Munoz P, Mendıvil MA (2013) Biomass logistics: financial & environmental costs. Case study: 2 MW electrical power plants. Biomass Bioenergy 56:260–267

Sallwey J, Hettiarachchi H, Hülsmann S (2017) Challenges and opportunities in municipal solid waste management in Mozambique: a review in the light of nexus thinking. AIMS Environ Sci 4:621–639

Schwärzel K, Hülsmann S, Ardakanian R (2015) Advancing a nexus approach for the sustainable management of water, soil and waste. In: White book: advancing a nexus approach to the sustainable management of water, soil and waste. United Nations University

Shafie SM, Masjuki HH, Mahlia TMI (2014) Life cycle assessment of rice straw-based power generation in Malaysia. Energy 70:401–410

UNEP (2010) Solid waste management in Nairobi: a situation analysis. Technical document accompanying the integrated solid waste management plan. Nairobi, Kenya

Weckenbrock P, Alabaster G (2015) Designing sustainable wastewater reuse systems: towards an agroecology of wastewater irrigation. In: Kurian M, Ardakanian R (eds) Governing the nexus. Springer, Cham

World Bank (2014) State and trends of carbon pricing. Washington, DC. https://doi.org/10.1596/978-1-4648-0268-3

Sustainable and Safe Use of Wastewater for Food Production in Peri-urban Areas of Karnataka, India

Girija Ramakrishna and Matti Hanisch

Abstract

Agriculture in India is facing a deep crisis due to deficient rainfall and the increasing cost of inputs such as fertilizers and irrigation water. The use of wastewater for food production in peri-urban areas plays an important role in feeding cities and ensuring urban resilience. However, the main concern is the health risks associated with the usage of raw wastewater. This paper reviews existing wastewater use practices in food production in the peri-urban areas of the Hubli–Dharwad region (Karnataka, India). The main challenge faced by these twin cities is the lack of adequate sewerage infrastructure. The farmers located in the periphery of Dharward have been using wastewater for irrigation for more than three decades, and they are well aware of the benefits. In this study, the risks associated with untreated wastewater use for agriculture are assessed through literature review, farmer and practitioner knowledge and field experiences. The major risks associated with wastewater use in agriculture in the study area were found to be medical waste followed by pesticides, weedicides usage and risks implied by mosquitoes and flies. There was limited knowledge among the farmers on the safety measures to be taken during the application of wastewater to the farmlands. The study documents the best practices of farmers and suggests the safety measures based on literature review and field experiences, concluding minimal and cost-effective on-farm treatment technologies reduce the pathogen load and minimize health risks. The study also recommends conducting a quantitative risk assessment for the safe use of wastewater in agriculture.

Keywords

Wastewater irrigation · Nexus approach · Safety measures · On-site measures

1 Introduction

Wastewater is increasingly being used for agriculture, especially in areas of water scarcity. There is also a growing recognition of the value of wastewater as a source of both water and nutrients. Wastewater is considered a reliable source of water for irrigation throughout the year (Drechsel et al. 2010). However, there are risks associated with wastewater usage. Hazards

G. Ramakrishna (✉)
Consortium for DEWATS Dissemination (CDD) Society, Bengaluru, Karnataka, India
e-mail: girija.r@cddindia.org

M. Hanisch
CDD-BORDA (Bremen Overseas Research and Development Association), Bremen, Germany

S. Hülsmann and M. Jampani (eds.), *A Nexus Approach for Sustainable Development*,
https://doi.org/10.1007/978-3-030-57530-4_4

associated with wastewater-irrigated products include excreta-related pathogens, toxic chemicals (Ensink et al. 2008; Fuhrimann et al. 2016) concerning (mainly) human health as well as nitrate and salinization concerning environmental health (e.g., Jampani et al. 2018). It has been estimated that in 2016, approximately 93.3% of the total fecal sludge and wastewater generated in India was released into the environment without treatment, resulting in pollution of water bodies and other environmental and health disasters. It is estimated that in 2025, 80.5% of the fecal sludge and wastewater generated will be unsafely disposed into the environment and thus only 19.5% will be effectively treated. There is no substantial reduction in unsafe disposal forecasted for 2025 when compared with 2016 data (Borda and CDD Society 2017).

Currently, irrigation accounts for more than 80% of India's water use. However, growing demand for other applications, such as municipal and industrial uses, is putting increasing pressure on water availability, especially in and around urban and peri-urban areas. The projections for 2050 show aggregate water demand increasing to 1,447 m^3 and while agriculture retains its relative dominance, other uses are projected to increase their relative share. With rising water stress and water exploitation, a 50% deficit between water demand and supply has been predicted by 2030. Thus, the overall analysis of water resources indicates that in the coming years, there will be a double-edged problem—the reduced availability of freshwater and increased generation of wastewater due to the increasing population, food production, and industrialization. In such a scenario, the use of wastewater for irrigation would become an inevitable and sustainable option, especially in arid, semi-arid, and poor peri-urban areas, primarily because of its high nutrient content and the scarcity of freshwater sources—as already practiced in many regions worldwide (Hettiarachchi and Ardakanian 2016). In the case of vegetable cultivation with wastewater, there can be a considerable aggregate benefit for society in terms of year-round production for farmers and a more balanced diet for consumers. However, both these stakeholders are potentially at risk of adverse health effects from untreated wastewater (Jiménez and Navarro 2013). Irrigation with untreated wastewater is associated with major health risks such as intestinal nematodes. While treatment reduces this risk, there is no single best strategy and each situation requires its own specific approach to health protection (Strauss 1991).

It is against this background that the CDD Society, a not-for-profit organization working in the space of sanitation, decided to conduct a study in the Hubli and Dharwad region where farmers have been practicing wastewater irrigation for decades, under a lighthouse project called Nexus. The main purpose of the study was to understand the existing practices of wastewater irrigation and assess the benefits and risks associated with its usage. The project aimed to demonstrate the interlinkages between sanitation, food, water, energy, and health. This nexus of sanitation and agriculture can provide solutions for food production in poor peri-urban areas through the safe use of nutrient-rich wastewater in agriculture. On the one hand, this approach improves the sanitation situation by minimizing the health risks imposed by untreated wastewater, so that a favourable environment for food production is created. On the other hand, it aims to increase the supply of food by using nutrient-rich wastewater for food production. The main objective of the current study was to improve the sanitation situation, provide better health, increase food supply, and reduce environmental degradation. There is a great need and potential for the scaling up of safe practices of wastewater use in agriculture to reduce stress on diminishing freshwater resources, improve soil quality, and reduce food miles in urban and peri-urban areas.

The main objectives of the study are: (1) to review existing wastewater use practices and assess the health hazards associated with wastewater irrigation, (2) to identify the existing control measures and document the best practices to mitigate the risks, and (3) to recommend contextual safety measures for the use of wastewater in agriculture. This study documents the best practices followed by farmers in the peri-urban areas of Karnataka while recommending some safety measures to crop handlers and consumers. It also elaborates on how we can effectively and safely recover the nutrients present in human waste/wastewater while simultaneously mitigating the risks through locally available solutions.

2 Materials and Methods

2.1 Study Area: Hubli–Dharwad, Karnataka

Hubli and Dharwad (Hubli–Dharwad) are twin cities in the Indian state of Karnataka and the second-largest conurbation in Karnataka after Bengaluru (Villageinfo nd). They are located in North Karnataka, about 430 km north of Bangalore. While Dharwad is the administrative headquarters, the city of Hubli, situated about 20 km southeast of Dharwad, is the commercial center and business hub of North Karnataka. The cities have a single municipal corporation called Hubli–Dharwad Municipal Corporation (HDMC). Hubli is located at an altitude of 2,200 ft. and Dharwad at an altitude of 2,500 ft. above mean sea level, the terrain sloping from Dharwad towards Hubli. Dharwad has more favourable climatic conditions and is colder than Hubli because of the altitude difference of 300 ft. The temperature in Hubli reaches up to 40 °C (April), the average high is around 30 °C. The average rainfall is around 675 mm for Hubli and 812 mm for Dharwad. The Hubli–Dharwad urban area lies on the Deccan Plateau. The soil towards the western side of the national highway is mostly black cotton soil and that on the eastern side of the highway is red and gravelly, which is also taken into consideration in the prospective growth of the city; development proposals are made with due consideration to preserving the agriculturally fertile region (HDMC, Census 2011).

The main challenge faced by these twin cities is the lack of adequate sewerage infrastructure and thus the wastewater generated in the twin cities is not completely treated. In Dharwad, the main wastewater channel flows into the peri-urban areas of the city. The farmers located in the peripheries of the city have been using wastewater for irrigation for more than three decades (HDMC; Bradford et al. 2003).

2.2 Methodology

The methodology followed to conduct the study was as follows:

- A pre-feasibility study to identify the event site (wastewater village) and control site (freshwater village).
- Qualitative assessment of health risks associated with wastewater use at the identified sites.
- Extensive field surveys, inspections, and interviews to identify existing control measures and documentation of the best practices of wastewater use through literature review.

2.2.1 Pre-feasibility Study

A pre-feasibility study was conducted in the Hubli–Dharwad region to review the existing practices of wastewater use and to identify potential event and control sites. Farmers of rural Dharwad have been using wastewater for irrigation for more than three decades. In Dharwad, the main wastewater canal flows to Madihal, which was once an outlying village but now incorporated as a suburb because of the expansion of the city. From Madihal, the canal passes the peripheries of Govankoppa, Gongadikoppa, and Maradagi. A focused group discussion (FGD) conducted with the farmers in the month of May 2016 revealed that there are 60 farmers in total in Madihal and the total farmland is around 200 acres. Wastewater use in agriculture is a very old practice, especially in places like Govankoppa and Madihal, which are located on the periphery of Dharwad. This is because of the early assessment of high nutrient content and a non-chargeable continuous supply of water for irrigation followed by the continuous year-round production of vegetables, fruits, crops, and fodder (Hunshal et al. 1997). The proximity to urban areas (i.e., the source of sewage) ensures a reliable supply of wastewater for irrigation during the dry season

(February–May). These crop production systems were predominantly found at Madihal in Dharwad (Fig. 1) and at Bidanal on the outskirts of Hubli. The ease of access to local urban markets and the high urban demand ensures a secure market for vegetable production, particularly during the dry season when vegetable market prices increase three-to-five-fold (Hunshal et al. 1997).

Farmers on either side of the canal have been practicing wastewater agriculture for more than 20 years and the main crops grown with wastewater are leafy vegetables (palak and rajgiri), groundnut, soybean, and root vegetables. After keeping some products for their own consumption, farmers sell the rest at the local market. They also grow fodder crops such as jowar and maize. The other two villages identified were Govankoppa (Fig. 2) and Gongadikoppa, where the wastewater canal flows. Gongadikoppa is a small village of around 40 households and 100 acres of land, formed long back during the flood when the people from Govankoppa migrated to Gongadikoppa because of flooding (according to information collected during FGD).

Based on field inspections and discussions with the University of Agricultural Sciences, Dharwad, Municipal Corporation and from the FGD with the farmers, the wastewater-irrigated villages shortlisted for the study were: Madihal, Lakumnahalli, Hosayellapur, Govankoppa, Gongadikoppa, Maradgi, Kankur, Kavalgiri, Hebbali, and Shivalli, which include areas on both sides of the wastewater channel. Freshwater-irrigated villages were Mansoor, Mangundi, Kogiligere, Kumbarkoppa, Kamalapura, Malapura, and Annavara. Out of these villages, Govankoppa was finalized as the event site/wastewater village and Kamalapura as the control site/freshwater village. The criteria for selection were topography of the location, distance (approx. 5–10 km) from the wastewater canal (preferably on both the sides of the canal and the tail end of the stream), wastewater irrigation in practice for more than two decades, sample size, and proximity of local

Fig. 1 Wastewater-irrigated agricultural field at Madihal, Dharwad district, Karnataka; *picture credit* Ms. Susmita Banerjee

Fig. 2 Wastewater stream at Govankoppa (Event site), Dharwad district, Karnataka; *picture credit* Ms. Susmita Banerjee

health care centers, using a considerable section of the wastewater-irrigated area for growing food crops, and consuming part of the produce (i.e., what remains after sending to market) locally within the village.

Event site: Govankoppa, Dharwad district, Karnataka

In the wastewater-irrigated village of Govankoppa, wastewater irrigation has been practiced for more than three decades. The cropping pattern followed is vegetable production, field crops, and agroforestry. Vegetables, especially green leafy vegetables, are grown in almost all places where wastewater is used for agriculture. Wastewater is especially used for growing chili, brinjal, ridge gourd, cucumber, spinach, and tomato. Along with vegetables, field crops such as rice, ragi, jowar, maize, pulses, oilseeds, cotton, groundnut, green gram, pea, wheat, and Bengal gram are also grown. Cultivation of fruit crops such as banana, guava, and sapota are also occasionally grown. The cropping calendar of Govankoppa has revealed that crops are grown intensively throughout the year. The land is left fallow for one or two months a year. Generally, three crops are grown because of the free available wastewater in all seasons. Vegetable farming is preferred, especially in the dry season.

According to respondents, the installed sprinkler system is effective and its yield is high as it uses an optimum quantity of wastewater and distributes it uniformly throughout the farm, especially during the dry season. Using wastewater directly from the pump without filtering or screening was the most commonly observed method of irrigation in Govankoppa during field visits. In many cases, it was observed that pipes were directly put into the channel, and then water was pumped out. This pumped water was directly poured onto fields without passing through any kind of filter, which is particularly damaging because all kinds of solid waste such as plastic bottles, polythene bags, used syringes, droppers, and napkins also end up on the fields. No settling tank or filtering or screening arrangements were observed on this site. The key advantage of wastewater irrigation for farmers was that they could grow vegetables in summer when there is a great demand in the market and they, therefore, did not need to wait for the rains, unlike in the rain-fed situation.

Control site: Kamalapur, Dharwad district, Karnataka

Freshwater villages identified in Dharwad were Kamalapur and Malapura as these villages are close to the identified wastewater villages. Kamalapur is a small village/hamlet in Dharwad Taluk in Dharwad District of Karnataka State, India. It comes under the village council of Kamalapur Panchayath (Sindhindia nd). It is located 1 km north from district headquarters in Dharwad and 459 km from the state capital of Bangalore. Kamalapur is surrounded by Hubli Taluk (a Taluk is an administrative district for taxation purposes, typically comprising a number of villages), Haliyal Taluk, Kalghatgi Taluk, and Kundgol Taluk. The total geographical area of the village is 684.45 ha. Dharwad, Hubli, Saundatti-Yellamma, and Navalgund are the closest nearby urban areas. Local health care centers, and private and Government hospitals are present in Kamalapur. Table 1 summarizes the number of farmers and cultivable land area in Kamalapur.

Agriculture in Kamalapur is mainly rain-fed and farmers with borewells apply freshwater irrigation. As there is no wastewater source in Kamalapur, the entire agricultural land is freshwater-irrigated. The quality of freshwater is good and farmers get an optimum and continuous supply. The main crops grown are groundnut, soybean, potato, green gram, Bengal gram, wheat, cotton, peas, and green chili. The crops grown for fodder are jowar, maize, and the dried straw of groundnut and soybean. Medium- and small-scale farms mostly follow furrow and ridge irrigation (30%). However, the sprinkler system is very common amongst large-scale farms (40%).

Table 1 Land holdings details of Kamalapur village, Dharwad (*Source* Own elaboration based on oral communication by Krishi Vignyana Kendra, KVK)

	Number of farmers	Cultivable land (ha)
Marginal farmers	193	109.72
Small farmers	158	215.38
Big farmers	119	183.49

2.2.2 Qualitative Assessment of Health Risks Associated with Wastewater Use

Based on the pre-feasibility study conducted in the Hubli–Dharwad area, a further study was proposed to assess the health risks associated with wastewater use at the two identified sites (impact of wastewater irrigation on farmers' health). The study was initiated by the International Water Management Institute (IWMI), and conducted in collaboration with CDD-BORDA (Bremen Overseas Research and Development Association). The study aimed to quantify the health impacts in this village through a survey and comparison with the results for the neighboring control village, Kamalapur. The selected control site was expected to nullify the background effects to obtain a clear investigation result after root-cause analysis.

The study comprised four important steps—screening, scoping, risk assessment, and recommendation. An overview of the methodology is as given in Fig. 3.

Screening

Screening is an initiation of the study that involves communication needs assessment, identification of stakeholders, question sets, desk study based on secondary data, and a pre-feasibility study. The screening involved selecting stakeholders and communication need assessment, closed-ended questions interview, and finally the pre-feasibility study report review. Identifying the questions to be asked in the assessment process was the most important part of the screening process, which covered scope, range, duration, variation, and outcomes of existing practices directly or indirectly related to wastewater use in irrigation. The FGD results and personal interviews were taken into consideration. For screening and scoping, a gradient-based questionnaire was used to outline the possible hazards. Around 25 respondents from each of the control (Kamalapura) and event (Govankoppa) sites were interviewed. A total of

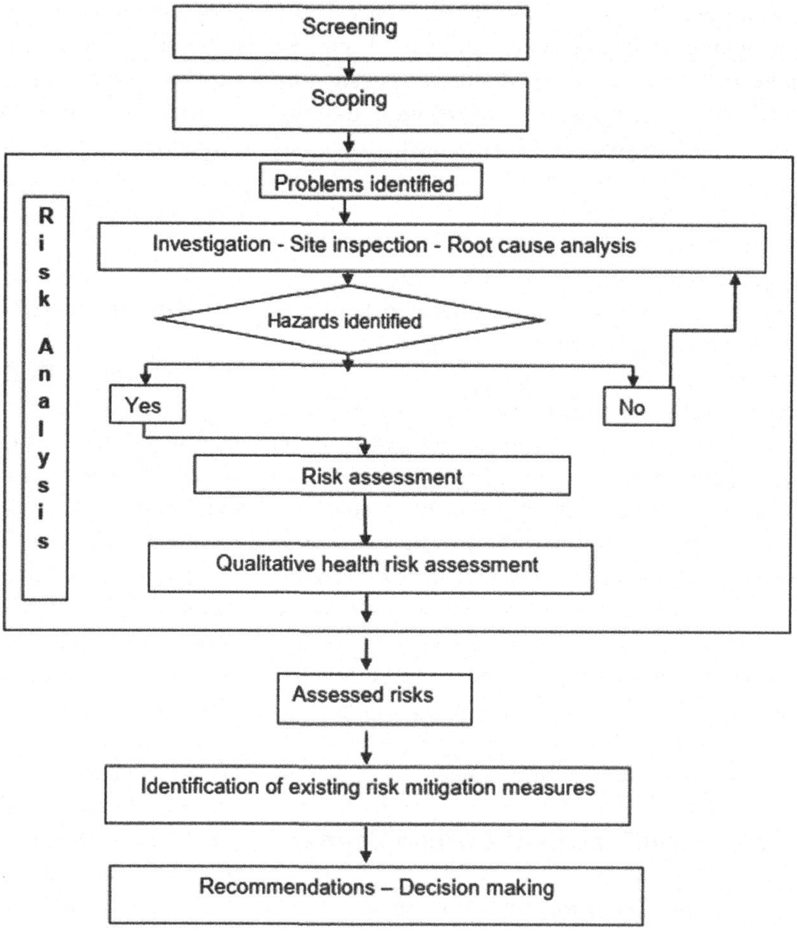

Fig. 3 Process flow: health risk assessment study by Ms. Susmita Banerjee, IWMI, and CDD-BORDA

60 closed-ended and 50 open-ended questions were structured for the event site and 42 closed-ended questions and 18 open-ended questions were asked from control site respondents. Gradient-based questionnaires were adopted for the quantification of impact and nature of risk. The questions included in the survey covered farmers' perception of wastewater irrigation, health issues relating to wastewater usage, final product quality, yield parameters, and existing control measures.

Scoping

This involves problem identification through multi-layered interviews and FGD. The following steps were involved: open-ended questions in the FGD, gradient-based questionnaire, and opinion poll with associated data analysis. Risk assessment, while playing a central role in guidelines for safe wastewater use for agriculture, was covered in this study as health impact assessment only, but not as quantitative risk assessment due to the lack of data.

The gradient-based FGD and opinion poll were based on an environmental management and auditing system, where every stakeholder gives his/her opinion on various issues directly or indirectly related to the identified/probable risk-prone areas. Finally, all the opinion polls from all the respondents were collected to assess the overall conditions or to evaluate a current scenario, which can be a process, practice, or procedure. A total of 27 risk areas were identified and analyzed for overall impact, magnitude, intensity, and occurrence frequency. Once risk areas were identified through FGDs, interviews and surveys, estimation was carried out by quantifying the probability of occurrence based on opinion poll and secondary data. Then, the potential risk areas were shortlisted considering medical waste, pesticides, weedicides, mosquitoes, flies breeding, etc. Based on the obtained results, an investigation was initiated, followed by a root-cause analysis (RCA).

Root Cause Analysis (RCA)

In this study, potential hazards were identified through investigation and probable root causes were further assessed. A comparative account between reference, test, and control was expected to reveal the magnitude of deviation (if any), and further associated risks were estimated. Data verification was conducted by visiting, interviewing, and verifying documents at hospitals, primary health centers, the sewage treatment plant, local bodies, and the revenue office. Some of the root causes identified were poor sanitation, exposure to sewage, infrastructural issues, chemical exposure, etc. Based on these findings and considering existing control measures, mitigation measures were recommended for the safe use of wastewater in agriculture.

2.2.3 Identification of Existing Control Measures and Documentation of Best Practices

Agricultural workers and their families, crop handlers, consumers, and those living in areas irrigated with wastewater are considered as the exposure groups for wastewater (Pescod 1992). These groups are exposed to wastewater during farming practices while selling produce in markets and during consumption. There are some minimal and cost-effective on-farm treatment methods like the low cost drip irrigation systems using suitable local materials that could reduce pathogen load and minimize health risks (WHO nd). The existing control measures to mitigate the risks associated with wastewater use were identified through extensive field inspections, surveys, and interviews with the farmers, consumers, related government officials, and other stakeholders.

The best practices were documented through literature review and through extensive field surveys. From the perspective of the health risk mitigation, for each of the above-mentioned exposure groups and risk factors, practical multicarrier control approaches as relevant and elaborated in the World Health Organization (WHO) Sanitation Safety Plan module were explored (WHO 2016).

3 Results and Discussions

3.1 Opinion Poll Results

During the event site and control site surveys, almost 25 opinion polls were collected (Fig. 4) in the identified 27 risk areas. Here, opinions are represented as positive and negative for the given questions. A positive opinion means the satisfaction level is high and the identified area is in good condition/status. The negative opinion is one, where the satisfaction level is low, and the identified area is in bad condition/status.

The findings of the event site- wastewater (WW) irrigation, survey is as shown in the below Fig. 4. The areas like nutrient load, product yield, wastewater availability and income through wastewater irrigation have received positive opinions. These results are attributed to the fertigation (fertilization and irrigation) nature of wastewater. Wastewater is an inevitable option for farmers for irrigation due to water scarcity in the study area. Also, the rich nutrient content of wastewater will lead to increased yield and thereby increased income to the farmers.

However, the areas like weed menace, sewage sickness, mosquitoes and flies breeding and overall health impact have received negative opinions. This is due to many uncontrollable factors in wastewater like nutrient content. Unlike the chemical fertilizers, where the nutrient application is as per crop requirement, the wastewater nutrient content is unknown and hence, the application will lead to excess or deficit of one or major nutrients. This in turn will lead to the growth of unwanted plants like weeds, long term continuous application will affect the soil properties. The organic matter content of wastewater will give rise to mosquitoes breeding.

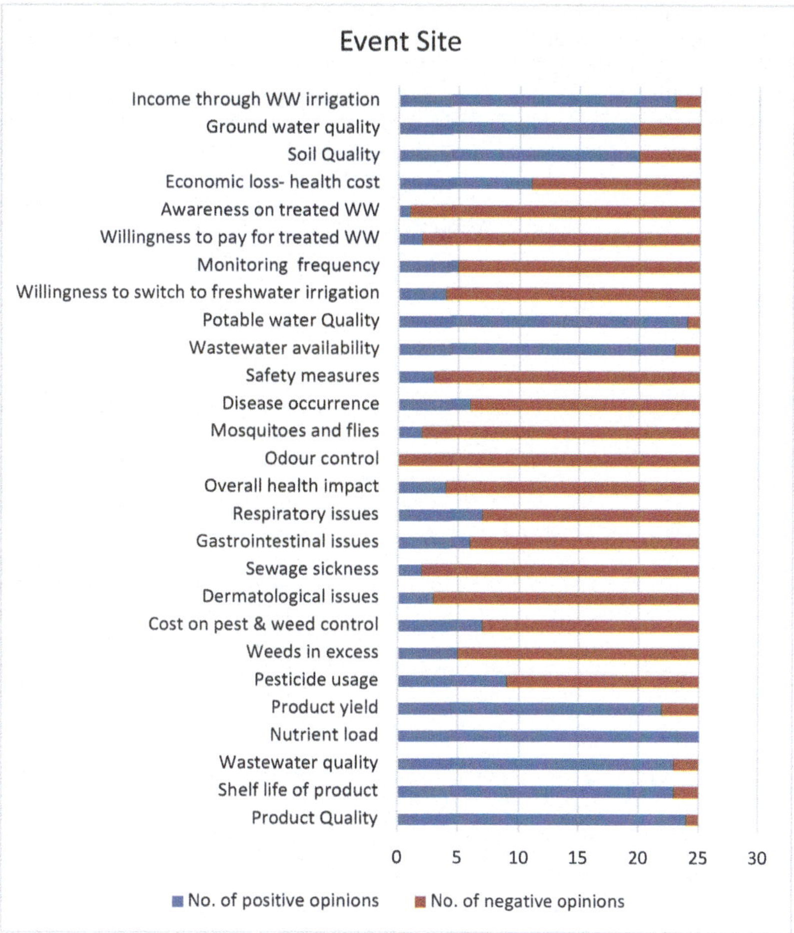

Fig. 4 Graphical representation of opinion poll result at the wastewater-irrigated event site, Govankoppa

In the control site survey (Freshwater irrigation), almost 25 opinion polls were collected and the results obtained are given in Fig. 5. In this study, results obtained from the control site were used to estimate background effects and to identify the potential hazards caused by the current practice of irrigation by untreated wastewater in the event site after the investigation process. The findings of the survey conducted in the control site showed positive opinions for areas like cost on weed control, overall health impact, product quality, etc. However, a few areas like willingness to switch from freshwater to wastewater irrigation, nutrient load and crop yield have received negative opinions. This is because the farmers in these areas were doing freshwater irrigation for a long time and were not willing to shift to wastewater irrigation. Also, there was a reduced cost of inputs (weedicide) to control weed menace unlike conditions in the wastewater irrigated site. But they were also concerned about the low yield of freshwater grown crops compared to wastewater grown crops.

3.2 Field Investigations

Findings from field inspections related to the risks of wastewater usage were as follows:

- Hospital waste (syringes, gloves, PET bottles, tablet foils, and droppers) were found in 12 of the 16 inspected fields, so the probability of such medical waste on the field was 75%, hence the risk is high.

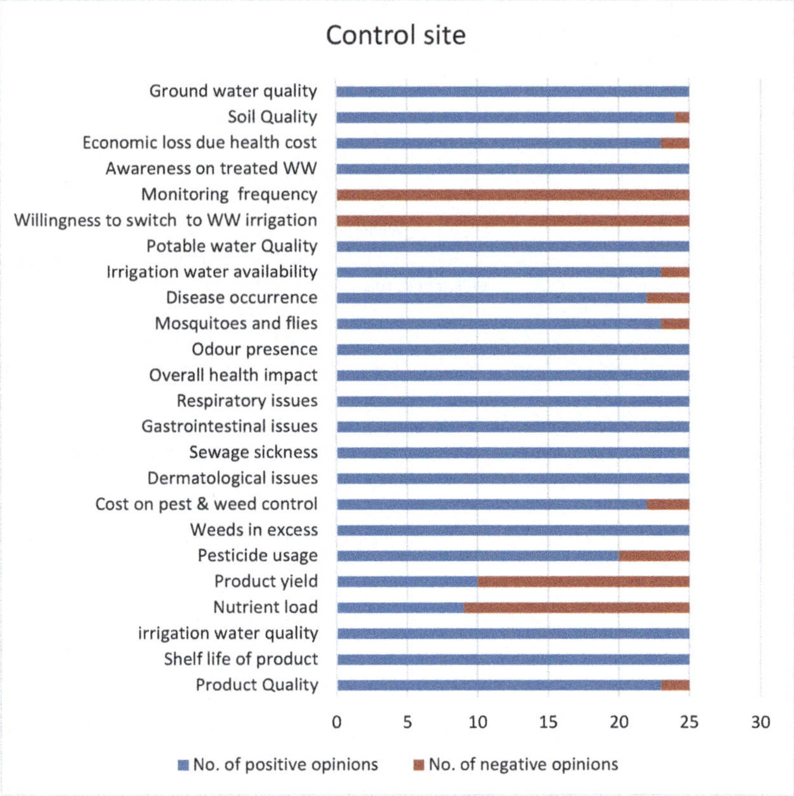

Fig. 5 Graphical representation of the opinion poll result at the freshwater-irrigated control site, Kamalapur

- Continuous bad odour was found and thus chances of contamination through inhalation (causing sickness to farmers) are high.
- 80% of farmers work without any safety measures, which means they were directly exposed to the wastewater.
- Non-judicious application of pesticides by farmers. It was found that there was a high dependence on manufacturers and private agencies for chemical pest controls.
- No preliminary treatment of wastewater, direct usage by pumping onto farmland (Fig. 6).
- The majority of the farmers were using sprinklers for irrigation, which would increase the risks through aerosol generation.

Field inspections related to operational issues at the field level revealed the following:

- Chemical fertilizers applied to wastewater-irrigated fields resulted in the burning of crops (withering and drying) because the high nutrient content of previously added wastewater becomes toxic to plants following the addition of fertilizers. The nutrients might be in excess or deficient in wastewater as the nutrient content of wastewater is unknown.
- Weeds are a major problem in wastewater farming since agricultural runoff with seeds and high nutrient content in wastewater facilitate weed growth. Many farmers were using weedicides to control weeds.
- There was clogging of drip irrigation pipes and other plumbing devices became blocked with algal growth when wastewater was pumped through drip irrigation.
- Wastewater farming promoted the breeding of mosquitoes in some of the fields. This can result in mosquito-borne diseases in addition to water-borne diseases, causing health issues such as diarrhoea, enteric fever, hepatitis, polio, and worm infestations.

Fig. 6 Pumping wastewater directly to the field at Govankoppa, Dharwad district; *picture credit* Ms. Susmita Banerjee

3.3 Root Cause Analysis (RCA)

Investigation results from the RCA are provided in Table 2. During the study, it was found that most of the respondents from the event site had almost zero control or no safety measures, neither in terms of personal protective equipment (e.g., gloves and boots) nor in terms of wastewater treatment techniques such as filtration and sedimentation. The biggest risks identified during field investigation were medical waste followed by pesticide usage and risks implied by mosquitoes and flies. According to the field investigation, the presence of medical waste in the wastewater stream can be considered as highly hazardous and dangerous. Usually, these wastes were handpicked and manually sorted by farmers, imposing even greater risks to farmer health. Along with this, municipal solid waste and weedicide usage were the major risk areas identified during FGD. Awareness about the current environmental impacts, mitigation measures, and existing hazards among users was very low, followed by a lower amount of willingness for treated wastewater irrigation and use. Most of the respondents were found to apply minimal safety measures. Also regarding the adverse effects of different pesticides and weedicides the awareness level was low, especially about which chemicals are the most toxic or carcinogenic in nature. There was a low awareness, in general, of the cumulative effects of pesticide poisoning, which often manifests itself in the gradual failure of the immune system, making it less detectable by health workers and epidemiologists (Satterthwaite 2003). Further study is still in progress to assess the quantitative risk of event sites through data collection and evaluation.

3.4 Field Inspections—Implications of Wastewater Irrigation

The risks associated with wastewater use can be due to microbiological and chemical contamination. The use of untreated wastewater in agriculture has the risk of pathogenic microorganisms such as bacteria, viruses, protozoa, and helminths, which are harmful to humans. These microorganisms can enter the human body through broken skin, direct ingestion, ingestion following inadequate handwashing practices, and inhalation. All excreted pathogens present in wastewater can survive in soil, causing potential risk to farmworkers. "Pathogens survive for a lesser time on the crops, due to sunlight. However, it still can pose risk to crop handlers and consumers" (FAO 1997).

Micropollutants are an emerging threat to the safety of water bodies across the world. These include diverse classes of chemicals, such as pesticides, pharmaceuticals, personal care products, household chemicals, and surfactants (Daughton and Thomas 1999). Some of these are endocrine-disrupting compounds and natural estrogens. The impact of these micropollutants on human health and the environment is not yet well understood (Cardenas Reyes et al. 2016).

Despite the presence of plant nutrients, pathogens, organic pollutants, biodegradable organics, and micropollutants in wastewater,

Table 2 Investigation results of control and event sites through RCA

S. no.	Probable root causes	Investigation results (comparative account of control and event site)	Remarks
01	Unsafe environmental conditions	Temperature, relative humidity, air quality, rainfall, dryness almost same as Kamalapur, not much variability	Cannot be the root cause
02	Chemical exposure	Usually with personal protective equipment, but could be because of excess pesticide/weedicide used without safety measures in place	Could be the root cause
03	Contaminated food	Almost the same food habits and the same conditions	Cannot be the root cause
04	Contaminated potable water	Potable water source is the same and quality is good for both of the sites, none of the respondents uses groundwater/borewell water	Cannot be the root cause
05	Exposure to sewage/wastewater	No such exposure at the control site but direct exposure at the event site	Could be the root cause
06	Infrastructural issues	All respondents live in their own house and their basic needs are fulfilled, most of them own land	Cannot be the root cause
07	Poor sanitation	Toilet available at every house and every member uses it	Cannot be the root cause
08	Poor hygiene practices	Interviewed and found satisfactory	Cannot be the root cause

wastewater continues to be used. The following factors determine the acceptability of wastewater as a sustainable source of irrigation:

- The ever-increasing demand for alternative sources of water for irrigation
- Uncontrolled urban growth increases the demand for potable water supply
- Increase in recognition of wastewater as a valuable source of nutrients
- Treatment of wastewater and adequate precautions while irrigating
- Socio-cultural acceptance of the practice of wastewater farming all over the world.

Wastewater is a reliable and inexpensive source of water supply. At times, the yield of crops might be increased with wastewater irrigation relative to freshwater irrigation, owing to the presence of nutrients such as nitrogen, phosphorus, and potassium. However, the disadvantage is that some of these nutrients might be present in excess or be deficient.

Fertilizers contaminate water bodies because of agricultural runoff, resulting in eutrophication. Eutrophication causes increased algal and weed growth, which poses a challenge to the use of wastewater for agriculture. This causes a serious threat to potable drinking water, fisheries, and recreational water bodies, causing plant die-off as a result of oxygen depletion (Chislock et al. 2013).

As per the field experience of the farmers, wastewater agriculture incurs fewer input costs, such as water and fertilizer, and yields higher income. However, the main operational issue is with high weed infestation. Under the freshwater-irrigated area, farmers grow more cereals and flowers and face less of a weed problem. However, this approach involves more input costs and produces a lower income. When it comes to the quality of the produce, wastewater-irrigated produce is dark in colour and looks fresh, but its shelf life is reduced. In contrast, freshwater-irrigated produce is light in colour and has a longer shelf life. Horticultural products brought to Hubli–Dharwad for sale were not traced according to where they came from and production figures for vegetables irrigated by wastewater were not available as there were no records of production levels within the city.

Farmers were of the opinion that whilst wastewater is available, they will practice farming; when there is no water source, they will sell off the land. Despite the risks associated with unsafe usage of wastewater, the FGD revealed that the benefits gained through wastewater irrigation are low cost of cultivation (as less fertilizer is required than under freshwater irrigation) and higher productivity since wastewater is rich in nutrients. However, the amount of weedicide applied is higher on account of prolific weed growth. The weeds increased because of the fertile nature of the sewage and the increased availability of water in the areas that were traditionally rain-fed. The increase in the number of weeds also had implications in terms of labour required, which adds to either the direct costs of hired labour or opportunity costs of household members. All these need to be further studied to better understand the cost-effectiveness of wastewater agriculture.

3.5 Risk Mitigation Measures

The fundamental global challenges of human development and environmental protection with respect to sanitation and agriculture can be tackled only if they are viewed as a nexus (Hettiarachchi and Ardakanian 2018). The nexus perspective is a driver of sustainable development goal (SDG) #3 (good health and wellbeing), #6 (clean water and sanitation), and #11 (sustainable cities and communities). To make the safe use of treated human waste socially, epidemiologically, environmentally, and economically feasible and acceptable, it is essential to ensure safety to public health and the environment.

3.5.1 Best Farmers' Practices

There are some minimal and cost-effective on-farm treatment methods that will reduce pathogen load and minimize health risks. There is a wide range of treatment options available to treat wastewater. Some of the best farmer practices for treating wastewater through locally available solutions are settling tanks and filtration or screening.

Settling tanks

Settling tanks or wastewater storage tanks are constructed by some farmers on their fields to store wastewater for one to two days before use on fields (Fig. 7). This will help to remove the settle-able suspended solids. This cost-effective partial treatment through sedimentation and retention improves wastewater quality by retaining most of the nutrients.

A sedimentation/settling tank allows suspended particles to settle out of water or wastewater as it flows slowly through the tank, thereby providing some degree of purification. A layer of accumulated solids (sludge), forms at the bottom of the tank and is periodically removed. This settling tank method is common in Maradgi village, which is very near to Govankoppa.

Filtration or screening

About 25% of total respondents were found to be using a simple screen or filter prior to applying the wastewater to the field. This filtration serves two purposes: it prevents debris entering the pump thereby reducing wear and tear, and it prevents the degradation of soil with solid waste (basically plastic, medical waste, and textiles) present in the sewage.

3.5.2 Recommended Safe Use Practices

A multiple-barrier approach is recommended to mitigate the risks of wastewater farming. This refers to the sequential combination of control measures to guide improvements for agricultural use. In combination, all controls should ideally achieve or exceed the target log reductions (WHO 2016).

The key concepts behind the WHO guidelines (WHO nd) are as follows:

- All exposure groups (farmers, crop handlers, and consumers) should be adequately protected
- It may not initially be feasible to meet target log reductions for farmers and consumers in all circumstances. Improvement plans should aim to incrementally improve the situation.

Fig. 7 Settling tank, Mugalur, Karnataka; *picture credit* Dr. B. R. Goud

(1) **Wastewater treatment systems**

Negative health effects can be alleviated by the treatment of wastewater for the effective removal of pathogens (FAO 1997). Sedimentation can be effective in the removal of protozoa and helminths, with pathogen die-away during long-term storage (Lam et al. 2015). Wastewater treatment systems are designed based on a set of treatment principles and used for both domestic and industrial wastewater. The CDD Society is promoting Decentralized Wastewater Treatment Systems (DEWATS) technology and has been successful in applying the technology in India for over 14 years with over 230 successfully implemented projects across the country. DEWATS is based on the treatment methods of sedimentation, activated sludge treatment, and aeration and filtration (Fig. 8).

DEWATS systems are effective, reliable, cost-efficient, and custom-made wastewater treatment systems that are perfectly suited for small to medium-sized systems at the community level and for institutions like schools, hospitals, or enterprises (Small and medium-sized enterprises). However, DEWATS solutions are not intended to replace but rather to complement centralized systems in applicable areas. At the community level, DEWATS can be integrated into a sanitation complex, which is operated on a pay-and-use basis, creating income opportunities for local people (community-based sanitation). The technical options within DEWATS are based on a modular and partly standardized design. DEWATS is based on basic technical treatment processes: mechanical treatment (sedimentation and flotation) and biological (anaerobic and aerobic) treatment. These systems can be designed based on needs and the most common DEWATS modules are: septic tanks, biogas digesters, anaerobic baffled reactors, anaerobic filters, planted gravel filters, and (if needed) polishing ponds.

DEWATS treatment methodology provides a treatment efficiency that allows for the use of the

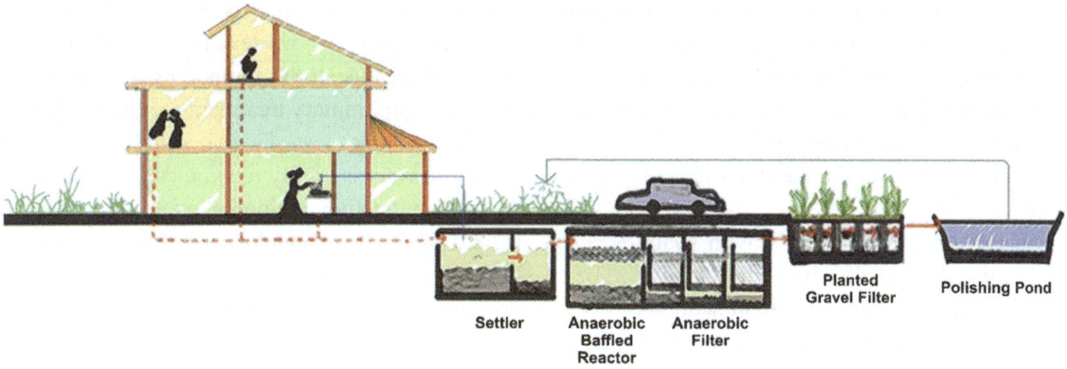

Fig. 8 DEWATS consisting of settler, anaerobic baffled reactors, anaerobic filters, planted gravel filters, and polishing pond; *source* modified after CDD Society

treated wastewater for gardening or irrigation as well as for safe disposal into a waterbody or natural drain. However, if further treatment is required to meet other water quality standards, then additional treatment units will be required. Due to the lack of appropriate disposal, the main purpose of the treatment unit is the safe disposal of wastewater and to improve deteriorating environmental and hygiene conditions.

(2) **Appropriate irrigation practices**
In the case of wastewater-irrigated villages, most of the farmers use overland irrigation (flood irrigation) and ridge and furrow irrigation, but farmers with larger farmlands mostly use sprinkler systems. However, it is recommended to follow proper irrigation practices to reduce pathogen load as follows:

- Localized irrigation techniques such as furrow or drip irrigation can protect farmers by reducing pathogen load
- Pre-harvest irrigation control such as cessation of irrigation before harvest
- UV (ultraviolet) exposure—Storing water in polishing ponds/lagoon lakes under sunlight for some period
- Dilution such as mixing of raw wastewater with freshwater can serve as a means to reduce pathogen load. Large dilution rates will, however, be required to eliminate pathogen contamination.
- Pathogen die-off before consumption can be achieved by keeping an interval between final irrigation and consumption.

(3) **Crop restrictions**
Certain crop restrictions are to be followed when using treated wastewater to irrigate food crops. The Sanitation Safety Planning by WHO (2016) recommends using wastewater to irrigate food crops where the fruit or leaves are above the ground and not in direct contact with the soil (e.g., root crops and leafy vegetables such as radish, carrot, and spinach). The latter crops and food crops that are eaten raw, such as salad leaves, must not be irrigated with wastewater as these are directly in contact with soil and wastewater. The recommended crops are non-food crops, such as cotton and flowers; food crops that have to be cooked before consumption; tall-growing crops, such as beans; and crops that are processed or canned before consumption, such as peas (WHO 2016).

(4) **Safety measures for crop handlers**
Hookworm and *Ascaris* infestation, enteric fever, cholera, *Helicobacter pylori* infection, and diarrheal diseases are some diseases associated with wastewater farming in the exposure groups (Fattal et al. 1986; Fuhrimann et al. 2016). Recommended measures to protect crop handlers are as follows (see also Alberta Government 2012):

- Use of personal protective equipment including masks for respiratory protection, goggles for eye protection, gloves for hand protection, boots for foot protection, and long-sleeved clothes for skin protection
- Cleaning of contaminated tools and clothes on-site
- Washing hands frequently with soap
- Keeping fingernails short and clean
- Cover and regularly disinfect wounds
- Avoid direct contact with wastewater
- Using a separate set of clothing for working with wastewater
- Safety instructions to labourers by farmers

(5) **Safety measures for consumers**
Pathogens survive for a low time on crops exposed to sunlight. However, they can still pose a risk to crop handlers and consumers (FAO 1997). Consumers who consume uncooked or inadequately cooked produce are also at risk of microbial infections. Hygienic food preparation measures are washing, cooking, and peeling of food produce.

A range of options to reduce the risk levels after prioritizing the potential risks to the exposed group such as operational methods (e.g., crop restrictions, longer retention time, and vector control) and behavioral methods (e.g., use of personal protective equipment, health education, and regular health check-ups) were chosen by farmers as measures for improvement. The factors should be considered in the selection of an appropriate safety measure: potential for improving the existing control, cost-effectiveness, technical feasibility, acceptability, and reliability of the measure in relation to the local cultural and behavioral habits.

4 Conclusions and Outlook

Direct usage of raw wastewater leads to health risks for farmers, crop handlers, and livestock. The risks can be minimized if treated wastewater is used for irrigation instead of raw sewage. In line with the recommendation that WHO guidelines need to be adapted to the specific conditions at the respective site of wastewater irrigation (Carr 2005). From this study, it emerges that the use of filters as a preliminary treatment measure and other locally available cost-effective solutions for treating wastewater is highly recommended. The use of personal protective equipment as a safety measure should be made mandatory for farmers working with wastewater. A good understanding of irrigation methods among farmers will lower health risks and ensure product quality and safety. Feasible control measures for mitigating risks associated with wastewater farming observed in the study area were the storage of irrigation water prior to application on fields, screening, filtration, and stopping irrigation (allowing pathogens to die off) before harvesting. Some measures for post-harvested produce include washing of produce in freshwater before selling at the market and ensuring hygienic cooking practices. There is a strong need for awareness, workshops, and training for farmers and other stakeholders regarding the adverse effects of untreated wastewater use for irrigation on the one hand and safe practices of wastewater use on the other hand.

There is a need to carry out many more epidemiological studies to better understand the health implications of the wastewater irrigation practice. Studies need to be conducted keeping in mind the role of medical waste and its impact on wastewater quality and more scientific data should be gathered focusing on the microbiological, chemical, and toxicological parameters of wastewater. A health impact assessment study on consumer health after consumption of wastewater-irrigated crops is also needed. Wastewater use, when adopted with best management practices, has huge potential to benefit farmers, improve public health by eliminating unsafe waste disposal, and enrich soil nutrient content. It is recommended to engage all stakeholders, from policymakers to sanitation experts, agriculture experts, health departments, and civil society organizations, to provide a culture of better understanding of wastewater use in a participatory approach, which had been successfully applied in the area before (Halkatti et al. 2016).

The strategy of nutrient and resource recovery has to be designed and implemented for a local area, based on local options and needs. The role of excreta and wastewater treatment infrastructure should be linked with city development and a food security agenda by urban planners and policymakers (Koné 2010). The use of wastewater in agriculture is an option that is increasingly being investigated and implemented since it enables freshwater to be exchanged for nutrient-rich wastewater (Hettiarachchi and Ardakanian 2016). This practice of agricultural use of wastewater after necessary treatment has many advantages such as reduced demand for freshwater, decreased pollution of surface and groundwater, increased soil fertility, reduced investment in chemical fertilizer, and efficient management of high volumes of urban wastewater. The safe use of wastewater for food production addresses the sanitation and agricultural issues faced by peri-urban areas while addressing also the malnutrition issue (Dasgupta et al. 2014). This explains the role of resource recovery and reuse through the sanitation–agriculture–public health nexus when building urban resilience.

Acknowledgements The first author (Girija R) would like to thank the CDD Society and Bremen Overseas Research and Development Association (BORDA) and International Water Management Institute (IWMI) for providing the opportunity to present the project findings. The authors would also like to thank the management and project team members for their valuable inputs and suggestions on this paper. Krishi Vignyana Kendra, Dharwad, provided land holdings details of Kamalapur village, Dharwad, Karnataka, India.

References

Alberta Government (2012) Occupational health and safety bulletin: workers exposure to sewage. Alberta Health Services

BORDA, CCD Society (2017) Scaling-up decentralized wastewater treatment. https://cdn.nimbu.io/s/gcj6927/channelentries/w3p3h2f/files/2.%20Edathoot%20BORDA%20GSTIC%20scalingup%20DEWT.pdf?eozwff7. Accessed 08 Jun 2020

Bradford A, Brook R, Hunshal CS (2003) Wastewater irrigation in Hubli-Dharwad, India: implications for health and livelihoods. Environ Urban 15:157–170. https://doi.org/10.1630/095624703101286600

Carr R (2005) Who guidelines for safe wastewater use—more than just numbers. Irrig Drain 54:S103–S111. https://doi.org/10.1002/ird.190

Cardenas Reyes MA, Ali I, Lai FY, Da L, Thier R, Rajapakse J (2016) Removal of micropollutants through a biological wastewater treatment plant in a subtropical climate, Queensland-Australia. J Environ Health Sci Eng 14(1). https://doi.org/10.1186/s40201-016-0257-8

Chislock MF, Doster E, Zitomer RA, Wilson AE (2013) Eutrophication: causes, consequences, and controls in aquatic ecosystems. Nat Educ Knowl 4(4):10

Dasgupta R, Sinha D, Yumnam V (2014) Programmatic response to malnutrition in India: room for more than one elephant? Indian Pediatr 51:863–868. https://doi.org/10.1007/s13312-014-0518-5

Daughton CG, Thomas AT (1999) Pharmaceuticals and personal care products in the environment: agents of subtle change? Environ Health Perspect 107:907. https://doi.org/10.2307/3434573

Drechsel P, Mara DD, Bartone C, Scheierling SM (2010) Improving wastewater use in agriculture: an emerging priority. The World Bank. https://elibrary.worldbank.org/doi/abs/10.1596/1813-9450-5412

Ensink JHJ, Blumenthal UJ, Brooker S (2008) Wastewater quality and the risk of intestinal nematode infection in sewage farming families in Hyderabad, India. Am J Trop Med Hyg 79(4):561–567. https://doi.org/10.4269/ajtmh.2008.79.561

FAO (1997) Chapter 2—Health risks associated with wastewater use. Rome. https://www.fao.org/docrep/w5367e/w5367e04.htm

Fattal B, Wax Y, Davies M, Shuval HI (1986) Health risks associated with wastewater irrigation: an epidemiological study. Am J Public Health 76:977–979. https://doi.org/10.2105/AJPH.76.8.977

Fuhrimann S, Winkler MS, Kabatereine NB, Tukahebwa EM, Halage AA, Rutebemberwa E, Medlicott K, Schindler C, Utzinger J, Cissé G (2016) Risk of intestinal parasitic infections in people with different exposures to wastewater and fecal sludge in Kampala, Uganda: a cross-sectional study. PLOS Negl Trop Dis 10(3):e0004469. https://doi.org/10.1371/journal.pntd.0004469

Halkatti M, Purushothaman S, Brook R (2016) Participatory action planning in the peri-urban interface: the twin city experience, Hubli–Dharwad, India: environment and urbanization. https://doi.org/10.1177/095624780301500112

HDMC Home | Hubballi-Dharwad City Corporation. https://www.hdmc.mrc.gov.in/en/home. Accessed 08 Jun 2020

Hettiarachchi H, Ardakanian R (eds) (2016) Safe use of wastewater in agriculture: good practice examples. UNU-FLORES, Dresden. https://collections.unu.edu/view/UNU:5764

Hettiarachchi H, Ardakanian R (eds) (2018) Safe use of wastewater in agriculture: from concept to implementation. Springer International Publishing

Hunshal C, Salakinkop S, Brook R (1997). Sewage irrigated vegetable production systems around Hubli-Dharwad, Karnataka, India. Kasetsart J (Nat Sci) 32(5):1–8

Jampani M, Hülsmann S, Liedl R, Sonkamble S, Ahmed S, Amerasinghe P (2018) Spatio-temporal distribution and chemical characterization of groundwater quality of a wastewater irrigated system: a case study. Sci Total Environ 636:1089–1098. https://doi.org/10.1016/j.scitotenv.2018.04.347

Jiménez B, Navarro I (2013) Wastewater use in agriculture: public health considerations. In: Encyclopedia of Water Science, 2nd edn. CRC Press, Boca Raton, p 1586. https://doi.org/10.1081/E-EEM-120046689

Koné D (2010) Making urban excreta and wastewater management contribute to cities' economic development: a paradigm shift. Water Policy 12:602–610. https://doi.org/10.2166/wp.2010.122

Lam S, Nguyen-Viet H, Tuyet-Hanh T, Nguyen-Mai H, Harper S (2015) Evidence for public health risks of wastewater and excreta management practices in Southeast Asia: a scoping review. Int J Environ Res Public Health 12(10):12863–12885. https://doi.org/10.3390/ijerph121012863

Pescod M (1992) Wastewater quality guidelines for agricultural use. In: Wastewater treatment and its use in agriculture—FAO irrigation and drainage paper. Chapter 2. FAO, Rome. https://www.fao.org/docrep/T0551E/t0551e04.htm#2.%20wastewater%20quality%20guidelines%20for%20agricultural%20use

Satterthwaite D (2003) Water and sanitation. In: Environment and urbanization, vol 15, p 168

Sindhindia (nd) Dharwad District, Aug 2018. Sindhindia.com. https://www.sindhindia.com/district.php?params=karnataka/dharwad/

Strauss M (1991) Human waste use: health protection practices and scheme monitoring. Water Sci Technol 24(9):67–79. https://doi.org/10.2166/wst.1991.0236

Villageinfo (nd) Dharwad village in Dharwad (Dharwad) Karnataka, Aug 2018. Villageinfo.in. https://villageinfo.in/karnataka/dharwad/dharwad/dharwad.html

WHO (nd) Guidelines for the safe use of wastewater, excreta and greywater—volume 4. In: WHO. https://www.who.int/water_sanitation_health/publications/gsuweg4/en/. Accessed 08 Jun 2020

WHO (2016) Sanitation safety planning. Manual for safe use and disposal of wastewater, grey water and excreta. WHO. ISBN 978 92 4 154924 0

Usefulness of Surface Water Retention Reservoirs Inspired by 'Permaculture Design': A Case Study in Southern Spain Using Bucket Modelling

Immo Fiebrig and Marco Van De Wiel

Abstract

Water harvesting (WH) techniques have experienced a renaissance within a grassroots sustainability concept and movement called 'permaculture'. Over the past decade, there has been a growing interest in the uptake of permaculture inspired solutions designed to restore the water cycle at a landscape level and to facilitate the delivery of ecosystem services as part of a holistic farming approach. In this study, we assessed four reservoirs built on a fruit farm in southern Spain in terms of usefulness. A simple hydrological model was developed utilising on-site data and two free calibration parameters, infiltration and evapotranspiration. The model matches the observations on the ground well, but indicates limited potential of WH within the boundaries of this particular farm. As the decade preceding the reservoir's inception received more rain (2000–2010), this may have led to a misjudgement for on-farm water harvesting potential in the planning phase. We conclude that the design of WH reservoirs or water landscapes as a contribution to reversing desertification processes and mitigating climate change would benefit from long-term studies on the ground. Moreover, modelling-based scenario analyses can help better understand the dynamics and extent of its potential in establishing an effective and economically viable land restoration process for the region, taking into account climate change projections of increasing desertification in the Mediterranean basin, thereby contributing to the water-soil-food nexus, addressing sustainable development goals (SDGs) related to food (2), water (6), responsible consumption and production (12), climate action (13) and life on land (15).

Keywords

Water harvesting · Water-soil nexus · Scenario analysis · Land restoration · Agroecology

1 Introduction

Land degradation is resulting in depletion of natural resources such as degradation of soil, reduction in vegetation and damage to water cycles on a global scale, thus putting world food

I. Fiebrig (✉)
NCMH, School of Biosciences, University of Nottingham, Sutton Bonington Campus, Loughborough L12 5RD, UK
e-mail: immo.fiebrig@nottingham.ac.uk

M. Van De Wiel
Centre of Agroecology, Water and Resilience, Coventry University, Wolston Lane, Garden Organic, Ryton-on-Dunsmore CV8 3LG, UK
e-mail: marco.vandewiel@coventry.ac.uk

S. Hülsmann and M. Jampani (eds.), *A Nexus Approach for Sustainable Development*,
https://doi.org/10.1007/978-3-030-57530-4_5

security at risk (McIntyre et al. 2009). This is largely due to human activity and significantly attributed to unsustainable farming practices, and in turn increases the impact of climate change by compromising the ability of the environment to mitigate climate disruption—with a rise in critical floods as well as droughts (Stringer 2008; Montanarella 2007; FAO 2011). Proposed strategies to improve unsustainable agricultural practices are intended to mitigate detrimental effects on the environment whilst improving outcomes in productivity, socio-economics, nutrition and health beyond current certified organic or biodynamic farming (IPES Food 2016) and in alignment with the sustainability concept People-Planet-Profits (United Nations 1987; Slaper and Hall 2011) and sustainable development goals (SDGs, United Nations 2015). Given the magnitude of worldwide land degradation (MEA 2005), the agricultural practice should not only change from 'unsustainable' systems to 'sustainable' systems that are 'able to last' or 'have the capacity to endure', but it should reach further towards 'restorative'[1] or 'regenerative'[2] processes that can be viewed as more in tune with natural cycling processes (Chabay et al. 2016; Rhodes 2015; Shepard 2013; Lyle 1994) and in line with SDGs, in particular SDG 15, life on land, supporting also SDG 2 (zero hunger) and SDG 6 (clean water and sanitation) (UN 2015).

1.1 Permaculture and Agroecology

Regenerative processes are supported by a concept known as permaculture. 'Permaculture' is a portmanteau word created from 'permanent' (for 'lasting') and 'agriculture' or 'culture'. It was coined in the late 1970s with the publication of the book *Permaculture One* by Bill Mollison and David Holmgren, advocating "…for an integrated, evolving system of perennial or self-perpetuating plant and animal species useful to man" (Mollison and Holmgren 1978). Permaculture is a holistic approach to designing farming systems and human settlement, starting with three ethical principles often referred to as Earth Care, People Care and Fair Shares (setting limits to growth and redistributing surplus). It focuses on a holistic agroecosystem design process that mimics nature (Mollison 1988). The framework has been playing a key role in grassroots engagement around social movements in transition to sustainability (Feola and Nunes 2014; Santiago et al. 2017). Grassroots networks of such social movements can be seen as agents of change who act against environmental degradation by implementing innovative solutions to pressing environmental problems and by influencing institutions to speed up this process (Ferguson and Lovell 2015). Such bottom-up initiatives may be important allies in the combined effort to address land degradation—in collaboration with governmental or intergovernmental approaches, thus making change processes more participatory and as such more regenerative.

In terms of scope, permaculture has developed to become possibly one of the most comprehensive sustainability frameworks devised so far and avoids recommending or imposing specific rules or practices, thus facilitating universal applicability irrespective of local conditions such as climate, geology or culture. Permaculture embraces many themes as important elements in 'the move beyond sustainability', within systems thinking approach as defined by the co-originator of the permaculture concept (Holmgren 2011)—for instance, land and nature stewardship, the built environment, tools and technology, education and culture, health and spiritual well-being, finances and economics or land tenure and community governance. However, permaculture is not a scientific discipline and existing literature is mostly written by non-scientists for a popular audience. It has only recently entered academic discourse as one of the various movements related to agroecology (Ferguson and Lovell 2013). The United Nations Convention to Combat Desertification (UNCCD) also lists

[1]'restore' (v.) from Latin *restaurare* "repair, rebuild, renew". (Harper nd).

[2]'regenerate' (v.) from Latin *regenerare* "bring into existence again" (ibid.).

'permaculture' as an *activity to prevent* land degradation within their framework to achieve Land Degradation Neutrality (LDN). LDN is defined as "a state whereby the amount and quality of land resources necessary to support ecosystem functions and services and enhance food security remain stable or increase within specified temporal and spatial scales and ecosystems". The framework clearly ascribes ecosystem services a value representing "stocks of natural capital associated with land resources" (Orr et al. 2017). The European Commission goes a step further and has recognized 'permaculture' as one of the several *prototypes of sustainable agro-ecology systems* within their Horizon 2020 research work programme (2016–2017) and a specific research challenge on productivity gains through functional biodiversity, linked to sustained delivery of natural habitats (EU Horizon 2020 2016). The Permaculture Worldwide Network (PWN nd) lists more than 2000 projects distributed across the globe. This internet-based network is run by the Australian Permaculture Research Institute (PRI), probably the most prominent non-academic permaculture teaching institution. A wider scientific analysis as to the social, environmental, economic or other favourable impacts of such projects is being established, both within the social as well as the natural sciences and across disciplines. The current study forms part of this research endeavour and puts permaculture as an integrative approach into perspective with a resources-focused nexus management approach (Hettiarachchi and Ardakanian 2016; Bleischwitz et al. 2018).

Agroecology, in turn, is an essential part of many permaculture practices. The term 'agroecology' was first used as an adjective—'agroecological'—by Bensin (1928, 1930) within scientific research. Today, the discipline represents the paradigm shift in sustainable agriculture, which promotes restoration processes at the level of soil, plot, farm and landscape (Gliessman 2015). Agroecology can be viewed as (1) a farming practice, (2) a socio-political movement and (3) a scientific discipline (Wezel et al. 2009). Permaculture, in turn, drawing on agroecological practices, can be considered (1) a design system, (2) a best practices framework, (3) a worldview and (4) a bottom-up movement (Ferguson and Lovell 2015).

1.2 Rainwater Harvesting

Within the agroecological practices of permaculture, water as a resource and on-site rainwater management plays an important role in relation to Holmgren's permaculture principle (2) 'catch and store energy' (2011) which refers to the closure of resource loops in the widest sense and on the farmland in particular. According to Mollison (1988), managing water flows on farmland, i.e. predominantly from rainfall, creates "long-term water and wildlife reserves in the total landscape" to improve "the global water cycle". Permaculture's recommended water management practices fall under the term of water harvesting (WH), defined by Critchley and Siegert (1991) as 'collection of run-off for its productive use', making deliberate use of surface run-off (Oweis and Hachum 2009) and shifting the timing of water availability (Richter 2014). They are 'low-external-input-techniques' or low impact nature-based solutions, in accordance with permaculture design principles such as 'use and value renewable resources and services' as well as 'use small and slow solutions'; building sustainably with natural, renewable materials that can also be considered aesthetically beautiful is preferred (Holmgren 2011) over large-scale concrete hydro dams primarily built for energy production or reservoirs for agricultural irrigation purposes, both of which have far-reaching environmental impacts (Ringler et al. nd; Keskinen et al. 2012; Biemans et al. 2011).

WH designs may include techniques that improve water infiltration into the soil, e.g. pitting, terracing or keyline channelling. Swales, *demi lunes* or semi-circular bunds can concentrate water near trees and other vegetation or crops, where it supports productivity. Such measures fall into the category of 'micro-catchment water harvesting' (Oweis and Hachum 2009). Macro-catchment WH techniques include the creation of reservoirs like lakes and ponds or 'dam storages' (ibid.) built from natural materials like clay or loam (Studer and Liniger 2013). Such macro-catchment measures tend to require heavy machinery like excavators and bulldozers and they involve considerably more planning and implementation costs, but they can also provide ecosystem services of a larger scale.

1.3 Water Harvesting by Building Earth Dams

Mountain farmer and permaculturist Sepp Holzer advocates for the creation of 'water landscapes' through a series of interconnected ponds and lakes as reservoirs, which he calls 'water retention spaces' designed to catch rainwater surface run-off from the actual farmland or adjacent land (Holzer nd). Water retention in the land as such is nothing innovative. As described by the USIDBR (2012), earth dams are said to have existed since the early days of civilisation with, e.g., an eleven miles long earth filled dam dating back to 504 B.C. in Sri Lanka. Later, around 250–900 A.D. (Maya Classic Period) the ancient Maya built larger scale reservoirs and dams as an adaptation strategy to changes in climate, whilst connecting water with ritual, ideology and control of power (Wyatt 2014). In the ancient city of Tikal in Guatemala, Mayan rulers monopolised reservoirs with a combined storage capacity of 100,000–250,000 m^3 within a catchment potential of 900,000 m^3, based on 1,500 mm of annual rainfall (Scarborough 1998).

With regard to current climate adaptation strategies, Ferrand and Cecunjanin (2014) state in their extensive review on ancient and traditional rainwater harvesting applications: "today, rainwater harvesting systems are mostly abandoned as a result of centralization of water resources".

The primary benefit of rainwater harvesting in agriculture is to prevent soil erosion and to keep the resource—water—under the farmer's control for subsequent benefits (Holzer 2008) such as e.g. on-farm microclimate regulation (temperature, moisture), habitat for waterfowl breeding and aquaculture, livestock drinking water, supplementary irrigation, wildlife habitat, recreation (swimming, bathing) and groundwater recharge (Power 2010; Biggs et al. 2016). Holzer gained empirical knowledge about this kind of holistic farm management on his 50 ha mountain farm Krameterhof in the Austrian Lungau (Salzburger Land) with more than 60 ponds within the top half of the land and pond sizes ranging from 10–80 m in length and 4–40 m in width (Holzer 2008). Köppen-Geiger climate classification for this region is Dfb (i.e. humid continental, with mild summer and long cold winter). Like many other gardeners and farmers who advocate for the principles of permaculture, Holzer acts as an independent consultant to land owners and farmers alike and advises on land restoration and management issues in other climatic zones (www.holzerpermaculture.us). Other experienced grassroots consultants specialise in teaching, training, planning and implementing large-scale earthworks to create water landscapes through swales, ponds and lakes. For example, Geoff Lawton is head of the PRI and known in the media for his project "Greening the Desert" in Jordan (Al Jazeera 2011), while Darren Doherty is known for his keyline water systems (TreeYo nd). Such consultants tend to capitalise on their empirical knowledge from many years of project experience. At the same time, the creation of water landscapes through ponds and lakes as a grassroots service provision has been emerging only recently.

1.4 Case Study: Commercial 'Permaculture' Farm

The current case study focuses on an EU organic and biodynamic (Demeter) certified fruit farm in southern Spain. According to Köppen-Geiger climate classification, the farm lies within the region of Csa (i.e. temperate Mediterranean, with dry and hot summer). In 2010, the farm owner decided to invest in diversifying his production system because he wanted to see productivity gains through a series of measures recommended by permaculture consultants and related popular permaculture literature (Holzer 2010, 2011) as well as recommended practices from regenerative agriculture (popular literature, Gras 2012). A 'permaculture design' including a series of water reservoirs (Holzer 2011, p. 43; design, p. 56; detailed section, p. 57) was created by an architect and was subsequently implemented with modifications. The intention was to reduce stormwater soil erosion as well as to store harvested rainwater for groundwater recharge and to create wildlife habitats. Bearing in mind the context of continuing controversies over large dams in particular (HLPE9 2015; Roy et al. 2011), the hope was to become more sustainable in terms of independency from costly water supplies piped from large artificial reservoirs in Spain, e.g. the nearby Andévalo reservoir.

A series of four reservoirs were built (Fig. 1). This was achieved either by simply digging a pit with an excavator (reservoirs 1B, 2 and 3; Fig. 1) and by combined pit digging and damming with an earthen dike, taking into account existing contour lines to keep water flow, geomorphology and visual appearance as natural as possible (reservoir 1A, Fig. 1). Typically, damming is performed at a narrow section of a valley by first excavating a core trench along the width of the valley. The core trench cuts into the hills on both sides and several meters below the future ground of the reservoir to become an aquifuge. As an impermeable barrier, the aquifuge should minimise seepage whilst impounding run-off water from upstream catchments. According to Holzer (2011), this 'impermeable core' should ideally be connected to the existing natural layer of clay or loam. The dam is subsequently built upwards by filling the trench with the same locally sourced material (clay/loam) which is compacted step-by-step either by a bulldozer or any compacting machine (e.g., trench roller, drum roller or vibratory rammer depending on the scale of the building site). Building material must be neither too dry nor too wet to ensure cohesiveness, plasticity and compressive resistance of the soil material (Röhlen and Ziegert 2014). Embankments on both sides of the dam may use soil comprising 20–30% of clay according to Studer and Liniger (2013) or any other soil material but not humus according to Holzer (2011); dam slopes should be no steeper than 1:2 (ibid., 69), spillways for stormwater run-off must be installed on natural ground away from the dam (ibid., 70) and the dam itself should be covered with a high diversity of shallow rooting plants to prevent wind and water erosion of the structure; deep rooting plants must be prevented from growing their roots into the aquifuge to avoid seepage. Trees like willows and alders may support the structure if planted at the downstream bottom of the dam, as described for a variety of pond and lake constructions in Europe (Holzer 2011). Reservoirs 1A and 1B are interconnected and allow any overflow from 1A to be stored in 1B, although interconnection disrupts the pre-reservoir natural drainage to the ephemeral south of reservoir 1A (Fig. 1).

1.5 Specific Challenge and Research Questions

Following the completion of reservoir 1A by the end of 2010 and remaining reservoirs 1B, 2 and 3 by the end of 2013, all reservoirs do catch water after rainfall events or rainfall on already saturated soil, but so far the reservoirs have only filled up to a fraction of their total capacity. Reservoirs 1B and 3 have failed to retain water beyond a few months and never retained water throughout the year. Although reservoirs 1A and 2 are constantly supplied with some water, this is only partly due to rainfall and mainly related to irrigation system maintenance. The required filters against particulate matter need to be flushed regularly by reversing flow direction. This rinsing water is then pumped into the respective reservoir for disposal several times a day (farm manager personal communication). The present research is aimed at indicating the root causes of potential performance limits of these reservoirs, in view of a considerable financial investment for land use conversion and reservoir construction.

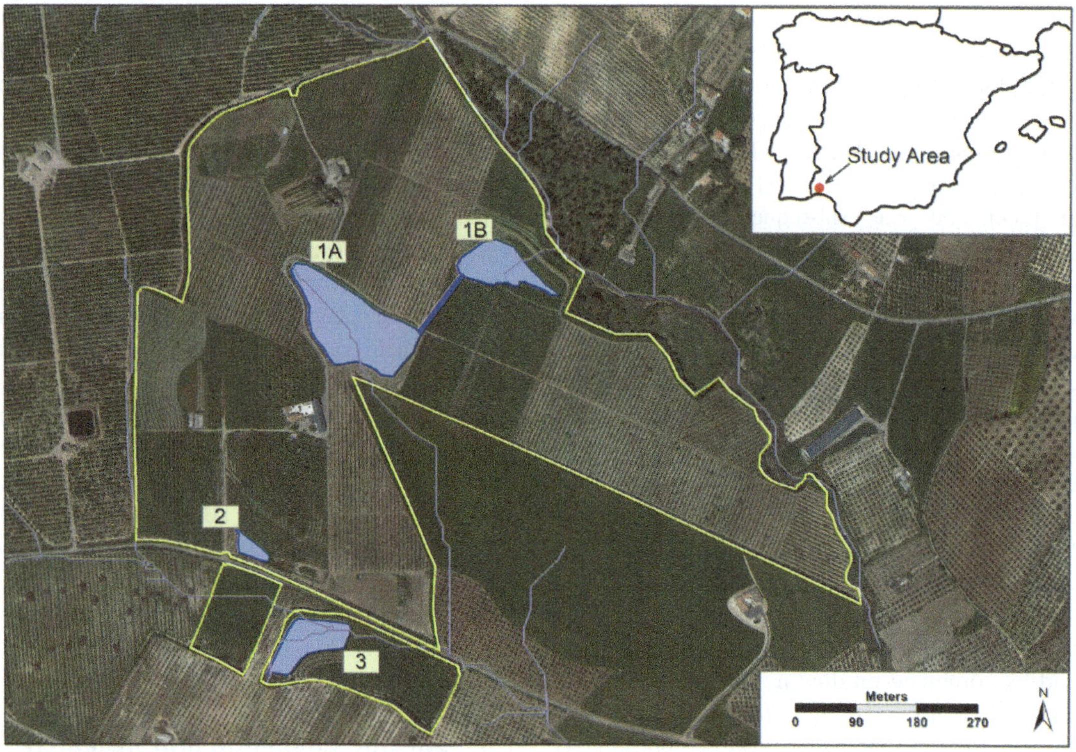

Fig. 1 Study area, with land use pattern and reservoirs 1A, 1B, 2 and 3. An interconnection allows any overflow from reservoir 1A to run into reservoir 1B (thick blue line, schematic). Yellow line marks farm boundary. Thin blue lines indicate pre-reservoir ephemeral channel drainage network. Inset shows location of the study area in Spain. (*Background image source* Google Earth, 37°23′08.5″N 7°01′19.7″W, April 2013)

Since reservoir inception, neither discharge nor water levels have been systematically monitored. A modelling approach is used for a retrospective performance and sustainability assessment of these four reservoirs.

2 Materials and Methods

2.1 Model Description

Hydrological models can vary in complexity, depending on the number of processes represented and the purpose of the modelling. Complex models are needed to understand complex feedbacks and interactions among different processes, but typically require many input data and extensive parameterization. Much simpler 'bucket' models only represent the most important processes (Zhang et al. 2002). They are water balance models that conceptualize the system of interest as a 'bucket', being filled up by rainfall and emptied by evapotranspiration and infiltration. When the bucket is full, extra water is assumed to overflow. Although their simplicity restricts their versatility and generalizability, bucket models can provide useful insights into the core behaviour of a system (ibid.).

In this study, a simple lumped 'bucket' model calculates reservoir water volumes as a function of inputs and losses to the reservoir (Fig. 2). Inputs are due to surface runoff from precipitation over a reservoir's catchment, losses are due to infiltration, evaporation, extraction and overflow. Subsurface hydrology, vegetation hydrology and spatial heterogeneity are not accounted for in this simple model.

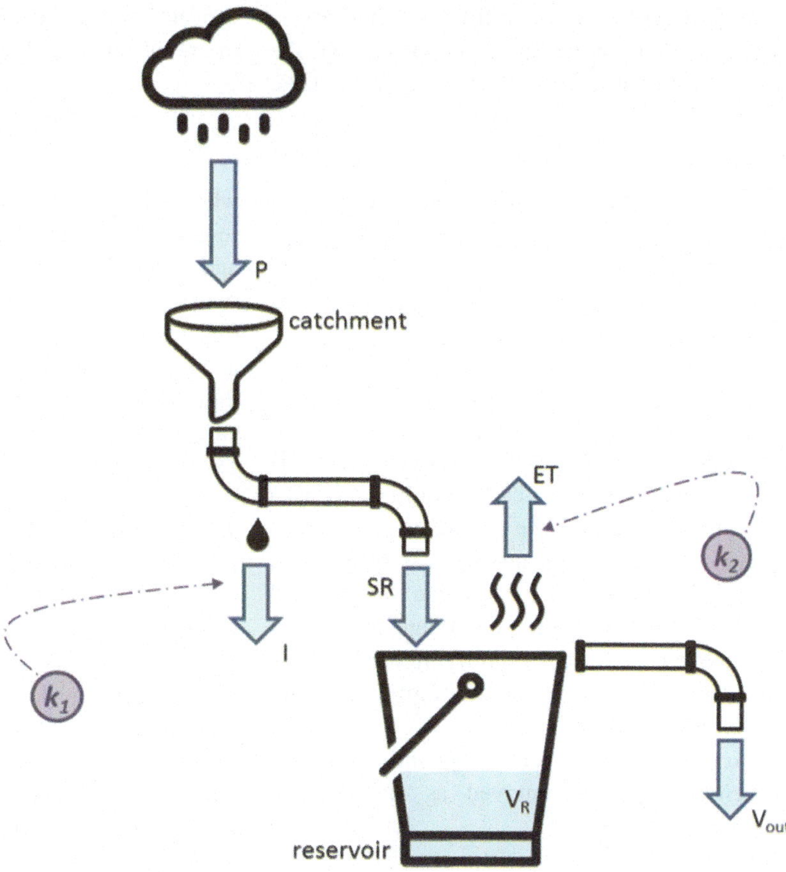

Fig. 2 Schematic representation of the 'bucket' hydrology model. The key processes affecting the reservoir's water volume, V_R, are indicated by P = precipitation, I = infiltration, ET = evapotranspiration, SR = surface runoff, and V_{out} = reservoir overflow. The circles, k_1 and k_2, indicate two main parameters in the model, affecting I and ET respectively

The water balance for a reservoir's water volume, V_R, over an interval Δt is given by:

$$V_{R,t+\Delta t} = V_{R,t} + \Delta V_R$$

with

$$\Delta V_R = V_{sr} - V_e - V_{out} + V_{in} - V_w$$

where V_{sr} is the influx from surface runoff, V_e denotes losses from the reservoir due to evaporation and reservoir seepage, V_{out} is outflux through overflow of the reservoir, V_{in} is influx from overflowing upstream reservoirs (i.e. where reservoirs are connected, a cascading hydrological system is created), and V_w denotes withdrawn volume (e.g. for irrigation).

The surface runoff volume, V_{sr}, contributing to a reservoir over a given interval Δt is obtained as:

$$V_{sr} = rA_c\Delta t$$

where A_c denotes the reservoir's catchment area and r is the surface run-off rate, which is calculated as a Hortonian infiltration-excess:

$$r = \{0p - i \quad p \leq ip > i$$

where p denotes precipitation rate and i denotes infiltration rate. Precipitation rate is temporally variable and is provided as model input for each time step. The infiltration rate is temporally variable and depends on preceding rainfall. Here, a simple linear decay in infiltration capacity is assumed, where the decay is proportional to the number of rain days, n_i, in the preceding 7 days:

$$i = k_1(7 - n_i)i_0$$

where i_0 is the time-invariable base infiltration rate for dry soil, and the multiplier k_1 is a calibration parameter.

Water losses, V_e, from a reservoir are calculated through a loss function:

$$V_e = k_2 e A_r \Delta t$$

where A_r denotes the reservoir surface area, e is the evaporation rate, and the multiplier k_2 is a calibration parameter. The calibration parameters k_1 and k_2 are free parameters that can be adjusted to optimize model output relative to observed data. In some sense, they are fudge parameters that compensate for the simplifying assumptions embedded in the representation of infiltration and evaporation process.

Excess water volume beyond reservoir capacity, V_{Rmax}, is removed as an overflow outflux:

$$V_{out} = \{0 V_R - V_{Rmax} \quad V_R \leq V_{Rmax} V_R > V_{Rmax}$$

where reservoirs are connected, the upstream overflow is added as an inflow to the downstream reservoir. Finally, any water withdrawals, V_w, are subtracted. In the simulations presented herein, however, there are no withdrawals ($V_w = 0$).

2.2 Model Setup

2.2.1 Reservoir and Catchment Properties

Three reservoir properties are needed in the calculations above: A_R, V_{Rmax}, and A_c. The exact dimensions of the reservoirs are not known. Although they could, in principle, be measured in a field survey, this was beyond the budget of this project. Instead, the reservoir surface area, A_R, is estimated by digitizing the contours from aerial photography (Google Earth). The reservoir capacity, V_{Rmax}, is estimated by approximating the reservoir as a shallow cone:

$$V_{Rmax} = \frac{1}{3} A_R d_R$$

where d_R denotes the maximum depth of the reservoir, estimated through personal observation and personal communication with the farm owner. The reservoir's catchment area, A_c, is obtained by calculating flow direction and flow accumulation area using ArcGIS on pre-reservoir digital topographic data (Fig. 3). Reservoir and catchment properties of the study area's four reservoirs are listed in Table 1.

2.2.2 Weather Data

Precipitation and evaporation data were derived from daily data from Gibraleón weather station, located 4.3 km to the northwest of the study area. The data cover the period from December 1999 until April 2017, with only 0.6% of missing values. Missing values have been filled where possible with data from El Tojalillo-Gibraleón weather station, located 7.5 km to the southwest of the study area. In the few instances where this was not possible, missing data were obtained by linear interpolation between available data from preceding and following days.

Precipitation mainly occurs in the autumn and winter months, while summers are mainly dry and hot (Fig. 4). The average precipitation is 631 mm/year but can vary markedly from year to year (Table 2; Fig. 5a). Temperature is more consistent throughout the observation period. Average daily precipitation (1.7 mm/day) is about half of the average daily potential evaporation rate (3.6 mm/day), indicating an overall tendency towards aridity. However, the maximum daily precipitation (62.1 mm/day) far exceeds the maximum daily potential evaporation rate (8.5 mm/day) meaning that rainfall can, in principle, accumulate in the reservoirs temporarily.

Since 2011, when the first reservoir was constructed, average annual rainfall has decreased by over 100 mm/year (Table 3; Fig. 5a) compared to

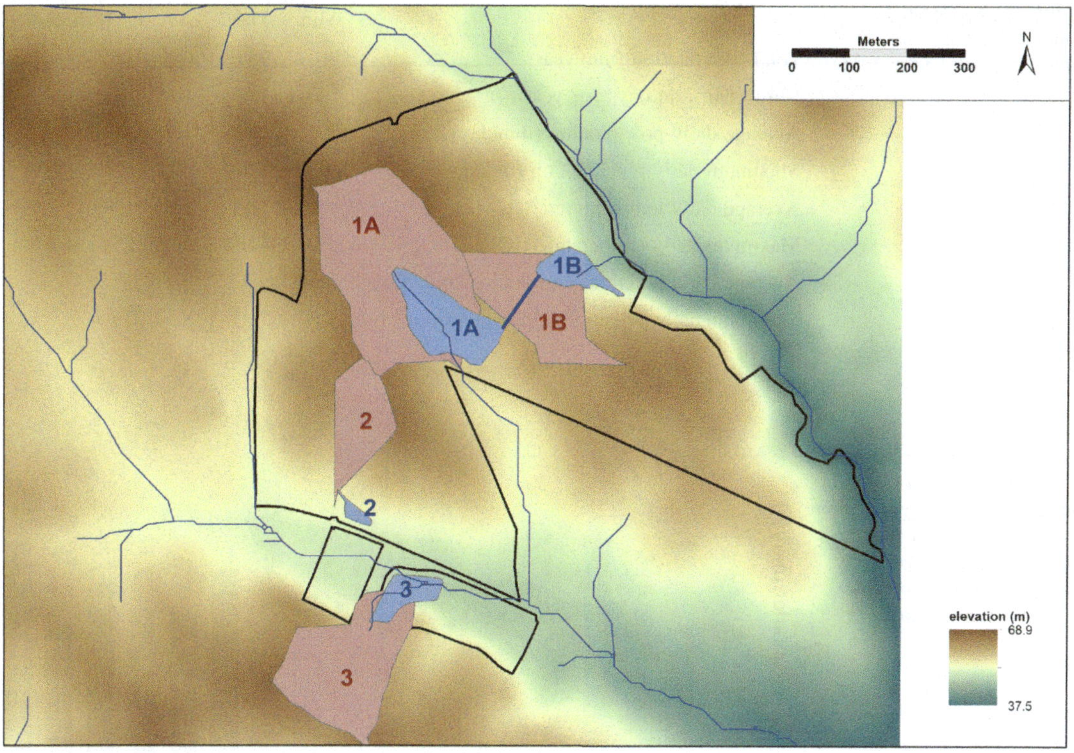

Fig. 3 Study area elevation map, with reservoirs (blue): 1A and 1B (interconnected; thick blue line, schematic), 2, 3 and their catchments (pink). Pre-reservoir natural drainage is indicated by thin blue lines. Farm boundary is indicated in black. (*DEM source* Instituto Geológico y Minero de España)

Table 1 Reservoir properties: surface area (A_R), depth (d), capacity (V_{Rmax}), catchment area (A_c)

Reservoir	A_R (m^2)	d (m)	V_{Rmax} (m^3)	A_c (km^2)
1A	16065	4	21420	0.068
1B	5910	3	5910	0.030
2	1155	2	770	0.016
3	5546	2	3697	0.038

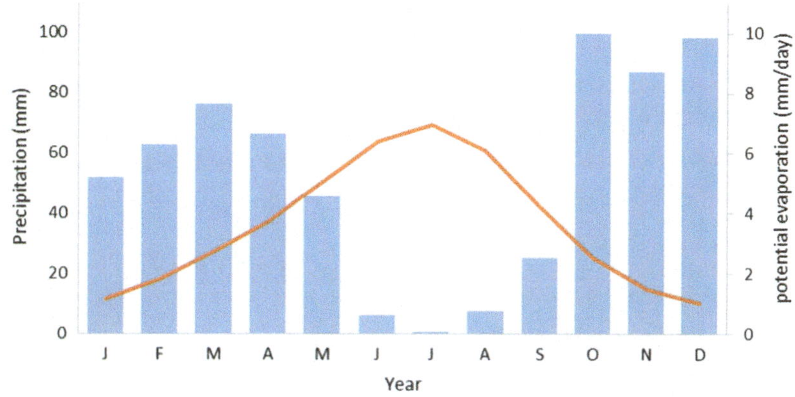

Fig. 4 Monthly rainfall (blue bars) and potential evaporation (orange) at the study site. Note different scales on the y-axes

Table 2 Aggregated annual weather characteristics (2000–2016)

	Min.	Max.	Average (Std. Dev.)
Total precipitation (mm/year)	407	954	631 (156)
Number of rain days per year (–)	81	124	100 (13)
Average daily precipitation (mm/day)	1.1	2.6	1.7 (0.4)
Maximum daily precipitation (mm/day)	36.0	110.8	62.1 (17.8)
Average daily temperature (°C)	17.0	18.1	17.4 (0.3)
Maximum daily temperature (°C)	37.4	41.8	40.0 (1.7)
Average daily potential evaporation (mm/day)	3.36	3.93	3.61 (0.15)
Maximum daily potential evaporation (mm/day)	9.82	7.31	8.52 (0.70)

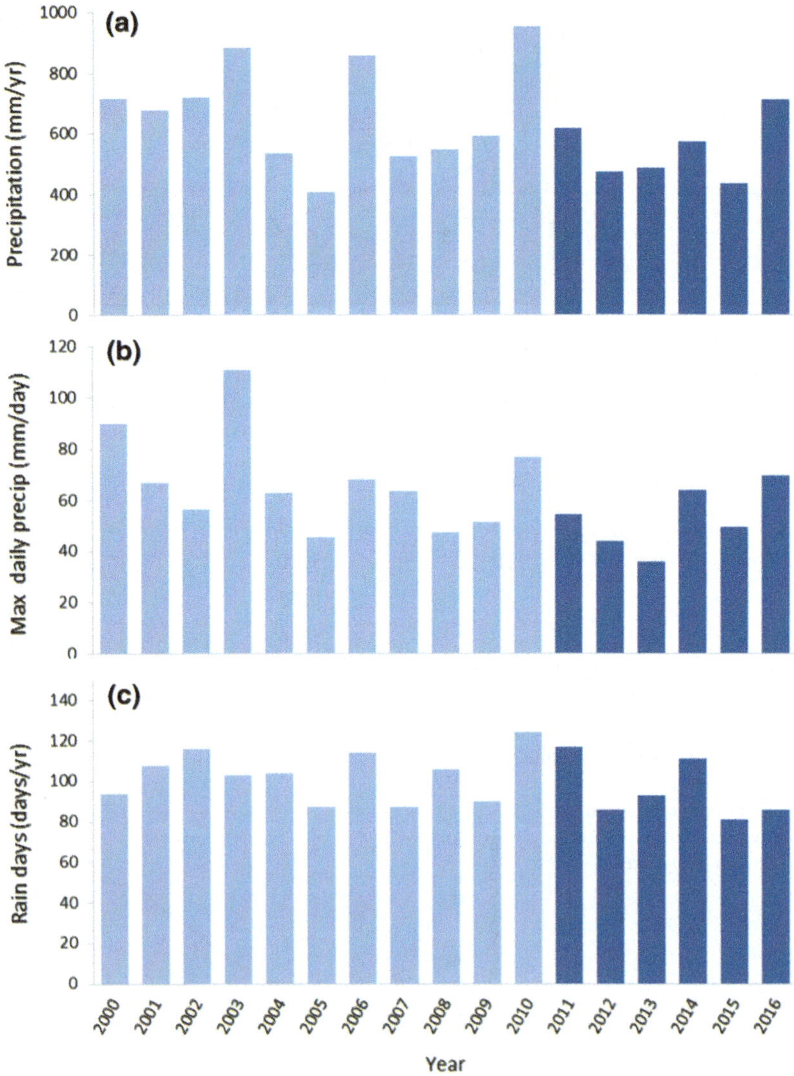

Fig. 5 Annual variation in total precipitation (**a**), maximum daily precipitation (**b**) and number of rain days (**c**). Light blue colours indicate the years prior to reservoir construction

Table 3 Precipitation differences in pre- and post-reservoir construction

	Average (Std. Dev.) 2000–2010	Average (Std. Dev.) 2011–2016	Trend (*p*-value)
Total precipitation (mm/year)	675 (164)	551 (96)	1.597 (0.131)
Max daily precipitation (mm/day)	67.2 (18.6)	52.8 (11.4)	1.622 (0.126)
Number of rain days per year (–)	103 (12)	96 (14)	1.092 (0.292)

the preceding decade. A similar observation holds for other rainfall characteristics, e.g. maximum daily rainfall and number of rain days per year (Fig. 5b, c). However, these trends are not statistically significant at p-value = 0.1 (Table 3) because of the large inter-annual variability.

2.2.3 Calibration and Validation

The model has two free parameters, i.e. the calibration parameters k_1 and k_2, which can be adjusted to optimize the model performance with respect to observed data. However, water levels or water volumes in the reservoirs at the study site have not been systematically monitored. There is, therefore, no direct data for calibrating or validating the model. However, a few aerial photographs (Google Earth) and on-site photographs (by farm manager and by first author) allow estimation of reservoir volumes on a limited number of occasions. Except for empty reservoirs ($V_R = 0$), these data are subject to some degree of uncertainty, as reservoir depths had to be approximated. Data for reservoirs 1B and 1A are used for calibration, whilst data for reservoir 3 are used for validation (Table 4).

Fit between observed and simulated data was measured using three different performance metrics: Pearson correlation, R^2, slope of the regression line, b, and Nash–Sutcliffe efficiency, E_{NS}. Although the optimal values for k_1 and k_2 could in principle vary for each of these metrics, they showed a remarkable consistency (Table 5). Nonetheless, the calibration is far from perfect, showing notable scatter of the individual points (Fig. 6a). Possible reasons for this are the incorrect estimation of observed water volumes, assuming the reservoirs are shallow cones, and oversimplification of physical processes in the model. However, plotting the observed data on the simulated reservoir hydrograph (Fig. 7b–d) suggests that the scatter is at least partly due to the model's tendency to under-predict the lag in reservoir volume peak volume and drying out, even though qualitatively the overall trends appear to be captured reasonably well. To test this idea, observed data, $V_{R,obs}$, are compared to a

Table 4 Observed[a] data used in calibration

Usage	Reservoir	Date	Source	Volume (m^3)
Calibration	1B	19/04/2013	Google Earth	1072
	1B	28/05/2016	Photograph	230
	1B	11/06/2016	Google Earth	42
	1B	23/09/2016	Photograph	0
	1A	19/04/2013	Google Earth	831
	1A	11/06/2016	Google Earth	26
Validation	3	19/04/2013	Google Earth	1304
	3	11/06/2016	Google Earth	0

[a]Water levels or water volumes in the reservoirs are unmonitored. For calibration purposes, volumes of water are estimated from aerial photographs (Google Earth) and on-site photographs

Table 5 Calibration results for different calibration metrics: Pearson correlation (R^2), slope of the regression line (b), Nash–Sutcliffe efficiency (E_{NS})

Metric	Value	k_1	k_2
R^2	0.514	7.81[a]	3.42[a]
b	0.707	7.81[a]	3.42[a]
E_{NS}	0.475	7.81[a]	3.81

[a]Value used in subsequent simulations

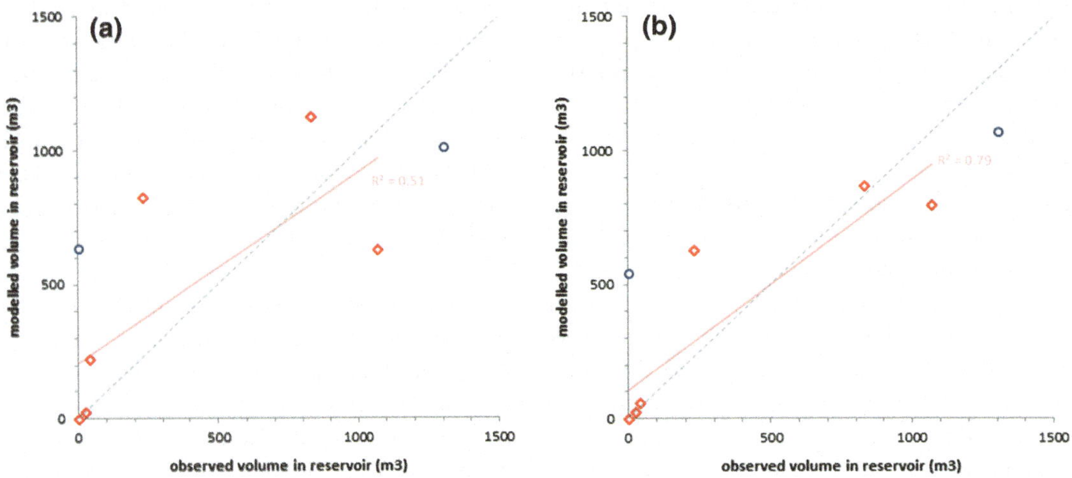

Fig. 6 Comparison of observed and simulated data after calibration, using direct comparison (**a**) and time-shifted comparison (**b**). A: Calibration data (red dots) are derived from reservoir 1A and 1B. Validation data (blue dots) are derived from reservoir 3. Grey dotted line indicates 1:1 correspondence. The thin red line is regression through calibration data

modified $V^*_{R,sim}$ which represents the closest fit to observed data within a 5-day kernel around the simulation date: $V^*_{R,sim,t} = V_{r,sim,t'}$, where $t - 5 \leq t' \leq t + 5$. The significantly better performance on all metrics of the model under these modified conditions confirms that the model captures the overall dynamic of the reservoir volume, albeit with a temporal error of plus or minus five days (Table 6; Fig. 6b).

3 Results

3.1 Sensitivity Analysis

Two sensitivity analyses are conducted to test the influence of reservoir input data (A_c, A_R) and model parameters (k_1, k_2). The reference scenario is the calibrated scenario, i.e. with input data as per Table 1 and parameter data as per Table 5. To test sensitivity each input and parameter are changed individually over a −20% to +20% range relative to its base value. Outputs are analysed in terms of average water volume in the reservoir, V_{Ravg}, in a five-year period (2012–2016) and number of reservoir dry days, n_{dry}, over the same period, as these are relevant properties for practical reservoir operation. The sensitivity to the model parameters k_1 and k_2 is also analysed in terms of the impact on the calibration metrics R^2, b, E_{NS}. This naïve analysis provides a first insight in the input data and parameter sensitivity but excludes the effects of parameter interaction.

With respect to V_{Ravg} and n_{dry}, the model shows very high sensitivity to A_c, high sensitivity

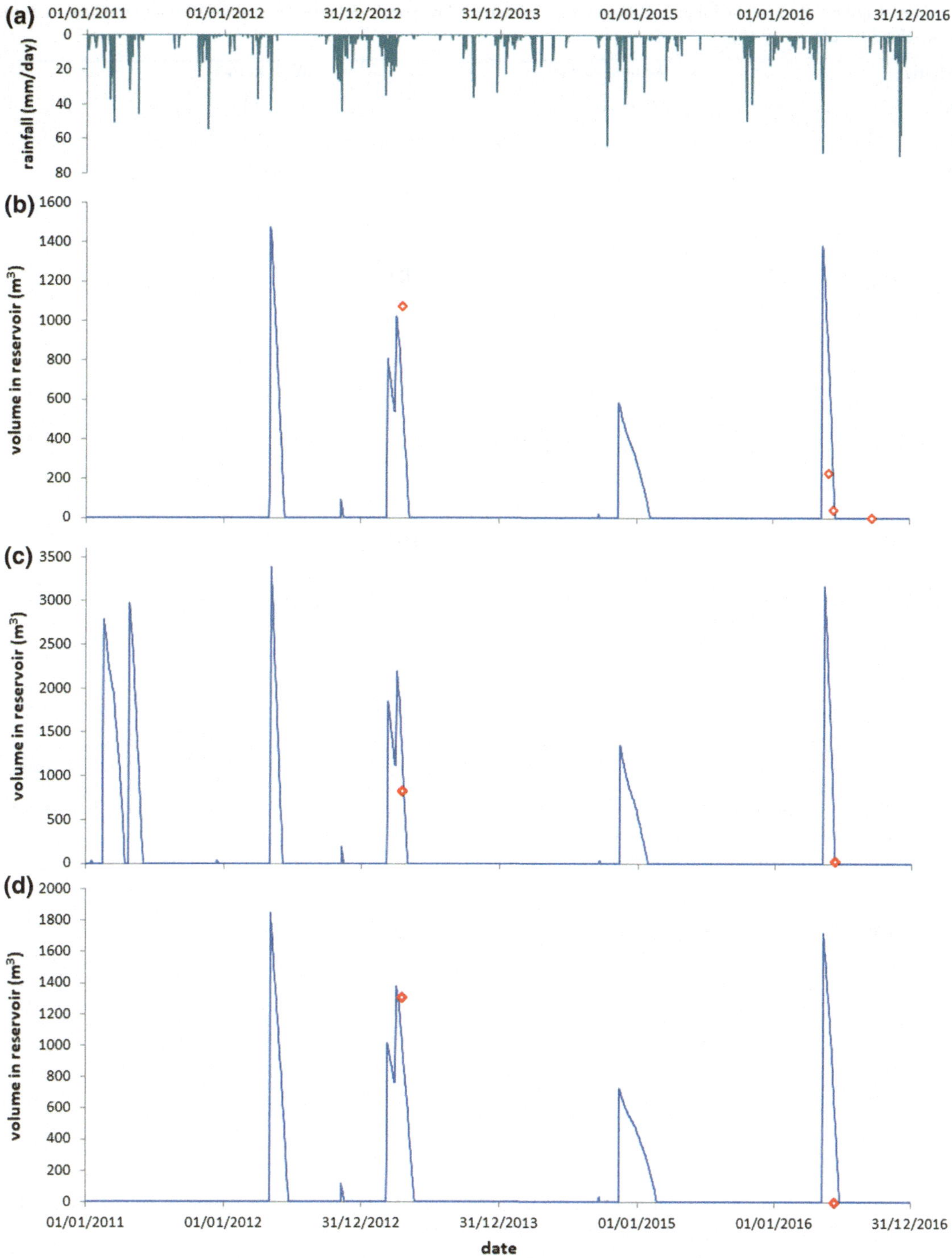

Fig. 7 Precipitation (**a**) and water volumes in reservoirs 1B (**b**), 1A (**c**) and 3 (**d**) over the period 2011–2016. Simulated water volumes are shown as blue lines, observed data as red dots. 1B and 3 were not operational until 2012. Data from 1B and 1A were used for calibration while data from reservoir 3 were used for validation. Note different scales on y-axes

Table 6 Calibration results with and without temporal kernel: Pearson correlation (R^2), slope of the regression line (b), Nash–Sutcliffe efficiency (E_{NS})

Metric	Without kernel	With kernel
R^2	0.514	0.813
b	0.707	0.786
E_{NS}	0.391[a]	0.760[a]

[a]Using parameters $k_1 = 7.81$, $k_2 = 3.42$

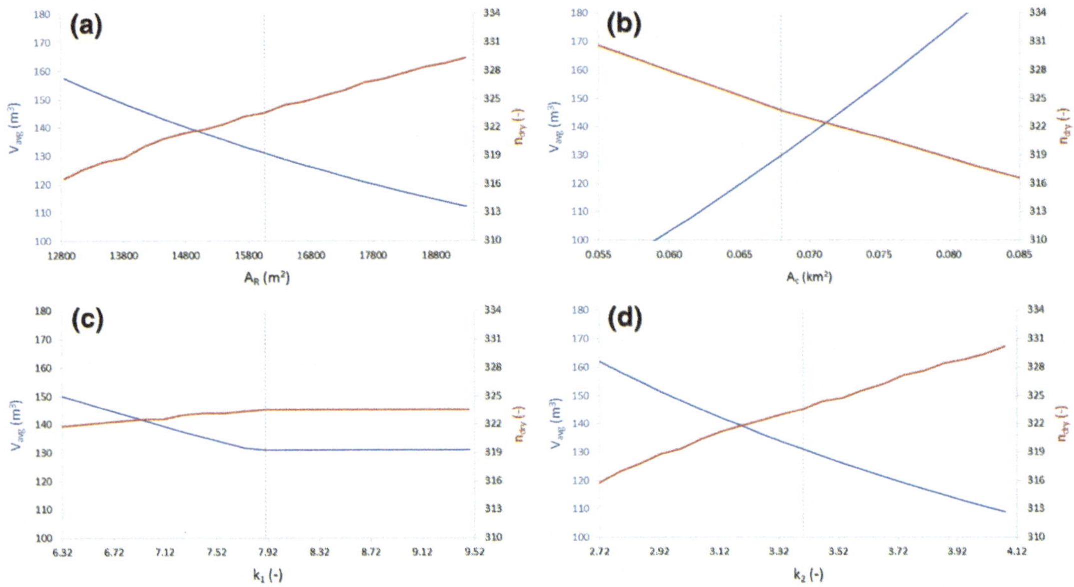

Fig. 8 Sensitivity of average reservoir volume V_{Ravg} (blue) and number of dry days n_{dry} (red) to reservoir surface area A_R (**a**) and catchment area A_C (**b**), and parameters k_1 (**c**) and k_2 (**d**). The dashed line indicates the value used in non-sensitivity simulations

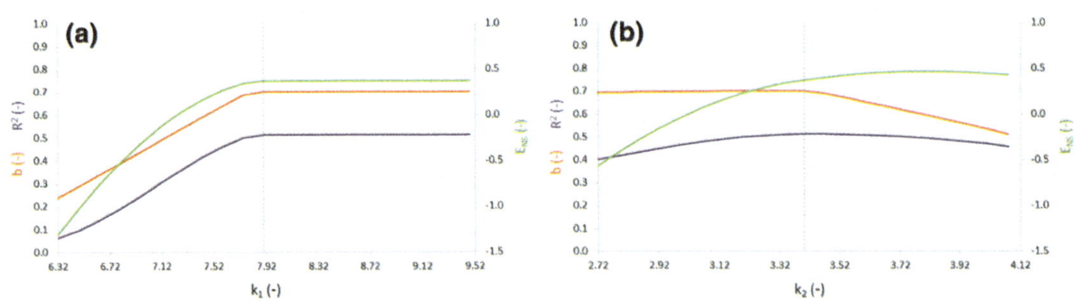

Fig. 9 Sensitivity of calibration metrics R^2 (purple), b (green) and E_{NS} (orange) to parameters k_1 (**a**) and k_2 (**b**). The dashed line indicates the value used in non-sensitivity simulations

to A_R and k_2, and low sensitivity to k_1 (Fig. 8). As A_c and A_R can be reliably obtained from the digital elevation data, this does not pose many problems, with the possible exception of k_2 which only can be obtained through calibration. Interestingly, the calibration metrics are far more sensitive to k_1 than to k_2 (Fig. 9). This suggests that k_1 has a greater impact on the temporal control of the hydrology, to which the calibration metrics are quite responsive, and that k_2 has a greater impact on the hydrology in terms of volumes of water.

3.2 Pre-construction Simulation

A separate set of simulations shows how the reservoirs would have worked with rainfall conditions prior to 2011, i.e. as if the reservoirs had been constructed earlier. This analysis allows to assess how much of the reservoirs' poor performance may be linked to the perceived change in climate conditions (see on 'weather data' above). Some of the markedly wet years (i.e. 2003, 2010) would have resulted in significantly higher water volumes, V_R, in the reservoirs (Fig. 10). Although not quite reaching full capacity, V_{Rmax}, these higher volumes persist for several months before evaporating during the summer. This gives hope that the reservoirs can function to a higher capacity in future years with higher rainfall. However, there also are several years with negligible contribution

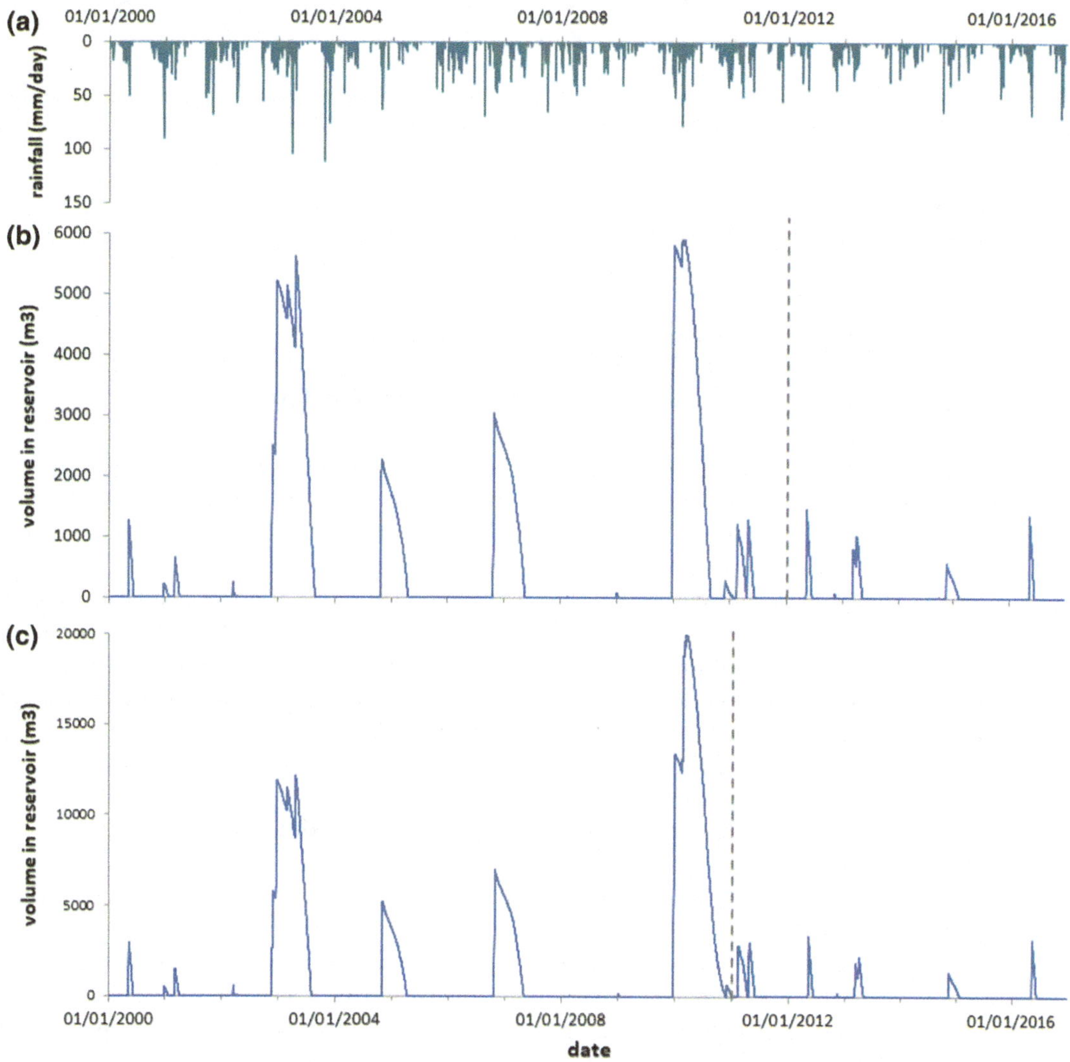

Fig. 10 Pre-construction rainfall (**a**) and associated hypothetical performance of reservoirs 1B (**b**) and 1A (**c**). The dashed line indicates the approximate construction of the reservoir

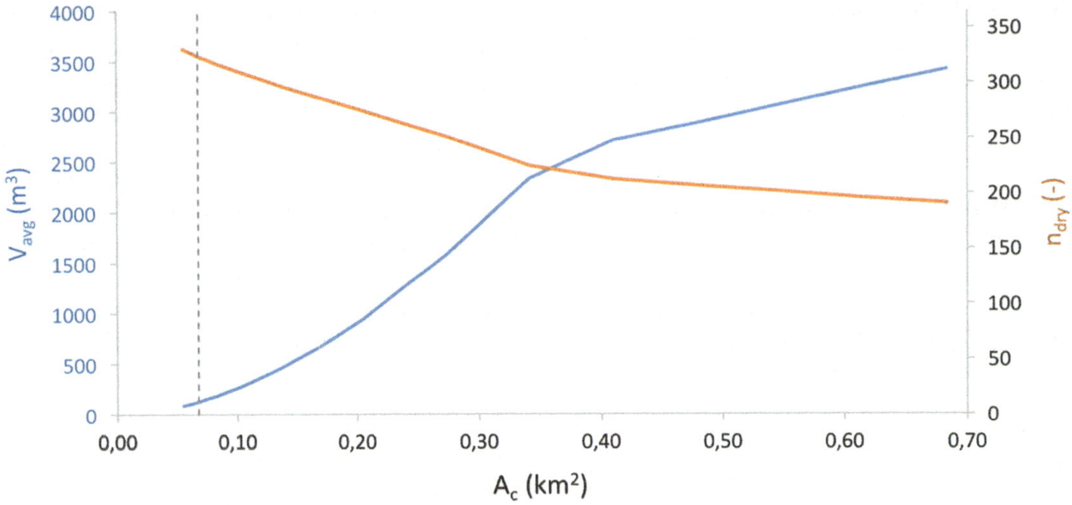

Fig. 11 Sensitivity of average reservoir 1A volume V_{Ravg} (blue) and number of dry days n_{dry} (red) to catchment area A_C. The dashed line indicates the value used in non-sensitivity simulations

to the reservoirs' water volume. Overall, the reservoirs appear to have limited potential for retaining water, except during years with notably high or prolonged precipitation.

The model shows limited potential of the reservoirs and their respective catchments to harvest rainwater in sufficient quantities between October and April when most rainfall occurs. Nonetheless, some rainfall can accumulate during several months. The overflow of reservoir 1A into reservoir 1B is not to be expected during the studied period (2010–2016), with all reservoirs being prone to drying up during the months of highest evaporation (June–August). The model represents the limited observations on the ground well, with a temporal error of plus or minus five days. Although statistically insignificant, the 2000–2010 decade may have appeared rainier, with the wet years 2003 and 2010 possibly having led to a misjudgement for on-farm WH potential during reservoir planning and implementation. Calibration metrics for infiltration (k_1) show high sensitivity in the model, which could support the assumption that the reservoirs themselves are not sufficiently impervious to infiltration. On the other hand, sensitivity analysis to catchment size (Fig. 11) indicates that larger catchments would significantly improve the reservoir's performance—due to given topographic limitations, however, only in theory.

4 Discussion

4.1 Model Findings

The applied model confirms observations on the ground, whereby none of the reservoirs fill and would only fully function under some unlikely weather scenarios. Furthermore, the creation of interconnected series of ponds and lakes across a landscape (e.g., Dregger 2010) may not be a viable contribution to combat desertification on-farm in the Gibraleón area and under the current or projected climatic conditions.

4.2 WH Potential

Assuming all four reservoirs at Gibraleón filled up to their maximum capacity of calculated 31,787 m^3 under rainier conditions (Table 1), the theoretical use of water for supplementary irrigation would allow water for only approx. 4 ha or 8% of the total fruit trees, without taking into account other factors like concurrent evaporation from reservoirs. This estimate is based on a model calculation of 400 fruit trees per ha and water consumption of 20 m^3 per tree and irrigation season (typical figures, farm consultant personal communication) on the approx. 50 ha of farmland (study site). As a consequence of irrigation, however, the reservoirs would limit their benefit for other ecosystem services such as wildlife habitat, microclimate regulation, recreational use or groundwater recharge. The calculation also illustrates how dependent agriculture in this area is on external water supplies.

4.3 Long-Term Weather Changes

Long-term average rainfall for Gibraleón (1901–2015) being 533 (151) mm/year, corresponds quite well with the recent record for 2011–2016 which is 551 (96) mm/year but is well below the previous decade's (2000–2010) which had more rain—675 (164) mm/year. Whether statistically significant or not, this might still have misled project developers. However, long-term average annual rainfall for Gibraleón is still slightly below current conditions, which suggest that dryer years may still come if the system is returning to its long-term average. This means that plenty of years with below-average rainfall can be expected for Gibraleón, without taking into account climate change projections.

4.4 Merits and Demerits of the Model

While some large-scale permaculture-inspired projects start on non-productive land without the need to produce commodities, an existing farm must remain productive and financially viable during any conversion process. In this context, the authors advocate for closer collaboration between producers interested in 'permaculture farming' and academic institutions as well as local authorities, given that many such conversions have strong experimental elements that may benefit from thorough interdisciplinary planning as well as systematic and ongoing monitoring. With this in mind, a simple WH computer modelling technique may help to improve planning and monitoring. The strengths and weaknesses of the model are summarised in Table 7. The merits of the model are a minimal amount of calculation processes due to few parameters, it is very fast and allows instant visualisation while running on widely available spreadsheet software, the input data for terrain (digital elevation model, soil properties) and weather are easily available. The demerits are linked to the merits. Only three and spatially lumped processes may not represent the complexity of reality well enough. The timing of rainfall events with reservoir filling levels is not captured very well (±5 days).

Table 7 Strengths and weaknesses of the simple 'bucket model'

Model merits	Model demerits
Simplicity—only 3 processes suffices: runoff, infiltration, evapotranspiration; spatially lumped	Only 3 processes may not be sufficient where more complexity needs to be embraced by the model; spatially lumped
Very fast: 17 year simulation for 3 reservoirs performed in less than 1 s	Accuracy: timing is not captured very well
Wide platform and instant visualisation; built in common spreadsheet software	Insufficiently tested; only evaluated on limited data points
Few parameters; only two: k_1, k_2	High sensitivity to parameters
Input data easily available: terrain and weather	

4.5 Comparison with Other European Projects

At this point, it also seems pertinent to compare the current WH study with other permaculture inspired WH projects like Holzer's Krameterhof and two sites within Csa climate, Tamera Ecovillage and Vivencia Dehesa for reference. Table 8 gives an overview of climatic conditions based on average annual rainfall data from 1901–2015 and a trending from 2000 onwards based on climate data until 2015.

Tamera is an ecovillage project in Portugal on 134 ha, strongly inspired by Holzer ('water retention landscape'; Dregger 2010), with a series of 13 reservoir elements. Kadura (2016) studied the impact of rainwater harvesting on the water balance, building a runoff model with field runoff measurements (HEC-HMS). He concludes: "there is a significant impact of the rainwater harvesting measures to the local water balance". Buffering and storing effects during rainfall events could be established. Agricultural production in the ecovillage is for subsistence and covers only parts of the food required by the community and their visitors, but is fully irrigated by harvested rainwater.

Vivencia Dehesa is a 250 ha property in the Extremadura region of Spain, known for their historical agrosilvopastoral landscapes (Schnabel and Ferrerira 2004). Holzer (2011) had planned a series of ponds and lakes to retain rainwater and regenerate land in an area prone to desertification due to general land overuse and subsequent abandonment. A series of 16 reservoirs covering 27 hectares has been established with noticeable greening in the vicinity of the water bodies (Google Earth). Various species of aquatic birds

Table 8 Climate data for various 'permaculture inspired' WH projects in Europe

Site	Average (Std. Dev.) 1901–2015	Average (Std. Dev.) post 2000	Post 2000 trend (*p*-value)
Gibraleón[a, c] (study site)			
Annual precipitation (mm)	533.4 (150.7)	552.0 (140.8)	−3.2 (0.19)
Annual temperature (°C)	17.6 (0.5)	17.5 (0.6)	+0.021 (0.03)
PET (mm)	1296.3 (46.0)	1289.9 (43.1)	+3.88 (0.003)
Vivencia Dehesa[a, c] (near Navalmoral de la Mata, Extremadura, Spain; Vivencia Dehesa 2016)			
Annual precipitation (mm)	491.0 (103.0)	488.9 (95.7)	−4.41; (0.19)
Annual temperature (°C)	14.1 (0.7)	14.1 (0.7)	+0.018 (0.01)
PET (mm)	1202.1 (58.3)	1203.6 (43.0)	+4.26; (0.04)
Tamera Ecovillage[a, c] (near Colos, Alentejo, Portugal; Dregger 2010)			
Annual precipitation (mm)	639.4 (167.4)	636.0 (155.0)	+5.60; (0.75)
Annual temperature (°C)	16.4 (0.6)	16.5 (0.6)	+0.024; (0.079)
PET (mm)	1105.8 (47.2)	1114.8 (42.3)	+4.71; (0.01)
Krameterhof[b, d] (near Tamsweg, Salzburger Land, Austria; Holzer 2010)			
Annual precipitation (mm)	1612.7 (175.2)	1605.9 (166.5)	+11.88; (0.34)
Annual temperature (°C)	3.8 (0.8)	3.8 (0.8)	+0.04 (0.22)
PET (mm)	526.2 (37.9)	532.7 (34.8)	−0.58 (0.96)

PET: potential evapotranspiration; Köppen-Geiger climate classification: [a]Csa and [b]Dfb; [c]negative water balance ([precipitation]—[PET]), [d]positive water balance. Climate data obtained from Climate Research Unit (CRU, East Anglia): gridded data with a resolution of 60 km rought (0.5° × 0.5°), cubic bilinear interpolation is used to estimate values at the specific locations. PET is calculated using the Penman–Monteith equation. Mann–Kendall trend test accounting for serial autocorrelation is used to estimate the trend since 2000. The intensity of the trend is determined through the Sen's slope have repopulated the area according to landowner's personal communication. In April 2016, the Spanish government nominated this project as the first 'private area of ecologic interest' in Extremadura (Vivencia Dehesa 2016). The landowner is developing regenerative agriculture (keyline design) and is reintroducing traditional animal husbandry (cattle and pork) whilst establishing bird watching as ecotourism.

Both Tamera and Vivencia Dehesa may be considered more successful in harvesting water and retaining it in the landscape than what can currently be said for Gibraleón, where so far WH has been limited. On the one hand, catchments at Gibraleón are relatively small, situated essentially on a hill rather than within a valley (Fig. 3), with no connection to catchments outside its boundaries. On the other hand, Tamera and Vivencia Dehesa may be benefiting from higher annual precipitation, lower annual temperature and reduced evapotranspiration (Table 8). While average annual precipitation (1901–2015) for Gibraleón (533 [151] mm/year) lies between Vivencia Dehesa (491 [103] mm/year) and Tamera (639 [167] mm/year), annual average temperature and thus potential evapotranspiration is highest for the Gibraleón site, indicating higher aridity here.

In terms of water balance, the Krameterhof clearly benefits from a humid and cool climate. Trending for both Gibraleón and Vivencia Dehesa gives an indication towards dryer climate and a negative water balance since 2000; Tamera's water balance, although negative, seems stable since that year.

4.6 Implications

Permaculture initiatives often start small and locally before spreading virtually within social networks as well as geographically with real land projects around the world. They intend to produce social, economic and environmental change to mitigate current and future challenges, such as land degradation and loss of natural capital (Lockyer and Veteto 2015). Given the increasing interest in more sustainable farming practices, permaculture has been considered for upscaling by commercial producers in Europe (Wezel et al. 2009; Fiebrig and Buley 2017), including WH in agreement with the European Commission's 'Blueprint to Safeguard Europe's Water Resources' (EC 2012). Assessments as to the sustainability and regenerative capacity of such farming systems are in their infancy. The current study intends to contribute to such assessments in a constructive manner. It is intended to help discern between the realistic and the idealistic, posing the overarching question about what practices are currently useful in a commercial setting to contribute to the ethics of Earth Care whilst taking into account the very real limits set by nature. Overall, producing crops that demand intensive irrigation within an arid climate can be considered as contentious in terms of sustainability, regeneration and the permaculture approach.

Although not dealt with in this study, upscaling permaculture should also adequately address the ethics regarding People Care and Fair Share (Holmgren 2011), implying changes in enterprise management within the food value chain together with a transformation of consumption patterns and economic as well as political power structures.

5 Conclusions and Outlook

Under the current climate conditions and farm management practices, two of the study site's reservoirs may fulfil a function as wildlife habitat as a result of the constant influx of rinsing waters, which can be viewed as beneficial wastewater cycling on the land (e.g. amphibians such as frogs have been sighted in the immediate vicinity of the reservoirs); the other two are partially filled during some months, even in the summer, and contribute to retaining rainwater on the land. All four reservoirs may play a future role in mitigating soil erosion during severe precipitation events in conjunction with keyline plantation design. However, the expectation of water being retained in the reservoirs to a considerable level throughout the year is currently not being met.

During the design phase of the reservoirs, potential precipitation may have been overestimated and infiltration of rainwater into the soil and evaporation as well as evapotranspiration may have been underestimated. If WH modelling had been included in the design process, it might have led to smaller sized reservoirs in relation to their respective catchments, thus improving functionality while reducing in-situ impact and costs. At least two of Holmgren's (2011) twelve permaculture design principles would have given valuable guidance during the design and planning process. Principle (9): "Use small and slow solutions"—smaller systems are easier to maintain than big ones, making better use of local resources and producing better outcomes; and principle (1): "Observe and interact"—by taking time to engage with nature, solutions that suit one's particular situation can be designed.

This would, e.g. translate to first identifying one reservoir area where the respective catchment would be expected to be most effective. A reduced reservoir size might suit the expected scenarios better regarding precipitation, infiltration and evapotranspiration. Allowing enough time to observe if the system actually works once installed, including ongoing monitoring and evaluation, would indicate if building more or larger reservoirs is reasonable.

The current study is focused on the analysis of reservoir functions. The permaculture design approach is able to produce nature-based solutions that can contribute to ecosystem services. Building on-farm WH reservoirs as for Gibraleón or water landscapes as in the case of Tamera or Vivencia Dehesa would benefit from long-term studies to better understand the dynamics and extent of their contribution in restoring groundwater levels, regenerating soil as well as biodiversity, sequestering carbon, reversing desertification processes, mitigating climate change and eventually in establishing an economically viable land restoration process that may translate to environmental stewardship schemes the land owner receives compensation for.

Further research into the application of the presented WH computer model may make modelling more accessible to farmers and land owners. This can improve planning processes, where well informed decisions on investments related to WH have to be taken. Moreover, modelling-based scenario analyses can help better understand the dynamics and extent of its potential in establishing an effective and economically viable land restoration process for the region, taking into account climate change projections of increasing desertification in the Mediterranean basin (Guiot and Cramer 2016). This would represent another initiative towards a nexus approach in the region (Hoff et al. 2019).

A collaboration with other experts, e.g. from geology, hydrology or ecology within the setting of a larger research project may increase the positive impact of WH initiatives, while a study that compares different available software tools with the current bucket modelling would help to establish a cost–benefit analysis. Some guidelines and standards related to earth dam building do exist and taking them into account may help to mitigate risks during and after construction (e.g., Stephens 2010; DPIW 2008).

Acknowledgements M. Buley, M. Báez Lozano, J. Schweikle and F. Lehmann for study site support; B. Dieppois for additional climate data, U. Schmutz, N. Giddings, D. Joshi and the external reviewers for critical review, Coventry University and its Centre for Agroecology, Water and Resilience (CAWR) for financial support to conduct the field research.

References

Al Jazeera (2011) Greening the desert, 28 Oct 2011. https://www.aljazeera.com/programmes/earthrise/2011/10/2011102793927343187.html. Accessed 19 Apr 2017

Bensin BM (1928) Agroecological characteristics description and classification of the local corn varieties chorotypes. Prague

Bensin BM (1930) Possibilities for international cooperation in agroecological investigations. Int Rev Agric Mon Bull Agric Sci Pract 21:277–284

Biemans H, Haddeland I, Kabat P, Ludwig F, Hutjes RWA, Heinke J, Von Bloh W, Gerten D (2011) Impact of reservoirs on river discharge and irrigation water supply during the 20th century. Water Resour Res 47(3):1–15. https://doi.org/10.1029/2009wr008929

Biggs J, Von Fumetti S, Kelly-Quinn M (2016) The importance of small waterbodies for biodiversity and ecosystem services: implications for policy makers. Hydrobiologia 793(1):3–39. https://doi.org/10.1007/s10750-016-3007-0
Bleischwitz R, Spataru C, VanDeveer SD, Obersteiner M, van der Voet E, Johnson C, Andrews-Speed P, Boersma T, Hoff H, van Vuuren DP (2018) Resource nexus perspectives towards the United Nations sustainable development goals. Nat Sustain 1:737. https://doi.org/10.1038/s41893-018-0173-2
Chabay I, Frick M, Helgeson J (2016) Land restoration: reclaiming landscapes for a sustainable future. Academic Press, Waltham MA
Critchley W, Siegert K (1991) Water harvesting: a manual for the design and construction of water harvesting schemes for plant production. FAO, Rome
DPIW (2008) Guidelines for the construction of earth-fill dams. Policy #2008/I. Department of Primary Industries and Water, Tasmania. https://dpipwe.tas.gov.au/Documents/Guidelines-for-earth-fill-dams.pdf. Accessed 19 Apr 2017
Dregger L (2010) Tamera—a model for the future. Verlag Meiga, Belzig
EC (2012) Communication from the commission to the European parliament, the council, the European economic and social committee and the committee of the regions. A blueprint to safeguard Europe's water resources. European Commission. https://eur-lex.europa.eu/legal-content/EN/TXT/PDF/?uri=CELEX:52012DC0673&from=EN. Accessed 29 Apr 2017
EU (2016) Horizon 2020. Work programme 2016–2017. Food security, sustainable agriculture and forestry, marine and maritime and inland water research and the bioeconomy. European Commission. https://ec.europa.eu/research/participants/data/ref/h2020/wp/2016_2017/main/h2020-wp1617-food_en.pdf. Accessed 5 Apr 2017
FAO (2011) The state of the world's land and water resources: managing systems at risk. Food and Agriculture Organization of the United Nations and Earthscan, Abingdon
Feola G, Nunes T (2014) Success and failure of grassroots innovations for addressing climate change: the case of the Transition Movement. Glob Environ Change 24:232–250. https://doi.org/10.1016/j.gloenvcha.2013.11.011
Ferguson RS, Lovell ST (2013) Permaculture for agroecology: design, movement, practice, and worldview. A review. Agron Sustain Dev 34(2):251–74. https://doi.org/10.1007/s13593-013-0181-6
Ferguson RS, Lovell ST (2015) Grassroots engagement with transition to sustainability: diversity and modes of participation in the international permaculture movement. Ecol Soc 20(4). https://doi.org/10.5751/es-08048-200439
Ferrand AE, Cecunjanin F (2014) Potential of rainwater harvesting in a thirsty world: a survey of ancient and traditional rainwater harvesting applications. Geogr Compass 8(6):395–413. https://doi.org/10.1111/gec3.12135
Fiebrig I, Buley M (2017) Permaculture and biodynamics. Do they have anything in common? Star Furrow 127:22–23
Gliessman SR (2015) Agroecology: the ecology of sustainable food systems. CRC Press, Boca Raton, FL
Gras E (2012) Cosecha de agua y tierra: diseño con permacultura [Harvesting water and soil: design through permaculture]. EcoHabitar, Olba, Teruel
Guiot J, Cramer W (2016) Climate change: the 2015 Paris Agreement thresholds and Mediterranean basin ecosystems. Science 354(6311):465–468. https://doi.org/10.1126/science.aah5015
Harper D (nd) Online etymology dictionary. https://www.etymonline.com/index.php?term=sustain=restore=regenerate. Accessed 10 Apr
Hettiarachchi H, Ardakanian R (2016) Managing water, soil, and waste in the context of global change. In: Hettiarachchi H, Ardakanian R (eds) Environmental resource management and the nexus approach: managing water, soil, and waste in the context of global change. Springer International Publishing, Cham, pp 1–7
HLPE9 (2015) Water for food security and nutrition. A report by the High Level Panel of Experts on Food Security and Nutrition of the Committee on World Food Security. FAO, Rome. https://www.fao.org/3/a-av045e.pdf. Accessed 20 Apr 2017
Hoff H, Alrahaife SA, El Hajj R, Lohr K, Mengoub FE, Farajalla N, Fritzsche K, Jobbins G, Özerol G, Schultz R, Ulrich A (2019) A nexus approach for the MENA region—from concept to knowledge to action. Front Environ Sci 7. https://doi.org/10.3389/fenvs.2019.00048
Holmgren D (2011) Permaculture: principles & pathways beyond sustainability. Permanent Publications, East Meon, Hampshire
Holzer S (2008) Sepp Holzers Permakultur. Leopold Stocker Verlag, Graz
Holzer S (2010) Sepp Holzer's permaculture. Permanent Publications, East Meon, Hampshire
Holzer S (2011) Desert or paradise: restoring endangered landscapes using water management, including lake and pond construction. Permanent Publications, East Meon, Hampshire
Holzer S (nd) Internationale Projekte 1998–2014 [International projects 1998–2014]. https://www.seppholzer.at/cms/fileadmin/user_upload/projekte/SeppHolzerProjekte2014.pdf. Accessed 19 Apr 2017
IPES Food (2016) From uniformity to diversity: a paradigm shift from industrial agriculture to diversified agroecoogical systems. International Panel of Experts on Sustainable Food Systems
Kadura J (2016) Impact of rainwater harvesting on the water balance in the area of Monte do Cerro, Portugal. BSc dissertation, Hochschule für Forstwirtschaft Rottenburg Schadenweilerhof
Keskinen M, Kummu M, Käkönen M, Varis O (2012) Mekong at the crossroads: next steps for impact

assessment of large dams. Ambio 41(3):319–324. https://doi.org/10.1007/s13280-012-0261-x
Lockyer J, Veteto JR (2015) Environmental anthropology engaging ecotopia. Bioregionalism, permaculture and ecovillages. Berghahn Books
Lyle JT (1994) Regenerative design for sustainable development. Wiley, New York, N.Y.
McIntyre BD, Herren HR, Wakhungu J, Watson RT (2009) Agriculture at a crossroads: global report. Island Press, Washington, DC
MEA (2005) Millenium ecosystem assessment. In: Ecosystems and human well-being: synthesis. Island Press, Washington, DC
Mollison BC (1988) Permaculture: a designer's manual. Permaculture Resources, Tyalgum, Australia
Mollison BC, Holmgren D (1978) Permaculture one: a perennial agricultural system for human settlements. Tagari Publications, Tasmania
Montanarella L (2007) Trends in land degradation in Europe. In: Climate and land degradation environmental science and engineering, pp 83–104, Oct 2007. https://doi.org/10.1007/978-3-540-72438-4_5
Orr BJ, Cowie AL, Castillo Sanchez VM, Chasek P, Crossman ND, Erlewein A, Louwagie G, Maron M, Metternicht GI, Minelli S, Tengberg AE, Walter S, Welton S (2017) Scientific conceptual framework for land degradation neutrality. A report of the science-policy interface. United Nations Convention to Combat Desertification (UNCCD), Bonn, Germany
Oweis TY, Hachum A (2009) Water harvesting for improved rainfed agriculture in the dry environments. In: Rainfed agriculture: unlocking the potential, pp 164–181. CABI, Wallingford, UK
Power AGE (2010) Ecosystem services and agriculture: trade-offs and synergies. Philos Trans R Soc B: Biol Sci 365(1554):2959–2971. https://doi.org/10.1098/rstb.2010.0143
PWN (nd) Worldwide permaculture projects. https://permacultureglobal.org/projects. Accessed 19 Apr 2017
Rhodes CJ (2015) Permaculture: regenerative—not merely sustainable. Sci Prog 2015:403–412. https://doi.org/10.3184/003685015X14467291596242
Richter BD (2014) Chasing water: a guide for moving from scarcity to sustainability. Island Press, Washington
Ringler C, Yang E, Bekchanov M (nd) Is hydropower for or against irrigation? In: CGIAR research program on water, land and ecosystems. https://wle.cgiar.org/thrive/2013/05/23/hydropower-or-against-irrigation. Accessed 16 Apr 2017
Röhlen U, Ziegert C (2014) Lehmbau-Praxis: Planung und Ausführung. Beuth, Berlin
Roy D, Barr J, Venema HD (2011) Ecosystem approaches in integrated water resources management (IWRM). IISD, Winnipeg, Canada
Santiago MP, Da Cunha LI, Meira Cartea PA (2017) Resilient communities to climate change. Environmental education and movement in transition connection. Procedia Soc Behav Sci 237:925–930. https://doi.org/10.1016/j.sbspro.2017.02.130
Scarborough VL (1998) Ecology and ritual: water management and the Maya. Lat Am Antiq 9(02):135–159. https://doi.org/10.2307/971991
Schnabel S, Ferrerira A (2004) Sustainability of agrosilvopastoral systems: dehesas, montados. Catena Verlag, Reiskirchen
Shepard M (2013) Restoration agriculture: real world permaculture for farmers. Acres U.S.A., Austin, Texas
Slaper TF, Hall TJ (2011) The triple bottom line: what is it and how does it work? Indiana Bus Rev
Stephens T (2010) Manual on small earth dams. Food and Agriculture Organization of the United Nations, Rome. https://www.fao.org/docrep/012/i1531e/i1531e.pdf. Accessed 23 Apr 2017
Stringer L (2008) Can the UN Convention to Combat Desertification guide sustainable use of the world's soils? Front Ecol Environ 6(3):138–144
Studer RM, Liniger H (2013) Water harvesting: guidelines to good practice. Centre for Development and Environment (CDE), Bern; Rainwater Harvesting Implementation Network (RAIN), Amsterdam; MetaMeta, Wageningen; The International Fund for Agricultural Development (IFAD), Rome
TreeYo (nd) TreeYo permaculture. https://treeyopermaculture.com/treeyo/previous-treeyo-courses/courses-attended/carbon-farming-course-the-farm-tennessee-usa/keyline-design-with-darren-daugherty/. Accessed 30 Apr 2017
United Nations (1987) UN world commission on environment and development. In: Our common future, from one earth to one world. https://www.un-documents.net/our-common-future.pdf. Accessed 13 Apr 2017
United Nations (2015) Transforming our World: the 2030 Agenda for Sustainable Development. UN, New York
USDIBR (2012) Design standards no. 13: embankment dams, chapter 2: embankment design. U.S. Department of the Interior Bureau of Reclamation, Washington, Dec 2012. https://www.usbr.gov/tsc/techreferences/designstandards-datacollectionguides/finalds-pdfs/DS13-2.pdf. Accessed 19 Apr 2017
Vivencia Dehesa (2016) Valdepajares de Tajo, primera área privada de interés ecológico de Extremadura [Valdepajares de Tajo, first private area of ecological interest in Extremadura] in Vivencia Dehesa. https://www.vivenciadehesa.es/valdepajares-de-tajo-area-privada-de-interes-ecologico-de-extremadura/. Accessed 24 Apr 2017
Wezel A, Bellon S, Doré T, Francis C, Vallod D, David C (2009) Agroecology as a science, a movement and a

practice. A review. Agron Sustain Dev 29(4):503–515. https://doi.org/10.1051/agro/2009004

Wyatt AR (2014) The scale and organization of ancient Maya water management. Wiley Interdiscip Rev: Water 1(5):449–467. https://doi.org/10.1002/wat2.1042

Zhang L, Walker GL, Dawes WR (2002) Water balance modelling: concepts and applications. In: McVicar, TR, Rui L, Walker J, Fitzpatrick RW, Changming L (eds), Regional water and soil assessment for managing sustainable agriculture in China and Australia, ACIAR Monograph No. 84, pp 31–47

Multifunctional Historical Data for Improved Management of Reservoirs

Joseph Sang and Caroline Maina

Abstract

Sedimentation surveys play a critical role in the management of reservoirs. They not only give information on current volumes but also records of past sedimentation status. Determining the sedimentation status of reservoirs provides a means of generating multifunctional historical datasets on sedimentation rates, thus explaining the nexus of water, soil and food and its impacts on water quantity and quality in reservoirs over time. Historical sedimentation rates, as demonstrated from assessments carried out in two reservoirs in Kenya, can inform management of reservoirs and watersheds where food production is carried out. This could be critical in Sub-Saharan Africa (SSA) where up to 58% of reservoirs are not monitored. A study in Ruiru Reservoir and Lake Sonachi in Kenya demonstrates how critical and missing information about the reservoirs and watersheds could be generated from sedimentation assessment. In the Ruiru Reservoir, which is one of the critical sources of water to Nairobi, information such as current volume and sedimentation rates were obtained. The assessment showed that the reservoir has lost 14% of its original volume at an average sediment accumulation rate of 1.75 cm/year. This information is key to the management of the reservoir and the upstream watershed.

Keywords

Reservoirs · Sedimentation · Bathymetric survey · Bathymetric survey system

1 Introduction

To tame the problems of water availability man has always resorted to building storage reservoirs. Reservoirs are used to control the unbalanced spatial and temporal distribution of water. In water scarce areas, reservoirs store water for the dry periods, whereas in flood prone areas reservoirs reduce peak flows and thus protect the downstream community from floods hazards. Worldwide over 40,000 large reservoirs (Tiğrek et al. 2009; Tigrek and Aras 2012) were constructed in the 20th century with a combined capacity of 6,000 km^3 (Kondolf et al. 2014). Approximately 2% of these large reservoirs are in Sub-Saharan Africa (SSA) and have a

J. Sang (✉)
Department of Soil Water and Environmental Engineering (SWEE), Jomo Kenyatta University of Agriculture and Technology (JKUAT), Juja, Kenya
e-mail: j.sang@jkuat.ac.ke

C. Maina
Agricultural Engineering Department, Egerton University, Njoro, Kenya
e-mail: caroline.maina@egerton.ac.ke

S. Hülsmann and M. Jampani (eds.), *A Nexus Approach for Sustainable Development*,
https://doi.org/10.1007/978-3-030-57530-4_6

combined storage capacity of about 575 km^3 (Tigrek and Aras 2012). The financial costs and human resources required to construct reservoirs are enormous. In addition, there is socio-economic and environmental disruption brought about by the construction of dams to the local populace lifestyles. Most countries, especially developing, rely on external funding in the form of grants or loans to offset the cost of developing their reservoir capacities.

Despite the benefits and enormous cost of constructing reservoirs, most natural lakes and reservoirs are currently threatened by sedimentation. Reservoirs are designed to have a given lifespan, where they offer useful services, after which they are supposed to be rehabilitated or decommissioned. However, sedimentation drastically reduces this lifespan. According to Wisser et al. (2013) sedimentation, which results in volume loss, controls the lifespan of reservoirs since it occurs faster than the loss of structural integrity of the dam wall. Globally, about 31 km^3 of reservoir volume is lost annually to sedimentation and this could deplete half of the current reservoir's capacity by 2100 (Sumi et al. 2004). According to Palmieri et al. (2003), the worldwide annual loss of reservoir storage capacity can be associated with the need for 45% additional reservoir capacity. This additional capacity is estimated to cost about US$ 13 billion per year (Palmieri et al. 2003). With capacity loss due to sedimentation, hydropower generation is also affected. On the other hand, sedimentation not only leads to capacity loss but also results in degradation of water quality, hence increases the costs of water treatment in the case of drinking water reservoirs.

The loss of volume in the reservoir is related to food production activities upstream—and in general with watershed management, which includes water, soil and land-use. At the source, mainly farmland, the costs are also discernible. The upstream farmers lose their fertile top soil to agents of erosion. The productive top layer is lost and consequently, the cost of food production goes up. This includes the expensive fertilizers applied to improve farm productivity. Consequently, farm income declines. If unchecked, it becomes a vicious cycle of loss, thus the more soil the farmer loses, the more fertilizer will be required to raise production. Therefore, it is imperative to quantify the nexus between food production and water quantity downstream and the ability of reservoirs to sustainably supply water and/or generate electricity. Similar nexus assessments have been done in other parts of SSA (e.g., Karlberg et al. 2015).

Controlling erosion upstream remains the most viable option for sedimentation control. Compared to the cost of dredging, scouring and trucking, upstream conservation remains the most viable measure (Kawashima et al. 2003). Conservation measures have been used to control erosion into major dams worldwide with or without reward schemes to the farmers. For instance, a study in Sasumua watershed, Kenya, showed that farmers under Payment for Environmental Services (PES) can significantly reduce sediment load into the reservoir as indicated by an observed decline of average Total Suspended Solids (TSS) from 71.05 mg/L to an average TSS of 42.73 mg/L in the inflow rivers (Nduhiu et al. 2016). The adoption of PES schemes has increased the chances of success in erosion control in Sasumua and other parts of the world (Dougill et al. 2012; Nduhiu et al. 2016). Unlike upstream erosion control, dredging has the capability of restoring the storage capacity of reservoirs where deposited sediment is removed. However, studies in the United States indicated that the cost of dredging might be twice as high as the cost of developing new reservoirs (Alan Plummer Associates et al. 2005). Thus, upstream erosion control remains one of the most viable options for sedimentation control in reservoirs and lakes in Kenya.

This paper gives information about the assessment of sedimentation rates of reservoirs in Kenya and identifies challenges and constraints to the assessment. The paper also evaluates how multifunctional historical data can be generated from sedimentation assessment via Bathymetric Survey System to inform the nexus of food production and water management.

1.1 Natural Lakes and Reservoirs in Kenya

Kenya's physiographic landscape, like the rest of the world, is dotted with both natural lakes and reservoirs. These lakes and reservoirs vary in their biophysical characteristics and socio-economic importance. Most natural lakes in Kenya are in the Rift valley physiographic province. From North to South of the Rift valley the lakes are outlined as: Turkana, Logipi, Baringo, Bogoria, Nakuru, Elementeita, Naivasha and Magadi (Becht et al. 2006). Lake Baringo and Lake Naivasha are the only freshwater lakes in the rift valley. Lake Victoria, Jipe, Chala Kamnarok, Solai, Chew Bahir are located outside the Rift Valley. The location of the major lakes in Kenya is shown in Fig. 1.

The lakes and reservoirs in Kenya vary in their depths and storage capacity. As shown in Table 1 the combined storage capacity of reservoirs in Kenya is about 4,100 M m^3 (FAO 2016; World Bank 2011). The locations of these reservoirs in Kenya are also shown in Fig. 1. The purposes of these reservoirs are mainly hydropower and water supply. Currently, it is estimated that 82% of Kenya's electricity supply is from hydropower (Oludhe 2010) and the bulk of this supply is from five reservoirs along the Tana River. Concerning water supply, the capital city, Nairobi, relies on Thika, Sasumua and Ruiru reservoirs, (Fig. 1), albeit insufficiently, especially during the dry season. Other cities within the country also rely on various reservoirs for their water supply. Despite the socio-economic importance of these reservoirs, the country has not fully explored its reservoir storage potential. In addition, a coherent management strategy, including adequate monitoring, would be critical (Hülsmann et al. 2020) for existing reservoirs but has not been implemented.

Like the rest of the world, sedimentation threatens the volume of natural lakes and reservoirs in Kenya. The scale of the sedimentation problem in Kenya is illustrated by Wooldridge (1984) who estimated a sediment inflow of 357 m^3/km^2/year to the Kamburu reservoir. In Kenya, this information is available for four reservoirs that have been surveyed. However, most of the lakes and reservoirs in Kenya have not been assessed to establish their current sedimentation status. This is further exacerbated by the lack of sediment load monitoring stations upstream of these reservoirs. For the management of these reservoirs and the watersheds upstream, it would be vital to know sedimentation rates in these reservoirs, identify the potential sources of the sediment, propose mitigation measures, and monitor the impacts of the proposed measures. Hence, there is a need to monitor the sediment load in upstream rivers or re-establish this from the rate of sedimentation in reservoirs.

1.2 The Need for Reservoir Sedimentation Assessment

Natural lakes and reservoirs support man's socio-economic activities and are hosts to a highly diverse flora and fauna. Reservoirs are critical for the spatial and temporal distribution of water for domestic and industrial use, food production and energy generation. In Kenya, without reservoirs, most cities would be experiencing poor access to water and sanitation. Reservoirs provide habitats for various flora and fauna. Furthermore, this flora and fauna support man's socio-economic activities like tourism, fishing, and sports. In Kenya, loss of reservoir capacity could also affect electricity generation and have a significant impact on the economy. Hence, sedimentation assessment plays a key role where regular monitoring of sediment load into the reservoir is not undertaken.

Reservoir sediment assessment provides multifunctional historical information about sedimentation rates. This multifunctional information includes the current storage capacity of the reservoirs, sediment thickness with its spatial location as well as physical-chemistry characteristics of sediment and water within the reservoir. Some of this crucial information is currently unknown, missing, or not up to date. When water managers know the current volume, they can adequately control the amount of water supply or decide whether they require alternative sources

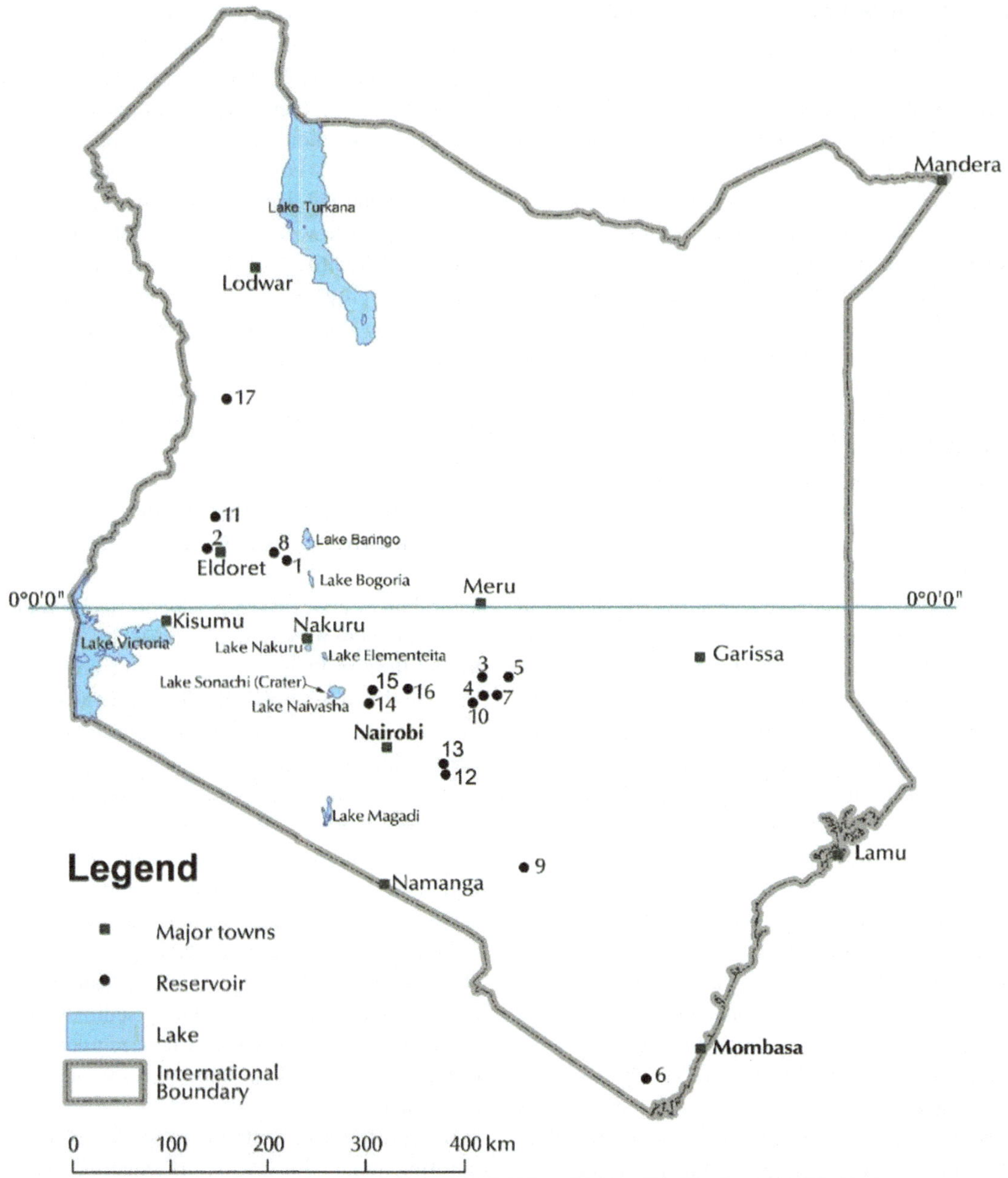

No	Name	No	Name	No	Name	No	Name
1	Chemeron	6	Kikoneni	11	Moiben	16	Thika
2	Ellegirini	7	Kindaruma	12	Mulima	17	Turkwel
3	Gitaru	8	Kirandich	13	Muoni		
4	Kamburu	9	Manooni	14	Ruiru		
5	Kiambere	10	Masinga	15	Sasumua		

Fig. 1 Location of major natural lakes and reservoirs in Kenya

Table 1 Characteristics and uses of the major reservoirs in Kenya

Name of reservoir	Nearest town	River	Start of operation	Capacity (10^6 m^3)	Purpose
Chemeron	Kabarnet	Chemeron	1984	4.60	Irrigation/Water supply
Ellegirini	Eldoret	Ellegirini	1987	2.00	Water supply
Gitaru	Embu	Tana	1978	20.00	Hydroelectricity
Kamburu	Embu	Tana	1974	150.00	Hydroelectricity
Kiambere	Embu	Tana	1987	585.00	Hydroelectricity
Kikoneni	Kwale	Mkanda	1981	1.26	Water supply
Kindaruma	Embu	Tana	1968	16.00	Hydroelectricity
Kirandich	Kabarnet	Kirandich	2000	3.00	Water supply
Manooni	Machakos	Manooni	1987	0.41	Water supply
Masinga	Embu	Tana	1980	1560.00	Flood control/Hydroelectricity
Moiben	Eldoret	Moiben	1991	6.20	Water supply
Mulima	Machakos	Mulima	1982	0.28	Water supply
Muoni	Machakos	Muoni	1987	0.83	Water supply
Ruiru	Nairobi	Ruiru	1949	2.98	Water supply
Sasumua	Nairobi	Sasumua	1956	13.25	Water supply
Thika	Thika	Thika	1994	70.00	Water supply
Turkwel	Kitale	Turkwel	1991	1645.00	Hydroelectricity

of water. In addition, information from reservoir surveys would inform managers on the sedimentation status of the reservoir. Physical–chemical analysis of the sediment samples could indicate if there is potential for water quality change due to contaminants inflow into the reservoir.

The results from surveys also give information on the impacts of anthropogenic activities upstream. This can be assessed from sediment cores and can aid in understanding the dynamics of sedimentation rates during a given time period. McAlister et al. (2013) reported that bathymetric surveys are useful in providing direct accurate assessments of sediment export within the watershed. When bathymetric surveys are combined with sediment cores to measure sediment thickness, accurate reservoir capacity and long-term water volume loss can be determined in a single survey. The advance in technology and survey techniques has made sedimentation assessment attractive for decision making.

1.3 Challenges to Reservoir Sediment Assessment in Kenya

Kenya has implemented various counter measures to curb sedimentation in natural lakes and reservoirs across the country. The main counter measure implemented aims at conserving the upstream catchment. These measures lower soil loss and thus reduce sediment inflow into the reservoirs. In some reservoirs, sediment removal by scouring within the reservoir is regularly undertaken. Even with the different counter measures to curb sedimentation of the reservoirs, little is known about their sedimentation rates. The country's ability to monitor sedimentation is limited. In Kenya, some of the limiting factors range from infrastructure to human resource capacity (Sang et al. 2017). Generally, little has been done on the training of technical expertise to undertake reservoir survey. Other factors that limit the execution of reservoir surveys include the high cost of external expertise, lack of equipment, and poor accessibility of various lakes and reservoirs (Dost and Mannaerts 2008). However, there are currently efforts, in collaboration with international partners, to address these limitations.

2 Reservoir Sedimentation Assessment Techniques

Various pre-and post-impoundments reservoir survey techniques can be used to determine the volume and sedimentation rate in reservoirs. The pre-impoundment techniques involve establishing the potential reservoir volume during the construction of the reservoir. Before the 20th century, water depth was measured manually from a boat using a sounding line and lead weights. At the tip of the weight, an adhesive was applied to indicate sediment level has been reached (Jakubauskas 2008). In some cases, where water is shallow, graduated poles were used. Another technique used to survey sediment distribution in the reservoir is the use of a spud survey. Spud survey estimates the thickness of sediment and it can even be used in areas where previous sediment thickness information was not available. In this method, a sounding weight is used to confirm the surface of the sediment (Ritchie and McHenry 1985; Morris and Fan 1998). The method works well in areas where sediment thickness is up to 4 m (Morris and Fan 1998). The spud is dropped into the water and allowed to fall vertically until it penetrates the deposited sediments. The spud has grooves where sediments enter, ensuring it is not washed away as the spud is being retrieved from the water. The method is easy, but the accuracy of data depends on the operator's experience. In addition, the method is more applicable to small reservoirs.

Another method that can be used for reservoir assessment is ranging. The method uses permanent ranges in the reservoir especially those that are re-surveyed at regular intervals. It is thus useful in computing the volume difference between the surveys. To get very accurate results the method can be used together with bathymetry survey or aerial photographs, where photogrammetry is used to define contour lines.

Photogrammetric techniques or airborne lasers are useful especially when the reservoirs are dry or where a high drop of water level is noted (Morris and Fan 1998). Initial reservoir volume is determined from pre-impoundment topographic maps and aerial surveys. To improve accuracy, the method is used together with ground-truthing activities. The use of aerial survey technique is illustrated by Wooldridge (1984) who used pre-impounded aerial photographs of 1965 and a sediment assessment of 1982 to determine sediment accumulation in the Kamburu reservoir, in Kenya. However, the aerial survey method is limited in its use since it can only be used before the impoundment of the reservoir or in cases where the reservoirs are completely empty. Finally, other options use Light Detection and Ranging (LIDAR). Although advances in remote sensing have enabled the use of LIDAR data in reservoir surveys, the method was reported to be very expensive and time consuming (Heyman et al. 2007), thus it cannot be easily used in regular surveys.

A need exists for a fast and cheap methodology that can be used for sedimentation assessment. This would benefit developing and some developed countries where over 60% of reservoirs do not have baseline sedimentation data. Modern technology that allows simultaneous operation of multiple transducers (i.e. collection of multiple transducer data separated by acoustic wavelength) is important. This makes it possible to collect spatially and temporally correlated acoustically independent data. An example of such a low cost methodology is the use of a Bathymetric Survey System (BSS). The BSS combines multi-frequency Acoustic Profiling System (APS) and a vibe-coring system (Fig. 2). The use of independent multi-frequency APS means that surveyors can utilize higher frequency acoustics to calculate water depth, while simultaneously utilizing the sediment-penetrating capability offered by lower frequency to map post-impoundment sediment thickness (Dunbar et al. 1999). The BSS works in three frequency ranges i.e. 200, 50 and 12 kHz (SDI 2017; Dunbar et al. 1999). High frequency is used to determine the current surface of the sediment and low frequency penetrates through the deposited sediment layer to pre-impoundment levels of the reservoir. Thus, BSS offers a technological means to verify historic sedimentation estimates and to quantify the effectiveness of conservation implementation (McAlister et al. 2013). As per McAlister et al. (2013), acoustic surveying techniques remain a superior methodology for accurately calculating reservoir volumes.

During a survey with BSS, a boat is driven in predetermined transects. The sounder which is mounted on the side of a boat aids in establishing the water depths and thickness of the post-impoundment deposited sediments. The depth of recently deposited sediments is further confirmed by core samples collected at selected locations using a vibe-coring device. The collected depth data are processed to generate water and sediment volume. Confirmation can also be derived from the spud bar method (McAlister et al. 2013). In addition, dating of the sediment core can provide a crucial historical timeline of deposition where no previous sedimentation records exist. Dating exploits the use of artificial radionuclides like ^{137}Cs and ^{210}Pb. However, the fallout radionuclides have rarely been exploited in Africa to assess soil erosion and reservoir sedimentation status (Maina et al. 2018).

3 Findings from a Reservoir and a Natural Lake in Kenya

BSS was used to survey two water bodies in Kenya. The aim of the survey was to determine the current volume and sedimentation status of Ruiru Reservoir and Lake Sonachi. Ruiru Reservoir was constructed and completed in 1949 (Table 1) and was the first source of water to the City of Nairobi. It is located 60 km north of Nairobi and currently supplies 22,700 m^3 per day of water to Kabete treatment works before being supplied to critical parts of the City. Given the age of the reservoir, its management has been under various government agencies, from the colonial government to various commissions governing Nairobi to the County government. Unfortunately, with this transition from one management organisation to the other most of it records were lost (Sang et al. 2017). Therefore, the management of the reservoir and its upstream watershed is dependent on routine practise instead of data-informed decision-making.

In this study, a natural lake that was surveyed is Lake Sonachi, (barren lake as per the local Maasai community), a small endorheic alkaline

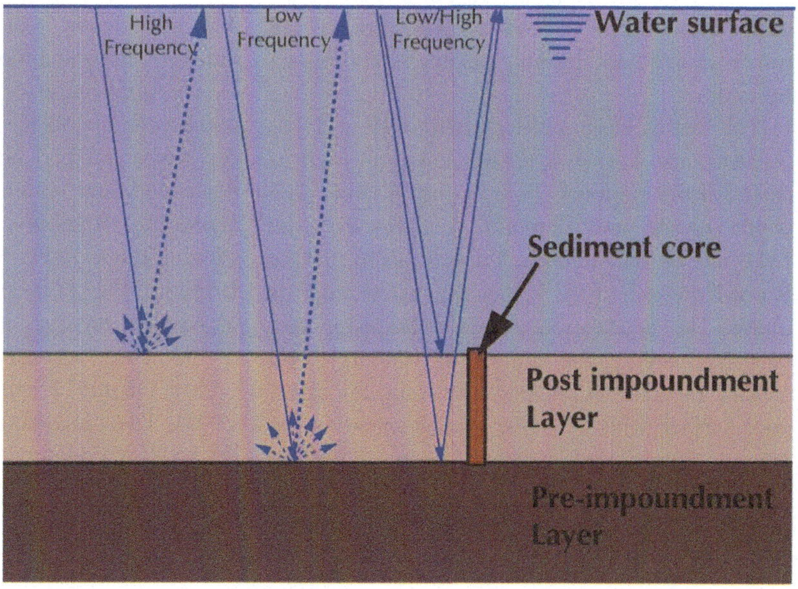

Fig. 2 Schematic view of the use of multi-frequency acoustic profiling system and vibro-coring techniques in reservoir sedimentation survey

lake located about 100 km northwest of Nairobi (Njuguna 1988; Verschuren 1999) and about 3 km from the main Lake Naivasha. Although the Lake is fully independent of the main Lake Naivasha, its levels are believed to oscillate in harmony with the main lake as a result of groundwater connection (Mavuti and Harper 2006). The lake's catchment is pristine. Unlike the surrounding catchment of the larger Lake Naivasha, clearance of natural vegetation is significantly limited, thus the sedimentation is probably very low.

In the Ruiru Reservoir, this study generated information about the reservoir that is critical for management but was missing before the study. Past undated records available show that the volume was 2,980,000 m^3, but this information was not sufficient for the management of the reservoir and water supply monitoring. However, sediment assessment showed information on current sediment distribution (Fig. 3), surface and volume-depth relationships (Fig. 4). The distribution of sediment in Ruiru Reservoir is shown by the Isopach in Fig. 3. Analysis of the sedimentation showed that the reservoir has lost approximately 14% of its recorded volume as reported also by Maloi et al. (2016) and Sang et al. (2017). This loss is equivalent to an increase in sediment thickness of 1.75 cm/year assuming a linear sedimentation rate. This is important for the company to be able to determine the volume of water available in the reservoir and even design plausible interventions upstream.

The bathymetric survey conducted in Lake Sonachi, also known as Crater Lake, allowed to establish the contours (Fig. 5); Fig. 6 shows the depth-volume and depth-area curves for the Lake. As aforementioned, Lake Sonachi, unlike Ruiru Reservoir (Fig. 3), has no visible inlet or outlet. The depth-surface area relationship has been compared to the findings of MacIntyre and Melack (1982). To compare this data, the reference surface with the same spatial extent as the current survey was assumed to be the point

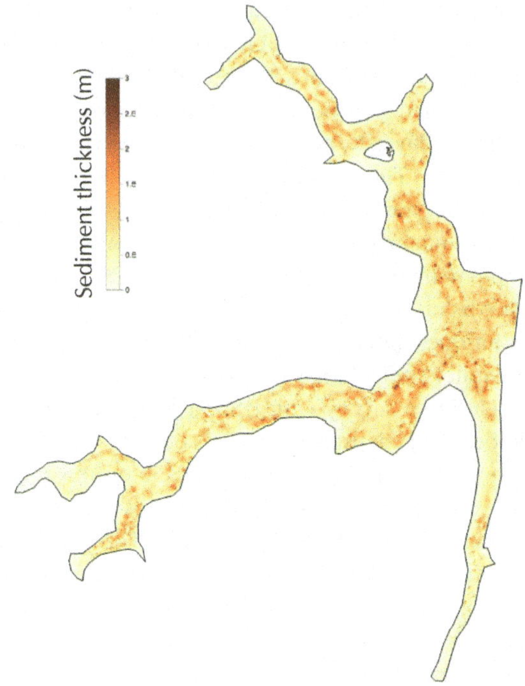

Fig. 3 Ruiru reservoir sediment isopach generated from analysing Bathymetric survey System (BSS) data

where the data overlap. Whereas this method might not be accurate, there was no other indication of the standard reference points such as height above sea level to aid in data comparison. However, comparison of the two graphs shows that there is a change in the area at the bottom and mid depth, which could be attributed to autochthonous sedimentation.

The application of this sedimentation assessment technique in Kenya is low. Other lakes and reservoirs that had been previously surveyed to determine their volumes are Lake Bogoria, Lake Naivasha, Kamburu, Masinga and Sasumua (World Bank 2011; Hunink et al. 2013). Lake Bogoria (Fig. 1) a saline lake in Eastern Rift Valley was surveyed in 2002 using a Lowrance X—15A chart recording echo sounder, coupled with a Garmin12 handheld GPS (Hickley et al. 2003). The processing of the data was manual, where X, Y coordinates were plotted and contour

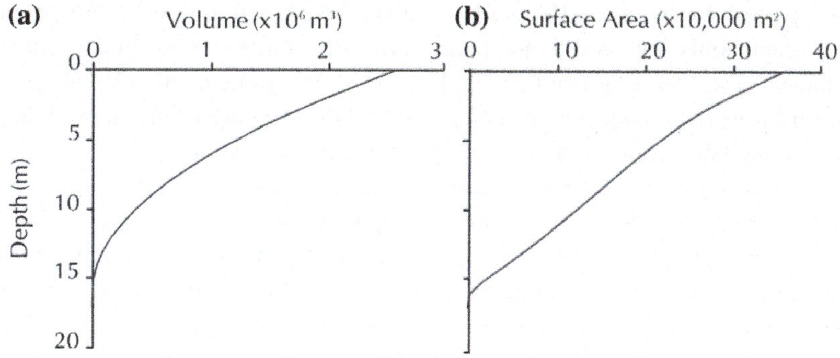

Fig. 4 The relationship between **a** depth and volume and **b** depth and surface area for Ruiru reservoir

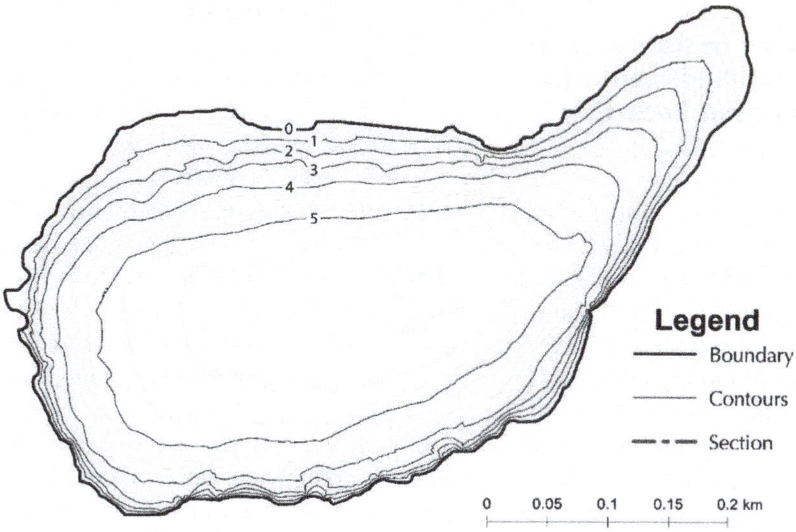

Fig. 5 Lake Sonachi (Crater Lake) contours generated from analysing bathymetric survey data

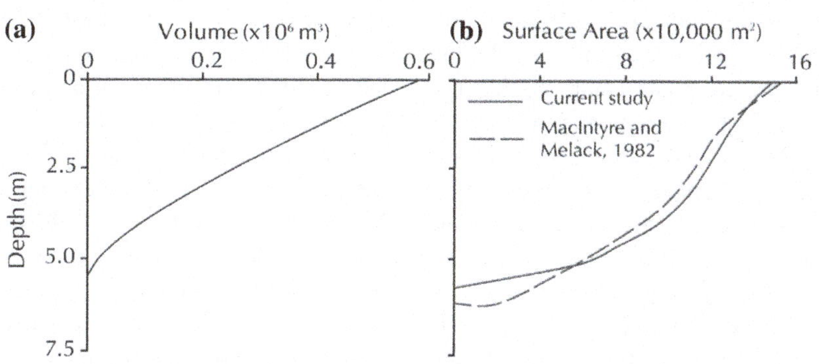

Fig. 6 Relationship between **a** depth and volume and **b** depth and surface area for Lake Sonachi (Crater Lake). The depth surface relationship from (MacIntyre and Melack 1982) is included for comparison

lines drawn using best fit by eye. The survey generated water depth only but could not map sediment thickness. The same procedure had previously been used in Lake Naivasha to determine the volume of the lake (Hickley et al. 2002). This indicates the need for a more robust method like the use of multi-frequency acoustic profiling system (APS). The multi-frequency survey would be more useful in the combined measurement of water depth and sediment thickness measurement. This method is important since it also gives an overview of the spatial distribution of sediment within the lakes and reservoirs.

4 Importance of Bathymetric Survey and Sedimentation Assessment in Reservoir Management

Developing these tools for reservoir management is important in informing water managers of the volume of water resources available. In successive surveys, change in depth-volume or depth-area relationships can be an indicator to reservoir managers of a need to change water abstraction. It also informs them of how much of the reservoir is affected by sedimentation. Successive surveys provide an indication of the sedimentation rate within the lakes or reservoirs. This approach was used by Wisser et al. (2013) to give an indication of sedimentation rates of 1,024 reservoirs in the contiguous US and of 191 reservoirs to give an indication of the global average. Changes in the depth-volume relationship or depth-surface relationship could also indicate a change in the sedimentation status of the reservoirs. This would call for intervention measures either upstream or within the dam to slow down the volume loss. Such measures could then be evaluated economically, at least for reservoirs used for hydropower generation as demonstrated by Ngondjeb and Ayuk (2021). In combination with dating, historical data generated can be used to analyse sediment yield dynamic and relate them with various events in the past such as extensive land use changes or fire events and propose mitigation measure in case of similar events in the future as demonstrated by Sang et al. (2015) and Allen et al. (2011) in the case of an increase in sediment rate after drought events.

The change in depth-volume relationship or depth-surface relationship could also be an indicator of the need to change the approach used in releasing water downstream or to the consumers. Since this change indicates a change in volume, a new approach to manage the available volume would be critical.

5 Conclusions

There are many lakes and reservoirs in Kenya, which are mainly used for water supply, irrigation, tourism, and hydropower generation. However, the reservoirs' functions are threatened by sedimentation which leads to loss of storage capacity, hence affecting the development and economy of the country. However, little has been done to understand the status of these lakes and reservoirs, since most of these water bodies do not have continuous monitoring of sediments other than the water level at a gauging station. In this study, BSS was successfully used to survey a reservoir and a natural lake in Kenya. The results from the sedimentation assessment showed that the Ruiru Reservoir has lost about 14% of its original volume due to sedimentation at an approximate sediment accumulation rate of 1.75 cm/year over the past 67 years. These results were used to provide critical tools and data for the management of the reservoirs, which were missing before these studies. The use of BSS in sedimentation assessment plays a key role in generating multifunctional historical data of these reservoirs. However, the application rate in Kenya is low where only three out of about 17 major reservoirs have been surveyed. Therefore, there is a need for technical capacity building on how these robust methods can be applied to the already threatened water bodies as part of a comprehensive monitoring strategy facilitating integrated resources management.

Acknowledgements The authors would like to thank Nairobi City Water and Sewerage Company (NCSWC) and the Jomo Kenyatta University of Agriculture and Technology (JKUAT) for funding parts of this project. The seed money for purchasing the survey equipment from Specialty Devices Inc. (SDI) was provided by the United States Department of Agriculture—Agriculture Research Service (USDA-ARS).

References

Alan Plummer Associates, Allen PM, Dunbar JA (2005) Dredging versus new reservoirs. Alan Plummer Associates, Fort Worth, Texas

Allen PM, Harmel RD, Dunbar JA, Arnold JG (2011) Upland contribution of sediment and runoff during extreme drought: a study of the 1947–1956 drought in the Blackland Prairie, Texas. J Hydrol 407(1–4):1–11. https://doi.org/10.1016/j.jhydrol.2011.04.039

Becht R, Mwango F, Amstrong Muno F (2006) Groundwater links between Kenyan Rift Valley lakes. In: Odada EO (ed) Water Kenya. Ministry of Irrigation and Committee International Lake Environment, Proceedings of the 11th World Lakes conference, Nairobi, Kenya, 31 October to 4th November 2005. International Lake Environment Committee, Kusatsu-shi, Shiga, Japan

Dost RJJ, Mannaerts CMM (2008) Generation of lake bathymetry using sonar, satellite imagery and GIS. In: ESRI 2008: proceedings of the 2008 ESRI international user conference

Dougill AJ, Stringer LC, Leventon J, Riddell M, Rueff H, Spracklen DV, Butt E (2012) Lessons from community-based payment for ecosystem service schemes: from forests to rangelands. Philos Trans Roy Soci. Series B—Biol Sci 367(1606):3178–3190. https://doi.org/10.1098/rstb.2011.0418

Dunbar JA, Allen PM, Higley PD (1999) Multifrequency acoustic profiling for water reservoir sedimentation studies. J Sediment Res 69(2):521–527

FAO (2016) AQUASTAT main database. Food and Agriculture Organization of the United Nations (FAO). Accessed 31 Jan 2016

Heyman WD, Ecochard J-LB, Biasi FB (2007) Low-cost bathymetric mapping for tropical marine conservation—a focus on reef fish spawning aggregation sites. Mar Geodesy 30(1–2):37–50. https://doi.org/10.1080/01490410701295996

Hickley P, Bailey R, Harper DM, Kundu R, Muchiri M, North R, Taylor A (2002) The status and future of the Lake Naivasha fishery, Kenya. Hydrobiologia 488 (1):181–190. https://doi.org/10.1023/a:1023334715893

Hickley P, Boar RR, Mavuti KM (2003) Bathymetry of Lake Bogoria, Kenya. J East Afr Nat History 92 (1):107–117. https://doi.org/10.2982/0012-8317(2003)92[107:BOLBK]2.0.CO;2

Hunink JE, Niadas IA, Antonaropoulos P, Droogers P, Vente J (2013) Targeting of intervention areas to reduce reservoir sedimentation in the Tana catchment (Kenya) using SWAT. Hydrol Sci J 58(3):600–614. https://doi.org/10.1080/02626667.2013.774090

Hülsmann S, Rinke K, Paul L, Diez Santos C (2020) Storage reservoir operation and management. in: Bogardi J et al. The springer handbook of water resources management, in press

Jakubauskas M (2008) Methods for assessing sedimentation in reservoirs. In: Hargrove WL (ed) Sedimentation in our reservoirs: causes and solutions, Kansas State University Agricultural Experiment Station and Cooperative Extension Service

Karlberg L, Hoff H, Amsalu T, Andersson K, Binnington T, Flores-Lopez F, de Bruin A, Gebrehiwot SG, Gedif B, zur Heide F, Johnson O, Osbeck M, Young C, (2015) Tackling complexity: understanding the food-energy-environment nexus in Ethiopia's Lake Tana sub-basin. Water Altern-An Interdisc J Water Polit Dev 8:710–734

Kawashima S, Johndrow TB, Annandale GW, Shah F (2003) Reservoir conservation vol ii: RESCON model and user manual. The World Bank, Washington, DC

Kondolf GM, Gao Y, Annandale GW, Morris GL, Jiang E, Zhang J, Cao Y, Carling P, Fu K, Guo Q, Hotchkiss R, Peteuil C, Sumi T, Wang HW, Wang Z, Wei Z, Wu B, Wu C, Yang CT (2014) Sustainable sediment management in reservoirs and regulated rivers: experiences from five continents. Earth's Future 2(5):256–280. https://doi.org/10.1002/2013ef000184

MacIntyre S, Melack JM (1982) Meromixis in an equatorial African soda lake. Limnol Oceanogr 27:595–609. https://doi.org/10.4319/lo.1982.27.4.0595

Maina CW, Sang JK, Mutua BM, Raude JM (2018) A review of radiometric analysis on soil erosion and deposition studies in Africa. Geochronometria 45 (1):10–19. https://doi.org/10.1515/geochr-2015-0085

Maloi SK, Sang JK, Raude JM, Mutwiwa UN, Mati BM, Maina CW (2016) Assessment of sedimentation status of Ruiru reservoir, central Kenya. Am J Water Resour 4(4):77–82

Mavuti KM, Harper D (2006) The ecological state of Lake Naivasha, Kenya, 2005: Turning 25 years research into an effective Ramsar monitoring programme. In: 11th World Lakes Conference, Nairobi, Kenya

McAlister JR, Fox WE, Wilcox B, Srinivasan R (2013) Reservoir volumetric and sedimentation survey data: a necessary tool for evaluating historic sediment flux and appropriate mitigation response. Lakes & Reservoirs: Res Manage 18(3):275–283

Morris GL, Fan J (1998) Reservoir sedimentation handbook: design and management of dams, reservoirs, and watersheds for sustainable use. McGraw-Hill, New York, N.Y.

Nduhiu C, Gathenya JM, Mwangi JK, Aman M, Mutisya T (2016) Assessment of the effectiveness of Payment for Ecosystem Services (PES) in the delivery of desired Ecosystem Services in Sasumua catchment, Kenya. Hydrol Earth Syst Sci Discuss 1–20. https://doi.org/10.5194/hess-2016-541

Njuguna SG (1988) Nutrient-phytoplankton relationships in a tropical meromictic soda lake. In: Melack JM (ed) Saline Lakes. Springer, Netherlands, Dordrecht, pp 15–28

Ngondjeb DY, Ayuk E (2021) Economic valuation of environmental services associated with agriculture in the watershed of Lake Lagdo, Cameroon. In: Hülsmann S, Jampani M (eds) A nexus approach for sustainable development. Springer Nature, Cham, Switzerland. https://doi.org/10.1007/978-3-030-57530-4_7

Oludhe C (2010) Impacts of climate variability on power genera on within the 7-Forks Dams. J Meteorol Related Sci (KMS 10th Conference Special Issue)

Palmieri A, Shah F, Annandale GW, Dinar A (2003) Reservoir conservation volume I: The RESCON approach economic and engineering evaluation of alternative strategies for managing sedimentation in storage reservoirs. World Bank, Washington, D.C.

Ritchie JC, McHenry R (1985) A comparison of three methods for measuring recent rates of sediment accumulation. J Am Water Resour Assoc 21(1):99–104. https://doi.org/10.1111/j.1752-1688.1985.tb05356.x

Sang JK, Allen PM, Dunbar JA, Arnold JG, White JD (2015) Sediment yield dynamics during the 1950s multi-year droughts from two ungauged basins in the Edwards Plateau, Texas. J Water Resour Prot 07 (16):1345–1362. https://doi.org/10.4236/jwarp.2015.716109

Sang JK, Raude JM, Mati BM, Mutwiwa UN, Ochieng F (2017) Dual echo sounder bathymetric survey for enhanced management of Ruiru reservoir, Kenya. J Sustain Res Eng 3(4):6

SDI (2017) Sediment mapping and sampling. Specialty Devices Inc. http://specialtydevices.com/index/survey-services/hydrographic-survey-services/sedimentation-mapping-and-sampling/. Accessed 12 August 2017

Sumi T, Okano M, Takata Y (2004) Reservoir sedimentation management with bypass tunnels in Japan. In: Proceedings of the ninth international symposium on river sedimentation

Tiğrek S, Aras T (2012) Reservoir sediment management, 1st edn. CRC Press/Balkema, London

Tiğrek Ş, Göbelez Ö, Aras T (eds) (2009) Sustainable management of reservoirs and preservation of water quality. In: Moujabber ME, Mandi L, Liuzzi GT, Martin I, Rabi A, Rodriguez R (eds) Options Méditerranéennes, Series A, No. 88 technological perspectives for rational use of water resources in the Mediterranean Region Bari, Centre International de Hautes Etudes Agronomiques Méditerranéennes (CIHEAM), Italy

Verschuren D (1999) Influence of depth and mixing regime on sedimentation in a small, fluctuating tropical soda lake. Limnol Oceanogr 44(4):1103–1113. https://doi.org/10.4319/lo.1999.44.4.1103

Wisser D, Frolking S, Hagen S, Bierkens MFP (2013) Beyond peak reservoir storage? A global estimate of declining water storage capacity in large reservoirs. Water Resour Res 49(9):5732–5739. https://doi.org/10.1002/wrcr.20452

Wooldridge R (1984) Sedimentation in reservoirs: Tana River basin, Kenya. Ill-Analysis of hydrographic surveys of three reservoirs in June/July 1983. In Report no. 6. Hydraulic Research, Wallingford, UK

World Bank (2011) Towards a strategic analysis of water resources investments in Kenya: hydrological, economic, and institutional assessment for storage development. World Bank, Washington DC, USA

Economic Valuation of Environmental Services Associated with Agriculture in the Watershed of Lake Lagdo, Cameroon

Dorothe Yong Ngondjeb and Elias Ayuk

Abstract

Sedimentation of the reservoir has been recognized as a major consequence of land degradation and erosion in the watershed. The objective of this paper was to determine the benefits of soil conservation practices on hydroelectric power generation at Lake Lagdo located in the northern part of Cameroon during the 1984–2009 period, using the damage function approach. The first step of the analysis was to estimate the physical effect of environmental change on soil erosion considering different land management practices. In the second step, the quantitative link between lake sedimentation and its effects on the storage capacity of the reservoir was established. Finally, the economic value of the impact was determined. The results show that the loss of direct storage capacity of the reservoir due to sedimentation, which increased the costs of electricity production. The annual average sedimentation of around 40.80 million cubic meters across the reservoir reduced total reservoir capacity by around 0.9%. Measures to reduce soil erosion include payments to landowners as powerful incentives for promoting environmentally friendly land-use practices that help to sustain ecosystem services (PES). The study illustrated very well the demand for a Nexus Approach looking at interdependencies, in this case between the agricultural and energy sectors.

Keywords

Ecosystem services · Land-use practices · Non-market valuation · Damage function · Sedimentation

1 Introduction

Land degradation in the form of soil erosion is one of the major environmental problems of developing countries (De Graaff 1996). Cameroon is one country that suffers from widespread and serious erosion induced by water. This problem is acute mainly in the Sudano-Sahelian Area (SSA), where agriculture is perceived as a major factor of environmental degradation and more particularly as promoting soil erosion (Abou et al. 2006; Ecam et al. 2008). In addition, the tropical ferruginous soils in SSA have limited intrinsic chemical and physical fertility; their reduced surface clay content, low organic matter and nitrogen content, limited water holding capacity and slightly acidic pH make them

D. Y. Ngondjeb (✉) · E. Ayuk
United Nations University Institute for Natural Resources in Africa, Accra, Ghana
e-mail: njedora@gmail.com

D. Y. Ngondjeb
Faculty of Economics and Management, University of Yaoundé II, P.O. Box 1365, Yaoundé, Cameroon

S. Hülsmann and M. Jampani (eds.), *A Nexus Approach for Sustainable Development*,
https://doi.org/10.1007/978-3-030-57530-4_7

susceptible to erosion (Baboule et al. 1993) This erosion is accentuated by a rapid increase in the agricultural population due to population growth and migration (M'biandoun and Olina 2007). Although biophysical and demographic factors are in part responsible for land degradation in SSA, inappropriate land-use practices mainly related to cotton development have accelerated the rate of harm (Ngondjeb et al. 2014).

On the other hand, for nearly 20 years, a series of projects have been promoting conservation agriculture with the support of donors in order to sustainably improve the agricultural production of vulnerable farms in the SSA (Abou et al. 2006). Given the known and potential negative effects of soil erosion, governments in Cameroon have been undertaking different policies and strategies to address soil degradation as part of poverty reduction and environmental conservation efforts. This is particularly the case of the 1994–2002 project "Développement paysanal et gestion des terroirs" (DPGT), followed by the Eau-Sol-Arbre (ESA) project since 2002. These projects promote agricultural conservation techniques including non-tillage practices, mulching, contour edges, alley farming and terrace construction that reduce the phenomenon of accelerated erosion and improve the organic status of soils (Thevoz 2000). Generally, however, efforts so far to promote the technologies have had insignificant impacts on increasing the adoption of different soil conservation practices. As a result soil erosion remains an intractable challenge to the area.

The limited success of various conservation practices suggests the need to explore the problem areas in implementing conservation activities. Most studies focus on the on-farm effects of the soil erosion but for public policy to address the challenge of soil erosion a comprehensive assessment of off-site impact is needed. This research aims at tackling this problem by assessing the off-site impacts of erosion and the value of avoiding this effect from society's point of view. Hence, our study provides the necessary inputs in designing possible intervention strategies.

The SSA of Cameroon is one of the most vulnerable areas for the off-site impacts of soil erosion, including proximate property damage, run-off and sedimentation of Lake Lagdo and Lagdo Dam, habitat degradation, and compromising rural road safety (Baboule et al. 1993). The presence of off-site costs/benefits leads to market failure or externalities (Barbier 1996; Boxall et al. 1996; Pretty et al. 2000; Hein 2007). These impacts are not reflected in market price since they affect other agents in the community who do not cause the action. The failures of the market to take account of this cost encourage actions that are not economically optimal for society as a whole. Bekele (2003) noted that this externality has implications for soil problems, especially in developing countries i.e. farmers lack sufficient incentive to consider the off-site impacts of their practice. This aggravates the problem further and leads to non-optimal resource use and allocation. The need to control the problem through appropriate mechanisms also requires an assessment of the value of the resource.

The change in land use practices generates multiple benefits as environmental, societal and cultural externalities (Zhang et al. 2007; Nijkamp et al. 2008). The economic assessment of the environmental benefits of agriculture requires a thorough understanding of the environmental services through the lens of environmental economics. When referring to environmental services, environmental economists do not rely on the functions that natural capital can fulfil, but rather on the environmental benefits that human activities can generate (Chevassus-au-Louis et al. 2009; Swinton et al. 2007). In this study, we frame positive externalities of production as service-externality, as opposed to service-function of the ecological approach (Gillis et al. 1998) and service-delivery of the services approach (Gadrey 2000; Aznar and Perrier-Cornet 2003).

The environmental service is therefore understood here as a positive externality of production resulting from human activity or also identified with an unintentional relationship between two agents without a specific production cost for the externality issuer and without compensation for the final receiver. This is the case, for example, with an agricultural practice that generates an unintended effect on the landscape. In this sense, Madelin (1995) identifies the positive action of agriculture on the environment as a positive externality of production and thus as an environmental service. Mahe (2001) does the same thing when he speaks of "environmental goods and services." Taking externalities into account and addressing the management of environmental resources in an integrative manner is the essence of the nexus approach (Hoff 2011).

The economic valuation of environmental goods and the environment is mainly based on methods that do not bring the environment back to the state of a simple commodity that we can buy or exhaust freely. It allows for providing an element of comparison to more classical economic variables mobilized in the analysis of decisions and political choices. Environmental economists propose to determine the economic value of the service provided to final users. They have developed at least a tacit understanding about the major categories of values to be considered in economic valuation (e.g., Brauer 2003). The total economic value is divided into three broad categories of values: use values, non-use values, and option values (Pearce and Markandya 1989; Pearce and Turner 1990; Tietenberg and Lewis 2008).

Using monetary value as a proxy for direct-use value is relatively straightforward and involves reliance on existing market prices. Non-market valuation can be used to quantify the benefits of different conservation practices with respect to off-site effects. The total economic value of a resource can be measured using either stated or revealed preference approaches. Given the objectives of this study, revealed preference techniques will be used to value the benefits of avoiding off-site impacts.

The revealed preference method used in this study relies on surrogate markets for environmental services to estimate monetary value based on indirect use values (Chevassus-au-Louis et al. 2009). These are the methods in which agents' choices and market prices serve as a basis for evaluation. In this study, we use the damage function approach (or dose–response function), which first evaluates the damage in physical terms and then monetizes it.

2 Materials and Methods

Exploring the interlinkages between resources and advancing an integrated management approach requires integrated modelling tools. For implementing such tools there is a pressing need for better disciplinary and interdisciplinary data. The damage function described below, highlights the interfaces, interactions, and fluxes between resources (soil–water) and sectors (water, energy, agriculture). This subsection aims to define data and data sources and to explain the methodology used to understand the interlinkages between environmental resources and sectors.

2.1 Study Area and Data Source

The Lagdo watershed was chosen as a geographically specific area to measure the off-site costs of erosion. The watershed covers approximately Cameroon's North and Far North provinces. It covers an area of 10.2 million hectares, of which 5.56 are cultivated. Its population of nearly 5 million is predominantly rural (77.6%). It is the cotton zone of Cameroon and while there is a great diversity of landscapes and economic activities, it is mainly focused on subsistence or agro-industrial agricultural production (CEDC 2002). Two reasons made this area an appealing choice. Firstly, the SSA contains many activities that are known to be impacted by erosion. Secondly, severe erosion problems in certain areas and under certain climatic conditions have already been documented (Abou et al. 2006; Ngondjeb et al. 2014). In the study area, soils are constituted of discordant or alkaline granites and alluvium with a sandy to loamy and sometimes loamy-sandy texture. Their reduced surface clay content, low organic matter and nitrogen content, limited water holding capacity and slightly acidic pH make them susceptible to erosion (Baboule et al. 1993). The climate is the sudano-sahelian type and it is generally characterized by a long dry season, which lasts seven to nine months (October to April) and a short rainy season of three to five months (mid-May to mid-October). The average rainfall amplitude is around 1,000 mm/year and temperatures oscillate around 28 °C. This rainfall plays a significant role in soil erosion, which itself favours the siltation of the lake during intense rainy episodes. In the watershed of Lake Lagdo, agricultural activity concerns 85,000 km^2 or almost 18% of the whole territory. The vegetation is the savannah type to thorny steppe characterised by the presence of species like Acacia albida (winter thorn), Balanites aegyptiaca (soapberry), Acacia spp. and Azadiratcha indica (Neem).

The Lagdo Dam is located on the upstream of the Benoue River, about 40 km from Garoua. This dam is of great importance to the northern region of the country. The benefits of Lake Lagdo include: electricity generation for users in Cameroon's three Northern regions, Adamaoua, the North and the Far North; agricultural activities within an irrigated area of about 1,000 hectares that have been developed downstream of the dam for the production of rice and other food crops; and fishing in the reservoir itself as well as in fish ponds developed downstream.

The lake is characterized by an elevation of 206 m above sea level (asl), the maximum water level is 218.18 m and the normal water level is 216 m. The commissioning of the hydroelectric plant in 1984 has stored approximately 7.7 billion cubic meters of water, covering an area of 700 square kilometres. The map of the SSA identifies the Lake Lagdo watershed and the location of the dam (Fig. 1).

Data used for this study are secondary data from both MEADEN (Mission d'Etudes pour l'Aménagement et le Développement de la province du Nord Cameroun) and the Lagdo Hydroelectric Plant. Data collection took place in February 2008. Data from the Lagdo Hydroelectric Plant include the capacity of dead and live storage, size and elevation of the storage, generated power, volume of turbine water, and prices of the kilowatt hour of electricity. Data from MEADEN include sedimentation volume in Lake Lagdo.

2.2 Methodology

2.2.1 The Damage Function Approach

The particular off-site cost of soil erosion and the sedimentation of dam reservoirs are examined in detail in this study. According to Pearce and Markandya (1989), the damage function method is a method of estimating effects based on observing actual market behaviour, rather than observing individuals' preferences. It aims to assign monetary values to goods or services induced by an improvement in the quality of the environment (or conversely to evaluate the cost of degradation of the quality of the environment). It belongs to the methods of monetary evaluation of the physical effects of the degradation of the environment and assumes that there is physical damage to non-market goods and services, for which the monetary equivalent is sought (Garrabé 1994).

Its objective is to assess the cost of deteriorating environmental quality. The use of the damage function method requires knowledge of the physical and ecological relationships between the cause and its impact, necessary for the prior specification of dose–response relationships. Pearce and Markandya (1989), quoted by Njomgang (2003) have developed a function whose simplest form is the following:

$$R = R(P, Z) \quad (1)$$

where R denotes impact (effect), P (dose) pollution, and Z other factors (e.g. a palliative to the R effect). The approach to measuring the monetary value of physical damage involves two steps. The first step calculates the variation of R with respect to P, i.e. the elasticity, or more simply ⊗R / ⊗P. The second step multiplies this quantity by the unit cost of the damage, denoted V * (⊗P/P), to obtain V * (⊗R/R), which designates the "avoided damage." V represents the value per unit of physical damage borne, which

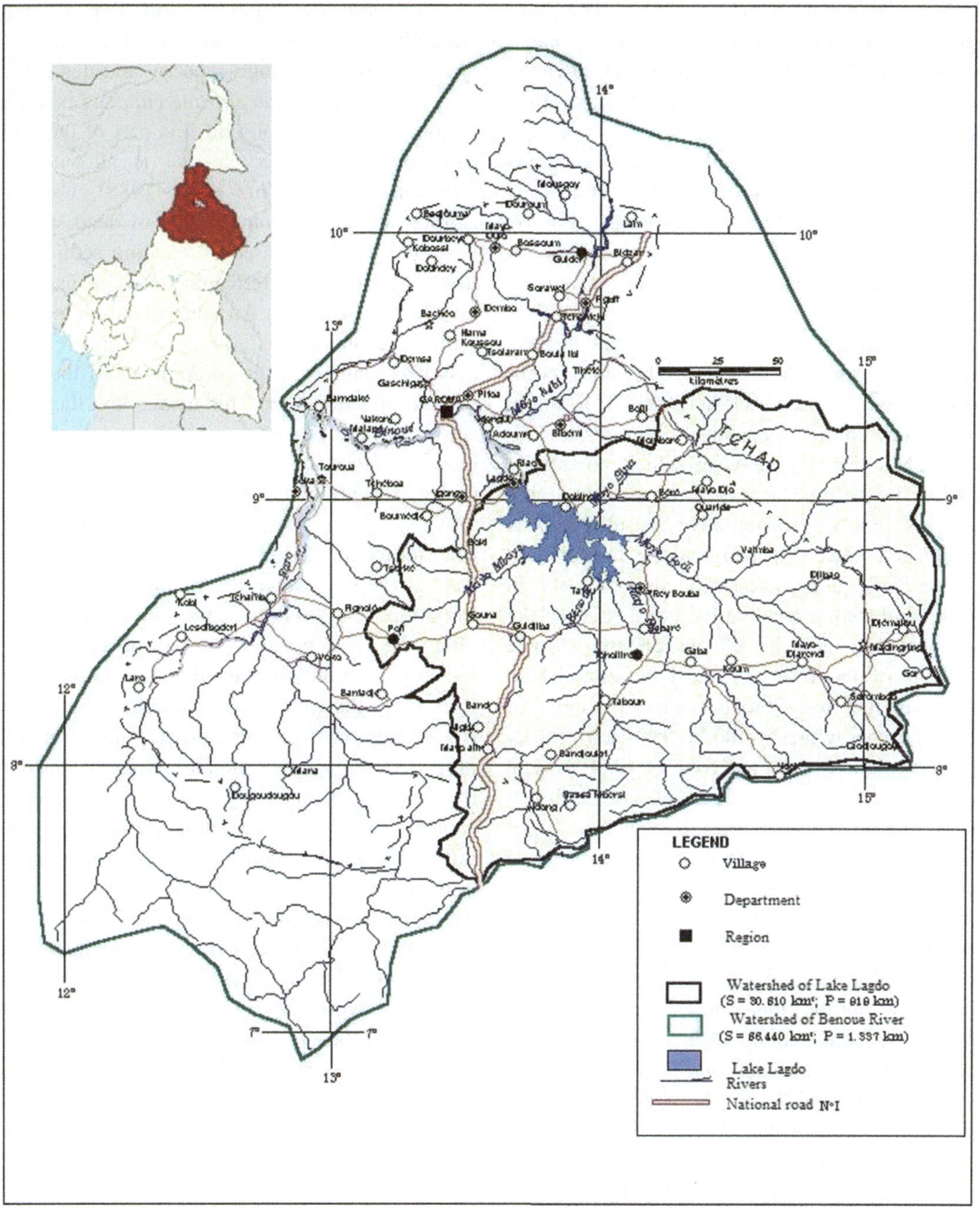

Fig. 1 Location of the reservoir dam and situation of Lagdo watershed

leads to a monetary damage function. The damage function is thus obtained from the estimated relationship between dose and effect after an economic price has been attributed to each degraded physical unit.

The damage function method, like all methods of monetary valuation of physical effects, is carried out in three fundamental stages. Firstly, the physical effect of environmental change on the receiver (real estate, natural resources, machinery or persons affected by the modification) is estimated. In the case of Lake Lagdo, this relationship was established by the study commissioned by MEADEN in 2005. This study revealed a sediment deposit of about 40.80 million cubic meters each year, which potentially implies the loss of a volume of water of the same order. Secondly, a quantitative causality link is established between a change in the environment and its consequences on production or costs. In the second step, a quantitative link between lake sedimentation and its effects on the storage capacity of the reservoir must be established. This step requires estimation of the amount of electrical energy that would be lost each year by the depreciation of the reservoir of 40.80 million cubic meters of water lost due to sediment deposition in the lake. Based on the study by (Cruz et al. 1988), which estimated the off-site costs of sedimentation of the dam reservoir in the Magat and Pantabangan watersheds in the Philippines, we can determine the volume of water needed to generate the kilowatt hour of electricity. Thirdly, the economic value of these impacts is estimated by estimating the market value of these impacts on production or costs. It is a question of associating a monetary value to the link ascertained in the second step. In other words, the quantified damage, in this case, the loss of electrical energy is evaluated by assigning a value to it based on the price of electricity on the market.

2.2.2 Description of the Variables

The description of the variables to be used for the estimation requires first an explanation of some notions that are relevant in understanding the impact of sedimentation in artificial lakes.

Dead storage capacity and live storage capacity: The storage capacity of the lake is divided into live storage capacity and dead storage capacity. *The dead storage capacity* is at the bottom of the reservoir and it is part of the total capacity of the reservoir that is allocated to sediment storage (Cruz et al. 1988). Usually, engineers plan a certain amount of dead storage in a dam reservoir based on existing sedimentation rates, with the remaining storage capacity assumed to be active. An unexpected increase in sedimentation due to soil erosion upstream is assumed to increase the components of the dead storage of a reservoir. This means that the more active storage becomes inactive and therefore, there is a loss of water available for hydraulic power, irrigation and other economic benefits. *The direct storage capacity* is above the tank. The storage of water in the direct reservoir can be used to supply water for hydroelectric power, irrigation, etc. Sedimentation can reduce the direct storage capacity of a reservoir resulting in the loss of economic benefits. The tank, therefore, plays the role of an accumulator of energy (Cruz et al. 1988). Consequently, any reduction in the live storage will impair the performance of the associated plant by reducing the energy capacity and the maximum power available during the most critical periods of consumption at the time of inadequate flows. With siltation of the tank, soil erosion reduces the hydroelectric potential and thus the potential benefits of the Lagdo power plant. This reduction in potential profits is expressed in terms of reduced tank capacity. Table 1 shows the capacity of these tanks for Lake Lagdo.

To cover its annual production, the plant needs a volume of water that is at the level of the normal coast (216 m asl), i.e., a volume of water equivalent to 4.57 billion cubic meters. The minimum water level is 206 m. If the volume of water falls short of this level, there will be no electricity production. If the maximum elevation of the lake level is 218.18 m asl, the Reliability Centered Maintenance (RCM) bathymetric measurements in 2004 show the lowest coasts of the current floor at 194 m asl, a maximum depth of 24.18 m. Also, turbine management is

Table 1 Characterization of Lake Lagdo reservoir

	Elevation (meter asl)	Size of the storage (meter)	Capacity (10^9 m^3)
Maximum water surface	218.18	–	–
Live storage	206–218.18	2.18	1.45
Useful storage	206–216	10	4.57
Dead storage	194–206	12	7.7
Dam bottom	194	–	–

parsimonious, and it is not possible to release water to empty the lake of its sediments. The tidal range, which is the annual variation of the lake level, is, in an average year, only 3.50 m, and it is the logical consequence of the area occupied by the lake. The lake is thus inexorably filled, and the power plant is condemned to close in the near future, with consequences for the economy of the country.

The rate of sedimentation of the lake: According to the study commissioned by MEA-DEN (2005), the silting effect of the Lagdo tank has the main effect of reducing the storage capacity of this structure. In 2004, this reduction was estimated at 23% of the total volume at the present stage of sedimentation. The results obtained from the various treatments carried out reveal that in 20 years, 815.9 million cubic meters of deposits, mostly of coarse sands, have accumulated in the reservoir. If we consider 1984, the year of optimum filling of the reservoir and the start-up of the Lagdo power station's number one, as the basis for calculating the rate of alluvial speed, the reservoir is being filled at the rate of 40.80 million cubic meters per year. This assessment could be verified by applying the method described by Sang and Maina (2021). If we assume that the dead reserve is empty at this time, this volume represents 0.9% of the useful reserve, which is 4.57 billion m^3.

Turbine water volume: These are water outlets that occur as discharges into the turbines during power generation (Magrath and Arens 1989). It is expressed in cubic meters.

Hydroelectric potential: The hydroelectric potential expresses the electrical power in kilowatt-hours at the output of the current generator. The electricity production of the hydroelectric power station in Lagdo is intended for users in the provinces of Adamaoua, the North and the Far North. There are four turbines with a total capacity of 72 megawatts to meet these needs.

3 Results and Discussion

Estimating the economic value of avoiding the off-site cost of erosion on the Lagdo Lake hydroelectric potential through the damage function method first requires an estimate of the volume of water needed to generate electrical energy.

3.1 Estimation of the Volume of Water Required to Generate Electrical Energy

The assessment of the effect of sedimentation on hydroelectric production requires, first of all, the estimation of the loss of direct storage capacity of the reservoir. Table 2 gives the results of estimating the volume of water needed to generate one kilowatt hour of electricity. This volume is obtained by dividing the total discharge or volume of turbine water by the corresponding amount of power generated per year (Cruz et al. 1988). It is observed that the change in the volume of turbine water in each period leads to a change in electricity production. The physical relationships that govern hydroelectric power generation suggest that the two factors

Table 2 Hydroelectric potential of Lagdo dam and volume of required water

Year	Generated power/year (millions of KWh)*	Total discharge or volume of turbine water/year (millions of m^3)*	Volume of required water (m^3/KWh)**
1984	99.901	2225	22.27
1985	90.725	1990	21.93
1986	120.009	2454	20.44
1987	118.242	2197	18.58
1988	114.487	1992	17.39
1989	124.298	2800	22.52
1990	113.721	2219	19.51
1991	117.342	1755	14.95
1992	124.314	1835	14.76
1993	131.102	1913	14.59
1994	130.255	1924	14.77
1995	139.511	1873	13.42
1996	151.158	2286	15.12
1997	155.320	2311	14.87
1998	162.345	2444	15.00
1999	162.838	2462	15.11
2000	173.567	2588	14.91
2001	186.820	2944	15.75
2002	192.246	2917	15.17
2003	198.617	3041	15.31
2004	199.414	3024	15.16
2005	206.736	3139	15.18
2006	212.011	3113	14.68
2007	217.670	3344	15.36
2008	221.675	3452	15.57
2009	235.830	3625	15.37
Total	4,100.124	63,954	423.23
Average	164.004	2558,16	16.92

Calculated source from (*) Lagdo Hydroelectric Power Station; (**) Authors calculation

determining the production function are: the amount of water discharged or used to drive turbines and the height of the landfill. However, as assumed by Aylward (1998) in the absence of a height effect, the variation in the electricity production that occurs is a function only of the variation in the discharge. Any change in the reservoir over a given period can be directly translated from units of water (cubic meters) to units of electricity (kWh) (Aylward 1998). An average discharged water volume of 16.92 cubic meters per kWh was calculated for 26 years after commissioning of the Lagdo hydroelectric plant.

3.2 Value of the Forgone Power Benefit

The quantitative relationship between sedimentation and the hydroelectric potential of Lake Lagdo

The accumulation of sediments in the direct storage of the reservoir leads to a reduction in the water storage capacity of the reservoir. The assumption is that all sediments go to direct storage. A loss of 16.92 m^3 of water due to an equivalent volume of deposited sediment would mean a loss of one kilowatt hour of electrical power. Thus, it is easy to establish that the 40.80 million cubic meters of water replaced each year by an identical volume of sediment has the potential for hydroelectric production loss of 2.4 million KWh per year.

Monetary valuation of the quantitative link

The aim here is to convert this potential energy lost every year into monetary value. The quantity of potential lost energy calculated above is multiplied by a price, in this case, the price established on the electricity market by Eneo Cameroon. It should be noted that Eneo Cameroon is distributing and pricing the energy produced from the Lagdo Dam according to a categorization of consumers (Table 3).

Table 3 Distribution of energy consumption at the plant outlet in 2011 (in million KWh)

Power generated	Industry (50%)	Non-domestic consumption (30%)	Domestic consumption (20%)
2. 4	1. 2	0.72	0.48

The energy produced is divided between industry, non-domestic consumption (petty trade, etc.) and domestic consumption (households). To determine the monetary value of the annual potential loss of electrical energy, different quantities of energy that can be consumed are valued at equivalent prices for each category of consumers. Table 4 shows the monetary value of the hydroelectric potential lost annually due to silting.

Table 4 Calculation of the value of hydroelectric potential lost each year

Consumed power	Tranche (kWh)	Price[a] (CFA francs[b])	Average Price[a] (CFA francs)	Monetary value (million CFA francs)
Industry	0–200	52	50	60. 2
	201–400	50		
	400 and more	48		
Non-domestic consumption	0–110	75	83.75	60. 5
	111–400	80		
	401–800	85		
	800 and more	92		
Domestic consumption	0–110	50	71.25	34. 3
	111–400	70		
	401–800	80		
	800 and more	85		
Total	Total monetary value on the 3 electricity range (155.2 million CFA s)			

[a]The prices used are late 2010 prices

[b]The CFA franc is the name of two currencies used in certain West and Central African countries, which are guaranteed by the French Treasury

The total value of electricity potentially lost due to the sedimentation of the lake is equivalent to the sum of the monetary value of the energy consumed for each category of consumer, i.e. CFA francs 155.2 million (USD 279,274.73[1]) per year. This amount represents the loss to Eneo Cameroon due to sedimentation. Based on the assumptions underlying this paper, the market value of the preservation of the hydroelectric potential of Lake Lagdo by agriculture is therefore also USD 279,274.73 per year. It is important to keep in mind that this value is still underestimating the true value of foregone benefits arising from sedimentation. Only lost power benefits were considered, though the dam and reservoir serve other functions such as irrigation, flood control, fisheries, and providing domestic water supply. Measurement and valuation of the impacts of watershed erosion on these other services require much more information than is currently available.

3.3 The Effects of Anti-Erosion Measures on the Reduction of the Sedimentation of Lake Lagdo Dam

Considering 1984, the year of optimum filling of the reservoir and the powering up of the Lagdo power station, as a basis for calculating the rate of alluvial flow, the reservoir fills at the rate of 40.80 million cubic meters per year. The calculation of sediment volumes accumulated in 2004 by the RCM is based on a relatively long time span of 20 years and the accumulated deposits for this period amount to 815.9 million cubic meters. According to the MEADEN study, if the current rate of dam filling remains constant and taking into account the quantities of sediment already accumulated in 2004, i.e. 815.9 million cubic meters over 20 years, the useful reserve on which electricity production basically depends can be seriously affected by 2020. This means that the accumulated sediments have reached the minimum level (206 m) with a capacity of 1,450 million cubic meters.

The amount of sediment accumulated since 1984 reduced the storage capacity to 634 million cubic meters. With this remaining capacity and a siltation rate of 40.7 million cubic meters per year, the lifetime of the structure is an additional 15 years, i.e. a lifetime of up to 2020.

In Table 5, if we assume scenarios for reducing the siltation speed by the anti-erosive installations of 15, 30 and 50%, we obtain results in terms of the lifetime of the structure. Thus, a reduction of 15, 30, and 50% increase the lifetime of the structure by three, seven, and 16 years, respectively, compared with the siltation rate of 2004. The monetary gain is equivalent to the number of additional years obtained as a result of sediment reduction multiplied by the total value of electricity potentially lost annually as a result of lake sedimentation.

A necessary condition for the adoption of these conservation practices in watershed farming is the allocation of secure claims over the land. However, farmers in the Lagdo watershed typically have no property rights to the land (Ngondjeb et al. 2014). As a consequence, there is no stake in ensuring its long-term productivity and the potential gain from reducing the loss of soil nutrients cannot be captured by the farmers. Therefore, it is not surprising that farmers exploit the land until its productivity declines and then move on to a new plot (Ngondjeb et al. 2014). With regard to the private incentives for conservation, it must be recognized that soil erosion does not necessarily impose current costs on the private land user as long as the topsoil layers are not completely depleted. Only with the removal of topsoil does the nutrient loss have a direct impact on the current productivity of the land. The substantial on-site benefits in terms of sustainable soil productivity will, in fact, result from the adoption of conservation-oriented farming and forestry practices. Watershed farmers, however, will adopt these practices (which are not costless) only if they can capture the long-term benefits that will accrue. This indicates that a necessary condition for conservation is for farmers to acquire a long-term stake in the land.

[1]Exchange rate (2018): 1 USD = 555.75 CFA francs.

Table 5 Effect of anti-erosion measures on the lifetime of the dam

Scenarios	Sedimentation speed (million cubic meters per year)	Length of life of the structure (year)	Gain in year	Monetary gain (in million CFA francs)
Sedimentation flow in 2004	40. 7	15.54	–	–
Reduction of 15% of the sedimentation flow	34. 6	18.28	3	465. 6
Reduction of 30% of the sedimentation flow	28. 5	22.20	7	1,086. 5
Reduction of 50% of the sedimentation flow	20. 3	31.08	16	2,483.5

At the same time, social benefits indicate that the government will benefit to actively subsidize conservation efforts as a sufficient condition for the abatement of soil erosion. In this light, the existing agroforestry program should be regarded as only a beginning, and the government must seriously look beyond it, towards a massive land reform program in the SSA of Cameroon supported by conservation-oriented subsidies.

In addition, taking into account the preservation of the hydroelectric potential of Lagdo Lake in deciding what agricultural activities to encourage is of considerable importance in terms of public decisions. While highlighting the existence of a positive territorial externality linked to land use, the environmental performance of agriculture will be strengthened further, which will encourage the implementation of agro-environmental policies in favour of the activity and its joint products.

Policies that allow landowners to capture the value of environmental services could provide powerful incentives for promoting environmentally-friendly land use practices (Pagiola et al. 2004). Therefore, it is important to develop mechanisms that pay land users for environmental services including conserving soil, so that the additional income stream makes environmentally-friendly land use practices privately profitable. While growing annual crops, incentives should also be given to adopting soil and water conservation technologies including non-tillage practices, mulching, contour and alley farming, and terrace construction, which can reduce soil erosion and other environmental costs substantially.

From a theoretical point of view, the application of the damage function is more common in the valorisation of the effects of pollution on health, and it is applied here in the valuation of environmental services rendered by agriculture.

The limits of this environmental service valuation are based on a number of important points. First of all, the basic hypothesis assumes that the anti-erosion installations can completely stop the sediment drain to the lake, which is not evident because the anti-erosion installations can only help to reduce the quantities of sediments that leak into the lake. Secondly, with natural reserves making up 68% of the total area of the watershed, agriculture can only occupy 32% of this area. This means that agriculture cannot totally stop the erosion that occurs in the watershed. Moreover, the study does not quantify the volume of sediments that can be stopped by each type of anti-erosive installation. To the extent that plots cannot be fully developed, the study also does not mention the total area required for the development of anti-erosive techniques in the watershed. In addition, the MEADEN study assumes that the volume of deposited sediments is constant from year to year and that the amount lost is exactly equal to the volume of sediment deposited in the lake. Finally, from the point of view of the consumption of the generated power, this paper does not provide the different proportions of consumption of the energy produced to which the specific prices are applied per consumption range, hence the use of average prices.

4 Conclusions

The objective of this study was to estimate the economic value of the benefit of avoiding the off-site costs imposed by erosion on the hydroelectric production of the Lagdo Dam, providing an example of a water-soil-energy nexus (Rasul 2014). Indeed, through erosion and sedimentation, upstream activities occurring in different biophysical components of the watershed generate downstream externalities. The damage function approach was carried out to measure the value of the hydroelectric potential loss (the marginal damage being the quantity of kilowatt hours lost) induced by the deposition of an additional cubic meter of sediment in the reservoir. The results show that due to erosion in the Lake Lagdo watershed, a sediment volume of about 40.80 million cubic meters is carried into the lower lake every year. On the basis of this observation made by MEADEN, the results show that the deposit of this quantity of sediment in the Lagdo Dam deprives the lake of an equivalent potential capacity of usable water for the production of electricity. Such a potential loss of water would cause a potential annual loss of hydroelectric power of about 2.4 million kWh per year. Monetary valuation based on kWh prices in the area shows that such annual loss of potential energy due to sedimentation is equivalent to 156 million CFA francs (USD 279,274.73) per year. In other words, the service provided by the adoption of anti-erosion measures by farmers such as non-tillage practices, mulching, contour and alley farming, and terrace construction while growing annual crops in the catchment area is estimated at 156 million CFA francs per year. This paper contributes to the monetary evaluation of an environmental service of agriculture in Northern Cameroon.

One very clear notion that comes from this research is the need to develop more damage estimation techniques to aid in the cost analysis. This would involve developing standardized tools for estimating average and marginal costs of erosion on certain activities. These tools could be employed by conservation agents with a limited amount of cost/damage information. Exploration of other erosion damages is also needed both for general and specific knowledge. Erosion impacts on irrigation systems, industrial water use, flood control, fish and wildlife populations, and recreation choices are currently undocumented. For the study area, a much better understanding of the distribution of erosion occurrences and damages is needed to help accurately project total benefits from conservation projects. Better reporting by producers of physical damages would also greatly help in the projection of specific off-site benefits from soil conservation projects.

Acknowledgements The authors thank the coordination of the Water-Soil-Tree Project in Garoua, for the financial support and the provision of materials. The authors also would like to thank the Study Mission for Planning and Development of the Northern Province without whose support this work could not have been completed.

References

Abou A, Hofs J-L, Mergeai G (2006) Relever les défis environnementaux pour les filières co-tonnières d'Afrique de l'Ouest et du Centre. Biotechnologie, Agronomie, Société Et Environ-Nement 10:351–359

Aylward B (1998) Economic valuation of the downstream hydrological effects of land use change: large hydroelectric reservoirs. Unpublished Ph.D.diss., Fletcher School of Law and Diplomacy, Medford

Aznar O, Perrier-Cornet P (2003) Les services environnementaux: diversité des conceptions et réalité des contours. Economie Rurale 273–274:153–168

Baboule ZB, Roose E, Sanon K, Waech F, Ziem BBA (1993) Effets des techniques culturales sur le ruissellement, l'érosion et la production de coton et de maïs sur un sol ferrugineux tropical sableux. Recherche de systèmes de culture intensifs et durable en région soudanienne au Nord Cameroun . Cah Orstom Sér Pédol 27:309–325

Barbier B (1996) The economics of soil erosion: theory, methodology and examples. Discussion papers in environmental economics and environmental management. University of York, Heslington, UK

Bekele W (2003) Economics of soil and water conservation: theory and empirical application to subsistence farming in the eastern ethiopian highlands. Doctoral

thesis Swedish University of Agricultural Sciences, Uppsala University
Boxall P, Adamowicz W, Williams M, Louviere J (1996) A comparison of stated preference methods for environmental valuation. Ecol Econ 18(3):243–253
Brauer I (2003) Money as an indicator: to make use of economic evaluation for biodiversity conservation. Agric Ecosyst Environ 98(1–3):483–491
CEDC, Centre d'études de l'environnement et du développement au Cameroun (2002) Etude d'impact des activités anthropiques sur la retenue d'eau de Lagdo. Centre d'études de l'environnement et du développement au Cameroun, Garoua, Cameroon
Chevassus-au-Louis B, Salles JM, Bielsa S, Richard D, Martin G, Pujol J (2009) Approche économique de la biodiversité et des services liés aux écosystèmes. Contribution à la décision publique
Cruz W, Herminia F, Tapawan-Conway Z (1988) The on-site and downstream costs of soil erosion in the Magat and Pantabangan watersheds. J Philippine Dev 15(26):85–111
De Graaff, J (1996) The price of soil erosion: an economic evaluation of soil conservation and watershed development. The Netherlands: Ph. D. thesis Wageningen Agricultural University
ECAM III, Auprès EC, des Ménages (2008) Troisième enquête camerounaise au-près des ménages: tendances, profil et déterminants de la pauvreté au Cameroun entre 2001–2007. Institut national de la statistique, Yaoundé, Cameroun
Gadrey J (2000) The characterisation of goods and services: an alternative approach. Rev Income Wealth 46(3):369–387
Garrabé M (1994) Ingenierie de l'évaluation économique. Edition Marketing
Gillis M, Perkins D, Roemer M, Snograss D (1998) Economie du développement. de Boeck and Larcer, Paris, p 784
Hein L (2007) Assessing the costs of land degradation: a case study for the Puentes catchment, south-east Spain. Land Degrad Dev 18(6):631–642
Hoff H (2011) Understanding the NEXUS, background paper for the Bonn 2011 conference: the water, energy and food security nexus. Stockholm Environment Institute, Stockholm
M'biandoun G, Olina JP (2007) Savoir paysan et fertilité des terres au Nord-Cameroun. Cahiers De l'Agriculture 16:185–197
Madelin V (1995) La rémunération des services environnementaux rendus par l'agriculture. Économie Et Prévision 117–118:77–88
Magrath W, Arens P (1989) The costs of soil erosion on Java: a natural resource accounting approach. Environment department working. World Bank, Washington
Mahe PL (2001) La multifonctionnalité en quête de légitimité. Problèmes économiques, 27 Juin 1–3
MEADEN, Mission d'Études pour l'Aménagement et le Développement de la Province Du Nord (2005) Evaluation de la sédimentation du lac de Lagdo et détermination de la fréquence de son cu-rage. Mission d'Études pour l'Aménagement et le Développement de la Province Du Nord, Yaoundé, Cameroun
Ngondjeb DY, Dia Kamgnia B, Njé P, Havard M (2014) Economic assessment of investment in soil conservation: the case of erosion control facilities in the catchment of Lake Lagdo in Cameroon. Can J Agric Econ 63(3):393–410
Nijkamp P, Vindigni G, Nunes P (2008) Economic valuation of biodiversity: a comparative study. Ecol Econ 67(2):217–231
Njomgang C (2003) Défaillances institutionnelles et externalités environnementales: le cas des programmes d'ajustement structurel. atelier international sur l'économie de l'environnement
Pagiola S, Agostini A, Gobbi J, de Haan C, Murgueitio I, Ramírez E, Rosales E, Ruíz M, Juan P (2004) Paying for biodiversity conservation services in agricultural landscapes. Environmental Department Papers no. 96. The World Bank
Pearce D, Markandya A (1989) L'évaluation monétaire des avantages des politiques de l'environnement. OCDE, Paris
Pearce D, Turner K (1990) Economics of natural resources and the environment. The Johns Hopkins University Press, p 378
Pretty J, Brett C, Gee D, Hine R, Mason C, Morison J, Rayment M, van der Bijl G, Dobbs T (2000) An assessment of the total external costs of UK agriculture. Agric Syst 65(2):113–136
Rasul G (2014) Food, water, and energy security in South Asia: a nexus perspective from the Hindu Kush Himalayan region. Environ Sci Policy 39:35–48. https://doi.org/10.1016/j.envsci.2014.01.010
Sang JK, Maina CW (2021) Multifunctional historical data for improved management of reservoirs. In: Hülsmann S, Jampani M (eds) A nexus approach for sustainable development. Springer Nature, Cham, Switzerland. https://doi.org/10.1007/978-3-030-57530-4_6
Swinton S, Lupia F, Robertson P, Hamilton S (2007) Ecosystem services and agriculture: cultivating agricultural ecosystems for diverse benefits. Ecol Econ 64 (2):245–252
Thevoz CM (2000) Le zaï ou les limites d'une stratégie de mise en culture des sols dégradés au Burkina Faso dans la sécurité alimentaire en question: dilemmes, constats et controversies, pp 217–229. Karthala, Paris, France
Tietenberg T, Lewis L (2008) Environmental and natural resource economics, 8th edn. Pearson International Edition
Zhang W, Ricketts T, Kremen C, Carney K, Swinton S (2007) Ecosystem services and disservices to agriculture. Ecol Econ 64(2):253–260

The Energy-Water Nexus in Iran: The Political Economy of Energy Subsidies for Groundwater Pumping

Tinoush Jamali Jaghdani
and Vasyl Kvartiuk

Abstract

The depletion of groundwater resources due to irrigation water pumping in Iran has become a serious problem that threatens both rural life and sustainable development in the country. The latest estimates show that 70% of groundwater resources have been overexploited over the last 15 years. The number of deep and shallow wells used for groundwater irrigation almost doubled in the last decade, reaching more than one million, which includes both permitted and non-permitted wells. Skyrocketing water consumption has become one of the primary reasons behind the devastation of groundwater resources. Cheap energy, resulting from energy subsidies, which have been provided for many years, made deep water pumping possible and huge investments in deepening and relocating wells feasible. This study focuses on the Iranian government's unsuccessful attempt to reform its subsidy policy in December of 2010 when they tried to phase out energy subsidies for groundwater pumping, however, they only did so for domestic and industrial consumers but failed to do so for the agricultural sector. We discuss the political economy of implementing these reforms using the World Bank's framework for policy reforms analysis. This overview study argues that the political overrepresentation of rural interests in Iranian politics impeded a successful reform implementation. Negative economic shocks to the Iranian rural economy may be counteracted by developing adjustment strategies that promote alternative livelihoods for residences of rural areas. Neglecting the political economy surrounding Iranian energy subsidies may further delay reform implementation and lead to further water depletion.

Keywords

Energy subsidies · Groundwater depletion · Irrigation water · Political economy · Price reform

1 Introduction

Adopted by the United Nations in 2015, the Sustainable Development Goals (SDGs) are a universal call to action to end poverty, protect the planet, and ensure that all people enjoy peace and

The original version of this chapter was revised: Incorrect sentence in this chapter has been corrected. The correction to this chapter is available at https://doi.org/10.1007/978-3-030-57530-4_16

T. Jamali Jaghdani (✉) · V. Kvartiuk
Leibniz Institute of Agricultural Development in Transition Economies (IAMO), Theodor-Lieser-Str. 2, 06120 Halle (Saale), Germany
e-mail: jaghdani@iamo.de

V. Kvartiuk
e-mail: kvartiuk@iamo.de

S. Hülsmann and M. Jampani (eds.), *A Nexus Approach for Sustainable Development*,
https://doi.org/10.1007/978-3-030-57530-4_8

prosperity by 2030 (UNDP 2018). All SDGs by design are an integrated set of global priorities and objectives that are fundamentally interdependent (ICSU 2017). Three SDGs are of special interest with regard to the water-energy-food (WEF) nexus: (1) end hunger, achieve food security and improved nutrition, and promote sustainable agriculture (SDG2); (2) ensure availability and sustainable management of water and sanitation for all (SDG6); and (3) ensure access to affordable, reliable, sustainable, and modern energy for all (SDG7) (Ringler et al. 2016). These goals are interlinked and cannot be dealt with separately (ICSU 2017). For instance, increases in food production require large additional water and energy resources. However, improved access to conventional or renewable energy depends on the use of larger amounts of water. In turn, access to safe water will require both increased amounts of water and more energy to treat this water to safe standards.

Competition over the same resources (land, water, energy) may result in trade-offs between all three of these SDGs. Consequently, all of these improvements require political will, significant investments, and institutional capacity (Ringler et al. 2016). There is no consensus among scholars on the definition of the nexus approach for analyzing the inter-linkage of resources. By considering the concepts of synergy and trade-offs, the nexus perspective emphasizes the inter-relatedness and interdependencies of environmental resources and their transitions and fluxes across spatial scales and between components of a system (Kurian et al. 2016). As UNU-FLORES (nd) summarizes: "Instead of just looking at individual components, the functioning, productivity, and management of a complex system are taken into consideration. In such complex systems, there are trade-offs as well as facilitation and amplification between the different components." The nexus perspective covers varying dimensions, such as (a) large dams and the nexus of hydropower generation, irrigation water provision, and environmental deterioration, (b) wastewater reuse for peri-urban agriculture, (c) the nexus of waste remediation, resource recovery, and water reuse, (d) renewable energy and the water-land nexus, and (e) biofuels and food trade-offs or complementarities. One of the important aspects of the nexus perspective is the groundwater irrigation power nexus (Scott et al. 2015), which is the focus of this study. The depletion of groundwater resources and the overexploitation of aquifers is a threat to the sustainable development of human communities and biodiversity. A NASA study in 2015 shows that 21 of the world's 37 largest aquifers have passed sustainability tipping points (Richey et al. 2015).

Modern groundwater extraction is energy intensive and commonly subsidized by governments. In contrast, practical old traditional groundwater extraction systems, such as qanats, which are common in the Middle East, use gravity to carry groundwater to the surface via underground tunnels without using energy (Jomehpour 2009). However, nowadays groundwater is mainly extracted with tube wells using electricity or diesel pumps to lift water to the surface. The relationship between energy and water is two-way: water can be used to produce energy and energy is used to pump groundwater (Zilberman et al. 2008). Energy is needed for pumping water from groundwater aquifers or rivers for domestic or industrial use, as well as food production. As water levels drop, more energy is required for pumping. These relationships represent an important central issue in the water-energy-food nexus (Coates et al. 2012). Instead of direct subsidies for water, farmers often receive electricity subsidies, which results in an overuse of both energy and water in groundwater-irrigated agriculture (Theesfeld 2010). The overexploitation of groundwater is a threat to the sustainable development of rural and urban areas, especially in arid and semi-arid regions. In order to avoid a possible resistance from rural communities receiving energy subsidies for irrigation, some scholars suggest other approaches to deal with groundwater depletion caused by supplying energy without changing energy prices. An alternative option to energy price increases could be the intelligent scheduling and management of a rationed power supply for agricultural consumers (Shah et al. 2004). This could be an alternative way to deal with local resistance against energy price changes. The effect of water and energy prices on the water-energy relationship is not minor; subsidy changes affiliated with these prices is a crucial factor. If the price of water does not accurately reflect the cost of energy required to obtain it, phasing out energy subsidies may improve the management of existing water resources by making them more costly. Getting the prices right is a key policy tool that can provide incentives to develop and adopt the most appropriate technologies for water management (OECD 2012, p. 147). Prices that don't reflect the true provision costs of water send the wrong signals to the water users and increase inefficient water use. Although water demand in agriculture is considered to be inelastic for lower irrigation water prices (de Fraiture and Perry 2007), a considerable reduction in water consumption because of price increases has been observed throughout different studies. For instance, in an Indian context, Shah et al. (2012) found that a 10% reduction in subsidies would reduce groundwater extraction by 5.4% and cost farmers 12% of their revenue. Similarly, in the high plains of the USA, a one dollar increase in energy prices (a 13.6% price increase) would reduce water pumping by 3.6% on average, thereby affecting crop selection and the use of technologies (Pfeiffer and Lin 2014). Importantly, investments in technologies that allow for cheaper water extraction and more efficient transportation can offset the effects of rising energy prices (Zilberman et al. 2008). Nevertheless, despite the clear economic arguments in favour of higher energy prices, the implementation of policies that affect voters' welfare negatively is challenging. Phasing out any type of subsidies is widely considered a complicated undertaking for politicians because they may risk facing opposition from the subsidy recipients. Policymakers all around the world have strong incentives to adopt, retain, and even expand consumer and producer subsidies because their constituents react to a redistribution of budgetary resources (Dixit and Londregan 1996). The logic of consumer subsidies is to shield the population from price shocks to basic commodities (food, gasoline, electricity, etc.) and other risks due to open economic policies (Garrett 1998). Without these protections, democratic or authoritarian regimes may be threatened by protests or even coups organized by dissatisfied recipients (Bates 1983). On the other hand, producer subsidies (such as energy subsidies) are to ensure the political loyalty of an industrial group that might otherwise threaten with redirecting political support to a competing party or parties (Acemoglu and Robinson 2001; Thomson 2014). In light of such high stakes for politicians themselves and possibly even entire countries, the political economy of reform processes must be carefully considered. Otherwise, any prospective policy change is unlikely to be effective, especially in the sphere of groundwater pumping subsidies (Scott and Shah 2004; Shah et al. 2012).

The depletion of groundwater resources due to irrigation water pumping in Iran is a clear problem that threatens rural life and sustainable rural development and sustainable agricultural production. The devastating effects of subsidies are recognized by experts and researchers in this area (Madani 2014). For instance, Moazedi et al. (2011) have demonstrated a non-linear relationship between groundwater levels and energy use. It can be concluded from this finding that higher energy subsidies should be provided as groundwater levels drop in order to keep the same level of production. As Dinar (2000) has presented, factors related to the political economy of irrigation water pricing can hinder any price reforms for surface water. Political factors may well influence the pricing of groundwater pumping. The experience of the Iranian government from 2010 until now highlights the consequences of disregarding the political pressures arising due to similar reforms. The following chapter reviews the political processes in Iran that may have contributed to the failure of phasing out subsidies for groundwater pumping. We first provide an overview of groundwater depletion and energy consumption in Iran. We then present a literature review on energy subsidy cuts and their respective effects on water pumping in Iran. Section 3 examines Iran's political economy surrounding its 2010 subsidy reform in detail and its interaction with rural energy subsidies. Finally, we present the outlook for the future and discuss possible solutions.

2 The Iranian Context

2.1 Groundwater Depletion and Energy Consumption Trends

The Iranian water crisis has become a hot public issue, especially since 2015 when high-level officials started to warn that the country risks experiencing a fate similar to Somalia over the next 20 years (for instance Kalantari (2015, announcement on 27.7.2015) and Nozaripur (2015, announcement on 9.2.2015)). However, the cause of Iran's water crisis has deeper roots in the country's water management patterns, which have especially transpired in recent years. At the moment, several controversies exist regarding available renewable water resources. Based on the latest information from Iran's Ministry of Energy (MOE), which is responsible for the country's water resource management, the average available annual precipitation has reduced from 225 to 205 mm, translating to a reduction in available renewable water resources from 130 billion cubic meters (BCM) to 100 BCM (Hajrasuliha 2015, IWRM[1] announcement on 27.9.2015). Based on the available estimations by the Iran Water Resources Management Company (IWRM), approximately 60 BCM of groundwater was extracted in 2014, of which more than 85% was used for irrigation (Water Resources Base Study[2]). The ministry of energy (MOE) announced that 92% of water resources were consumed by the agricultural sector (Chitchian 2014, Ministry announcement on 27.9.2014). The share of total water resources consumed by agriculture does not correspond to the share of agriculture in Iran's national economy. The agriculture sector accounts for approximately 10% of Iran's GDP and 18% of employment (Tehran Chamber of Commerce 2017). Groundwater consumption in Iran has been increasing in recent decades. Figure 1 shows the increase in the number of wells (deep and shallow) over the last 40 years. By 2013, over 200,000 wells had been refitted to take advantage of the electrical power network and farmers were encouraged to use electric pumps.

Table 1 shows the amount of water extracted and used by these wells and the share of the agricultural sector based on official data from the IWRM. From a total of 609 recognized aquifers in the country, the number of aquifers restricted for further groundwater extraction has increased from 46 in 1973 to 297 in 2013 (see Table 1). Based on the announcements in 2015, more than half (300–350) of the aquifers were in critical condition (Modaberi 2015, announcement on 28.12.2015). As we can see in Table 1, agriculture is the primary user of groundwater, with 90% of the extracted water consumed by the sector. Both Fig. 1 and Table 1 show that the number of wells with official permits has increased dramatically in the last 40 years. However, the deep wells running on subsidized energy (diesel or electricity) are mostly responsible for substantial increases in water consumption within the agriculture sector (see Table 1). The availability of illegal wells is another negative factor that has contributed to the depletion of groundwater resources. While there is no reliable estimation of the number of illegal wells, some official sources indicate that there are as many as 200,000 (Meidani 2015b, Ministry Deputy announcement on 6.8.2015), while others put the figure to nearly 300,000 (Meidani 2015a, Ministry Deputy announcement on 29.10.2015) or even 350,000 (Akbari 2018).

2.2 The Historical Trend of Energy Subsidies for Irrigation

The Iranian government has subsidized energy consumption (electricity and diesel) for the agricultural sector for the last 40 years. Since 1999, a special program for replacing diesel pumps with electric pumps has been a major focus. The major

[1]Iran Water Resources Management Co. (IWRM), website: http://www.wrm.ir/.

[2]Data source: http://wrbs.wrm.ir/ (last access 18.5.2016).

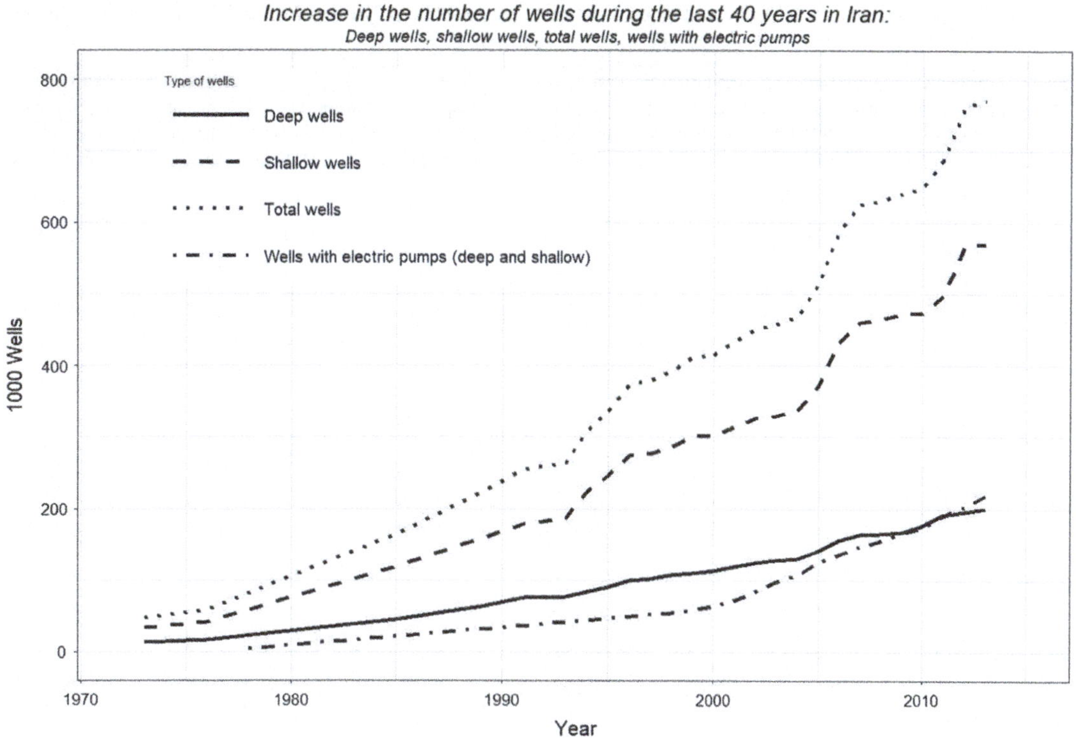

Fig. 1 The trend of well expansion in Iran during the last 40 years *Source* Iran Energy Balance Sheets 1973-2013 (MOE), Water Base Study Organization 2016 (IWRM)

Table 1 Number of wells, groundwater extraction, restricted aquifers for further well drilling, and the share of water used for agriculture in total

Year	No of confined wells	Groundwater extraction (million m^3)	No of unconfined wells	Groundwater extraction (million m^3)	Extraction from springs and qanats (million m^3)	Total groundwater extraction (million m^3)	No of restricted aquifers (from total 609)	Share of water used for agriculture (%)
1976	16656	7473922	42242	3883147	13007.88	24364.95		90
1985	45355	19114.83	119068	9486163	16905.45	45506.43	93	90
1991	75995	26152.26	179527	11882.66	15921.59	53956.51	146	90
1995	90008	27226.07	246258	11472.86	27195.14	65894.07	174	90
1996	100062	28895.79	273890	12975	29393.84	71264.63	176	90
1997	102032	29037.98	276230	12830.1	28430.32	70298.4		90
2000	112789	30655.29	301945	13327.37	28201.26	72183.92	195	90
2003	127800	31434.85	330269	13977.59	29201.21	74613.65	211	90
2013	200859	34544.85	569708	12163.87	18223.53	64932.25	297	88

Source Water Resources Base Studies Bureau of IWRM (WRBS), website: http://wrbs.wrm.ir/ (last access 10.8.2017)

goal of this program was to reduce fossil fuel consumption and its environmental effects. As a result, approximately 10,000 wells have been outfitted with electric pumps annually, and under a special tariff for agriculture, electricity consumption has dramatically increased over the last

Table 2 The average electricity tariffs paid by each sector between 1989 to 2014 (US cent/kWh) and the inflation rate in percent and the official annual exchange rate (rials for one USD)

Year	Domestic price (cent/kWh)	General use price (cent/kWh)	Agricultural price (cent/kWh)	Industrial price (cent/kWh)	Other usages price (cent/kWh)	Total average price (cent/kWh)	Official annual inflation rate (%)	Annual exchange rate (rial/USD)
1989	0.42	0.75	0.33	0.29	0.71	0.45	17.4	1200
1990	0.38	0.60	0.33	0.31	0.60	0.40	9	1410
1991	0.49	0.88	0.33	0.64	0.88	0.60	20.7	1420
1992	0.65	1.04	0.33	0.64	1.04	0.70	24.4	1490
1993	0.75	1.33	0.29	1.00	1.33	0.95	22.9	1800
1994	0.58	1.65	0.29	1.65	2.42	1.22	35.2	2660
1995	0.50	1.30	0.19	1.30	1.82	0.95	49.4	4070
1996	0.59	1.09	0.18	1.50	2.05	1.05	23.2	4440
1997	0.65	1.36	0.17	1.75	2.03	1.17	17.3	4780
1998	0.64	1.09	0.13	1.59	1.80	1.04	18.1	6460
1999	0.68	0.90	0.10	1.31	2.43	0.93	20.1	8630
2000	0.79	1.02	0.16	1.48	3.02	1.09	12.6	8190
2001	0.91	1.25	0.14	1.67	3.42	1.23	11.4	8000
2002	1.06	1.55	0.16	1.83	4.27	1.42	15.8	8010
2003	1.17	1.83	0.17	1.96	4.95	1.58	15.6	8320
2004	1.23	2.01	0.18	2.12	5.90	1.73	15.2	8740
2005	1.14	1.96	0.24	2.23	5.97	1.68	10.4	9040
2006	1.12	1.97	0.23	2.17	5.87	1.66	11.9	9220
2007	1.33	1.71	0.22	2.20	5.43	1.76	18.4	9350
2008	1.23	2.37	0.23	2.12	5.72	1.80	25.4	9660
2009	1.33	1.57	0.22	2.13	5.18	1.71	10.8	9670
2010	1.36	2.17	0.45	2.52	5.74	2.00	12.4	10440
2011	2.78	4.17	1.04	3.67	10.59	3.40	21.5	12040
2012	1.29	1.88	0.50	1.64	5.14	1.56	30.5	26070
2013	1.09	1.62	0.42	1.39	4.22	1.31	34.7	31830
2014	1.34	1.88	0.54	1.65	5.07	1.60	15.6	32800

Source Electricity prices: Statistical Centre of Iran; Annual inflation rate: Central Bank of Iran
Annual exchange rate: ISNA News Agency. https://www.isna.ir/news/97012006097/

30 years. Table 2 shows the average tariffs for each sector between 1989 and 2014. In order to make it comparable with international prices, the official exchange rate has been used to calculate the tariffs in USD cents. The agricultural sector shows the lowest levels of tariffs during this whole period, and this is even further lowered by the constant currency devaluation. Gasoline consumption remained relatively stable during the same period. The available data on Iran's Energy Balance Sheet[3] shows that gasoline consumption from agriculture has reduced from 4 billion litres in 1997 to 3.17 billion litres in 2015. There is no sufficient data showing the share of water pumping from this level of energy

[3]Iran's Energy Balance Sheet is a report that's published annually by MOE. This report presents a picture of energy consumption and production in Iran. It's prepared based on IEA and OECD standards and guidelines.

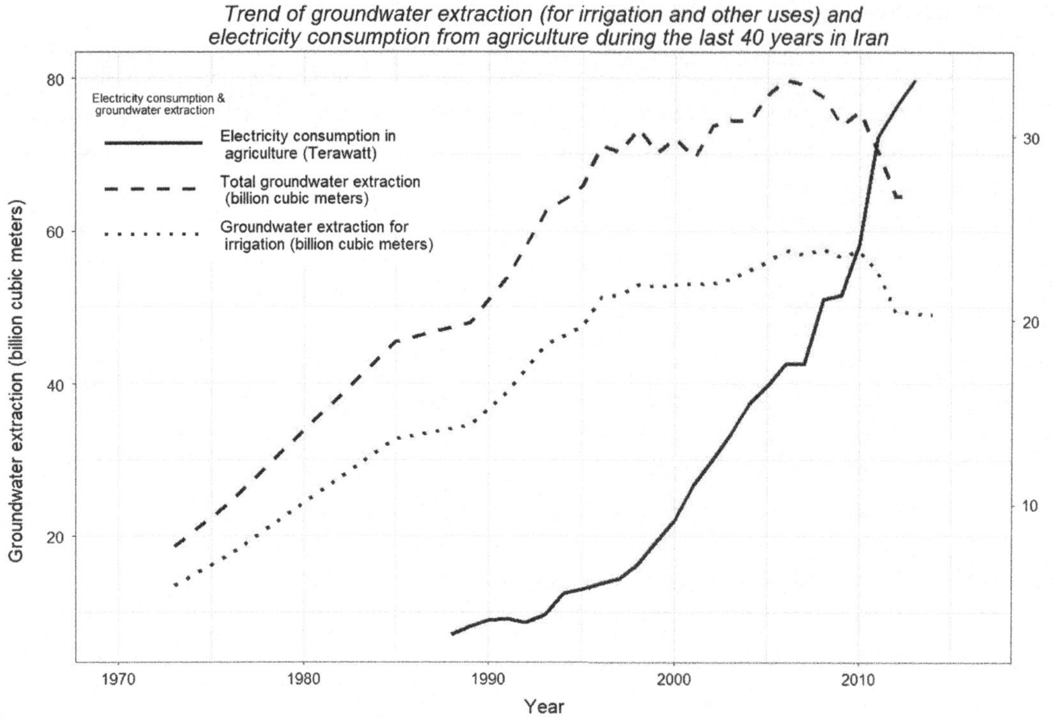

Fig. 2 Trend of groundwater extraction for irrigation and other uses and electricity consumption from agriculture during the last 40 years *Source* Iran Energy Balance Sheets 1973-2013 (MOE), Water Base Study Organization 2016 (IWRM)

consumption in agriculture (Morady et al. 2014).[4] Figure 2 displays electricity consumption and groundwater extraction in Iran during the last 40 years. As we can see, in spite of the increase in energy consumption and the number of wells (as shown in Fig. 1), the total volume of groundwater extraction has started to decline. This decline may be a reflection of the devastating condition of groundwater resources.

Few studies have analyzed the effects of energy subsidies on rural and urban households' welfare. These studies mainly focus on the reforms of pricing policy. Using 2001 data, Heidari and Parmeh (2011) anticipate a reduction in purchasing power between 16–19% for rural households and 18.5–22% for urban households due to energy subsidy elimination. In another study, by using a 2011 social accounting matrix, Mazaheri and Khiayabani (2017) foresee a 10.6–11.11% reduction in purchasing power for rural and urban households on average as a result of a 100% increase in energy prices.

One of the main consumers of energy in Iranian agriculture is irrigation water pumping in general, and groundwater pumping in particular. The latest estimations from 2014 show that water pumping accounts for 70% of electricity consumption in agriculture (Bagherzadeh 2017). During the last 40 years, different governments have implemented numerous support programs for the agricultural sector, with cheap energy provision being one of the central components. Prior to the attempt to phase out energy subsidies in 2010, electricity tariffs for agriculture followed a complicated pattern that was substantially simplified afterwards. The sector has been supported by a special tariff that is much lower than the electricity tariff for industrial and domestic consumption, the provision cost, or the price of electricity on the regional electricity market. Based on the announcements from Iran's MOE

[4]As the focus of this study is not gasoline, we mainly discuss electricity in this paper.

Table 3 Electricity consumption among different sectors in Iran (Gigawatt hours (GWh))

Year	Domestic	General	Commercial	Industrial	Transport	Agriculture	Other uses	Total	The share of agriculture (%)
1997	26523	6727	8160	23661		6009	2278	73358	8.2
1998	28686	7077	8484	24140		6782	2477	77646	8.7
1999	29754	10622	5567	26493	11	8019	4190	84656	9.5
2000	31266	11271	5991	28924	13	9147	3754	90366	10.1
2001	32891	11951	6394	30721	18	11079	4117	97171	11.4
2002	34946	12630	6925	33456	13	12435	4671	105076	11.8
2003	37967.1	13714	7461	36937.1	14.3	13858.6	4672.4	114909.5	12.1
2004	40563.9	15020.6	7862.7	40247.8	89.7	15489.1	5188	124461.8	12.4
2005	44108	16390	8542	42950	112	16469	4305	132876	12.4
2006	48085.5	18328.6	9319.5	46430.2	144.2	17666.2	4607.5	144581.7	12.2
2007	50776.7	19648	9952.6	49601.9	169.8	17670	4509.9	152329	11.6
2008	52896.1	20428	10741.8	51863.9	245.8	21178.7	4090.9	161445.1	13.1
2009	55629.6	21826.6	11015.3	54605.4	282.1	21405.1	3674.3	168438.3	12.7
2010	60907.7	21308.1	12726.8	61183.4	299.4	24188.8	3567.6	184181.8	13.1
2011	56773.7	16751.5	12663.6	63591.5	352.7	30020.3	3752.1	183905.4	16.3
2012	61350.9	17809.8	12598.8	66706.4	400.9	31646.6	3635.3	194148.5	16.3
2013	64379.2	17833.4	13377.9	70447.6	285.9	33125.8	3764.7	203214.6	16.3
2014	71162.7	19766.7	15404.4	73931.6	362.7	35187.9	3836.9	219652.8	16.0

Source Ministry of Energy. Energy Balance Sheet: https://bit.ly/2znFPXA last accessed (01.07.18)

in 2015, farmers pay 110 rials/kWh on average, which is much less than the provision price of 800 rials/kWh (Chitchian 2015b, Ministry announcement on 27.2.2015). In the same year, the average price of electricity for export was 8–11 cents per kWh (Mahsuli 2014, Ministry Deputy announcement on 15.11.2014). Based on Iran's 2014 Energy Balance Sheet, the average price of energy for domestic use in 2014 was 346.7 rials/kWh and 442.6 rials/kWh for industrial use. The average official USD-Iranian rials (IRR) exchange rate in 2014 was 32,800 IRR = 1 USD (according to the Central Bank of the Islamic Republic of Iran). The agricultural sector has the highest growth rate for electricity consumption (likely as a result of electric pump expansion) and consequently receives the highest subsidies for energy consumption. Table 3 shows the electricity consumption among different sectors in Iran. The share of the 2008 electricity consumption from agriculture was 13% of the total consumption (before the reform), which increased to 16% in 2014.

It is important to mention that Iran's electric power industry is mainly in the hands of state owned companies and organizations. Although privatization of some fossil fuel power stations started in 2006, the government still plays the main role in the production of energy from different sources (hydropower, nuclear, and fossil fuels) and the subsequent distribution (Falahatian 2015; Purihosseini 2015). Therefore, the cost of electricity generation and distribution is not determined by market forces.

2.3 Iranian Studies on the Effects of Subsidy Cuts on the WEF Nexus

The possible effects of energy subsidy elimination in Iran have been addressed by scholars. To our knowledge, only two studies have examined the trade-offs between reducing water pumping and phasing out energy subsidies while also considering income effects on the rural

Table 4 Previous studies on energy prices dependent on groundwater extraction

General	Paper	Balali et al. (2008)	Balali and Montashloo (2014)
General information of the studies	Data year	2008	2012–2013
	Model	Dynamic mathematical programming with tube-bath model of aquifer	Mathematical programming
	Crops	Multi-crop condition	Multi-crop condition
	Region	Hamadan-Bahar Plain	Garveh-Kordestan
	Energy type	Electricity and diesel	Electricity and diesel
	Scenario Structure	8 scenarios of energy price increases for 1 to 5 years periods are compared to the baseline scenario, no technological change assumptions for irrigation are made	8 scenarios of sudden increases in energy prices compared to a baseline scenario, technological change assumptions for irrigation are made for 5 scenarios
Scenarios from the two studies relevant for our discussion			
Scenario 1: "Small price increase"	Energy price increase	Annual increase in energy price for 5 years: 50% electricity and diesel	10–20% increase in energy price with no technological change
	Groundwater consumption decrease	5 years period: 3.04%	0.15%
	Gross income decrease	5 years period: 15.7%	–
Scenario 2: "Intermediate price increase"	Energy price increase	Annual increase in energy price for 5 years: 125% electricity and 67% diesel	50% increase in energy price with no technological change
	Groundwater consumption decrease	5 year period: 31.9%	2.04%
	Gross income decrease	5 year period: 53.7%	–
Scenario 3: "Large price increase"	Energy price increase	Annual increase in energy price for 5 years: 286% electricity and 211% diesel	–
	Groundwater consumption decrease	5 year period: 55.4%	–
	Gross income decrease	5 year period: 85%	–

population. Table 4 presents the major conclusions of the studies by Balali et al. (2008) and Balali and Montashloo (2014). Both studies rely on mathematical programming in their analysis and projections. Moreover, Balali et al. (2008) specifically consider the effects on the local population's incomes.

Both of the studies present a number of scenarios with different assumptions, from which we pick three of the most relevant ones for the topic of this chapter (Table 4). They are based on the assumptions of small, medium, and large electricity subsidies cuts. Interestingly, Balali et al. (2008) find that with small price increases of 50%, water consumption reduction will be very minor (just above 3%), but the income losses will be considerable (15.7%). In contrast, nearly tripling electricity and doubling diesel prices will have dramatic consequences for the rural economy as incomes were projected to drop by 85%.

However, water consumption would be more than halved. This would be a desirable outcome for irrigation water consumption but unacceptable for politicians, whose constituents would be affected by large negative income shocks. As a result, these negative income effects need to be directly addressed by the Iranian government. In the next chapter, we will examine the political processes surrounding the Iranian government's attempts to introduce these reforms.

3 The Political Economy of the 'Energy Subsidy-Groundwater Pumping' Nexus

In light of the discussions about Iran's water crisis, it is important to examine institutions that may have contributed to the overuse of groundwater for agriculture. This will enable the design of future policies for the efficient and sustainable use of groundwater resources. The 2010 attempt to phase out energy subsidies represents an opportunity to analyze the political economy of an unsuccessful reform in the realm of subsidy elimination for groundwater pumping. We utilize a World Bank framework for reform analysis from the political economy perspective (World Bank 2008). It provides a practical way to analyze economic reforms by focusing on several different aspects: (1) the reform context; (2) the reform arena that stipulates examining institutions, stakeholders, and involved economic-political interests; and (3) the reform process. Following this framework, we review the major regulations regarding water use and consumption within Iranian agriculture. Afterwards, we analyze the context, arena, and process of the Iranian reform attempt to phase out energy subsidies, with a particular focus on the agricultural context.[5]

[5]For instance, Iran spent 22.6% of its GDP on fuel subsidies in 2010.

3.1 Reform Context

As this subsidy reform was a law that was proposed by the government and was approved by the parliament, we review the rules and regulations that cover water-energy management using economic instruments. Table 5 shows the major rules and regulations that represent the economic instruments of groundwater resource management. Additionally, it displays the various organizations that are responsible for issuing a particular regulation and their effects on surface and groundwater use in agriculture. As we can see, the regulatory structure of the water/energy sector does not offer much guidance for the efficient and sustainable use of water resources. The only notable law on the list was part of the subsidy reform (the "Elimination of indirect subsidies", implemented in 2010), and it could have positive economic effects, as well as encourage more efficient groundwater use by eliminating energy subsidies for water pumping. However, this reform was not successful in its implementation. In the next subsections, we will discuss the potential factors that may have contributed to the unsuccessful implementation of this initiative in agriculture in comparison to other sectors. Although the economic value of water is emphasized in Iran's development plans,[6] the actual legislative framework does not contribute to the efficient use of water.

After providing an overview of the laws and regulation and their relationship, we analyze the context of subsidy reform from 2009–2010. The government of Iran announced its plan to eliminate its energy subsidies and some basic commodity subsidies, such as those for bread, for all types of consumers during the 5 year period starting in 2008. This plan was approved by the parliament after minor modifications in 2009 and was implemented at the end of 2010. As a compensation scheme for the reform's negative

[6]For instance, article 17 of the Fourth Five-Year Country Socio-Economic Development Plan (Majlis of Iran 2004b, Act no. 96/66,911) established an obligation to find, and to consider, the economic value of water of all river basins of the country. It also asked the government to prepare operational guidelines for this purpose.

Table 5 The main regulations affecting water or energy consumption for water pumping

Year of issue	Year of implementation	Issuing organization	Topic	Affected area: water/energy	Effects on efficiency	Source:
1999	1999	Parliament	Electric pumps for wells	Groundwater/energy	Negative; expansion of electricity consumption with cheap prices	(Majlis of Iran 1999)
2004	2005	Parliament	Elimination of observatory fees for groundwater pumps in agriculture	Groundwater	Negative; reduction of observatory power of MOE on groundwater consumption	(Majlis of Iran 2004a)
2010	2010	Parliament	Recognition the illegal wells	Groundwater/energy	Negative, many wells have received legal status and could be provided by subsidized energy	(Majlis of Iran 2010)
2009	2010	Parliament/government	Elimination of subsidies	Energy	Positive; if this law could be implemented, there was a chance of reducing groundwater pumping	(Majlis of Iran 2009)
2015	2015	Government	20 h free electricity for pumping groundwater	Groundwater/energy	Negative: this temporary plan has encouraged extensive groundwater pumping	(Chitchian 2015a, Announcement on 15.7.2015)
2015	Under review, not approved	Parliament	Free irrigation water	Surface water	Negative: if this will be law, it could have devastating effects on any efficient use of water	(Majlis of Iran 2015)
2018	2018	Government	20 h free electricity for pumping groundwater	Energy/groundwater	Negative: this temporary plan has eliminated monetary tools form water-energy management completely	(Ahmadi Yazdi 2018, announcement on 09.05.2018)

welfare effects on households, each affected person received 455,000 rials per month. The government also planned to compensate for the reform's adverse effects on industrial and agricultural consumers (Majlis of Iran 2009). The subsidy reform intended to increase the domestic fuel prices to the FOB[7] fuel price level in the Persian Gulf and to raise electricity tariffs to cost recovery levels over the five year period. However, this plan was not explicit at all with respect to the compensation scheme for consumers. Additionally, a clear instrument for compensating the industrial or agricultural sectors was not defined. Massive fiscal burdens generated by subsidies encouraged the government to pursue this large-scale reform. General aspects of the reform have been largely debated in the media. Moreover, the government recognized some resistance to different aspects of the reform during special parliamentary discussions. Potential welfare implications for different stakeholders were at the center of the debates, and different supporters and opponents paid special attention to those aspects that could affect an average household. We focus exclusively on the energy subsidies for pumping groundwater in our further discussion.

3.2 The Reform Arena

3.2.1 Government

The government of Iran is the major producer and the only distributor of electric power in Iran. With a large budgetary burden, it initiated the phasing out of fossil-fuel and electric power subsidies, resulting in over-consumption of energy. The government finalized the master plan of the reform and sent it to the parliament for ratification in December of 2008.[8] It should be noted that producing major cereals and food items self-sufficiently was a major goal of the Iranian government over the 30 year period prior to the reform. The reform did not explicitly deal with the self-sufficiency programs and their interaction with it. Agricultural producers of the major cereals and food items received another type of subsidy under the framework of the self-sufficiency plan. Energy subsidies and the provision of cheap or free irrigation water represented major elements of the support package. The reform did not clarify its approach towards the pre-existing support programs, potentially affecting millions of farmers and agriculture affiliated industries and communities in rural and urban areas. In particular, the reform intended to completely eliminate indirect subsidies for the agriculture sector, which must be compensated by direct assistance (Article 8 of the Subsidy Reform Law). The mechanism for this compensation was not clarified in the related legislation.

3.2.2 Farmers

Although the value of agricultural products is not a crucial part of Iran's GDP (10%),[9] there are more than 4 million agricultural holdings in the country (Statistical Center of Iran 2014, page 25). The agricultural sector mainly consists of smallholders and family farms. The rural population of Iran stands at around 21% of the total population, which is currently estimated to be 80 million.[10] Additionally, from the more than 16 million ha of agricultural land (fallow and cultivated areas), 54% is rainfed and 46% is irrigated (Statistical Center of Iran 2014, page 26). Table 6 shows that Iran's agricultural production is mainly dependent on irrigation.[11] Climatic conditions, together with the self-sufficiency programs for food products and inefficient water management, have made agriculture the primary consumer of groundwater and surface water. The cheapest energy is provided for pumping groundwater for the agricultural sector. Consequently, phasing out energy subsidies in the agricultural sector would dramatically increase

[7]FOB price stands for Free on Board (or Freight on Board) price level.

[8]Source: Parliament Research Center: https://rc.majlis.ir/fa/legal_draft/show/720654

[9]Source: Tehran Chamber of Commerce (2017).

[10]Source: https://www.amar.org.ir/Portals/0/Files/baravord/population75-95.xlsx, last accessed 05.12.2018.

[11]There are some inconsistencies between statistics from agricultural ministry and statistical center of Iran.

Table 6 Rainfed and irrigated agricultural production in Iran in 2014 and 2017

		Rainfed area (million ha)	Rainfed production (million ton)	Irrigated area (million ha)	Irrigated production (million ton)
Field crops	2008	4.42	2.98	6.03	50.29
	2012	5.8	5.88	6.57	59.63
	2017	5.05	5.84	5.97	76.36
Tree crops	2008	0.21	0.66	2.14	11.83
	2012	0.26	0.79	2.21	14.11
	2017	0.36	1.3	2.49	19.73

Source Ministry of Agriculture, yearbook statistics 2008, 2012 and 2017

water extraction costs. The proposed legislation did not address these issues, potentially jeopardizing agricultural production for millions of farms.

Theoretically, phasing out agricultural input subsidies could be considered as a negative shock for the rural economy. There are a few studies available that have analyzed the effects of subsidy elimination on rural life and agricultural activities in Iran. As agricultural subsidies used to support farmers for energy and other inputs, these studies cover different subsidy elimination conditions. These studies mainly used data available before the reform of 2010, and they are mostly based on economic theories and ad hoc hypotheses. Musavi et al. (2013) used a production function with the available data for the 1974–2005 period; their research showed a 100% increase in the energy price, which will increase production costs by 9%. Additionally, because of increases in the prices of agricultural products, the value of agricultural production will increase by 4.34%. Government expenditure also decreases by more than 20% due to a 100% increase in energy prices. Shahraki et al. (2016) studied the subsidy reduction on water prices and water abstraction costs. A computable general equilibrium model (CGE) was applied to data from 2001. They showed that a 100% increase in water extraction costs would reduce rural production by around 5.33% and rural consumption by more than 30%. Alijani et al. (2012) also used a CGE model to analyze the reduction of the whole production subsidy. The results show that a 100% decrease in the production subsidies would reduce the production of field crops and tree crops by 9.78% and 2.45%, respectively. The investment in the field and tree crop production and agricultural services would go down by 12.45%, 7.65%, and 24.45%, respectively. Additionally, employment in the field and tree crop production and agricultural services will drop by 17.9%, 14.46%, and 19.46%, respectively. They also conclude that importing agricultural products would increase dramatically, in turn jeopardizing food self-sufficiency goals. Overall, existing research suggests that the elimination of subsidies would make one of the key production factors more expensive and, as a result, negatively impact agricultural production.

3.2.3 Parliament

Another important actor in the legalization of the subsidy reform is the Iranian parliament, which is also known as the "Islamic Consultative Assembly", or "Iranian Majlis". Electoral districts are represented by one or more members who are elected in two-stage majoritarian elections. As a legislative body, its goal is to observe the activities of the government and to approve the legislation, which then has to be ratified by the Guardian Council. For instance, the parliament approved the subsidy reform. Up until 2018, there were 207 electoral district areas spread throughout 31 provinces and 290 members of parliament (MPs) should have been elected.[12] Five electoral districts belong to religious minorities. Elections have been based on

[12]Source: https://www.iribresearch.ir/entekhabat/danestani/hozehaye_entekhabi.pdf, last access 06.12.2018.

individual competition and not on political party list-based competition.[13]

The structure of the Iranian parliament appears to enhance the influence of rural areas. The MPs who represent non-metropolitan rural areas are likely to deal with and care about agricultural issues more than their counterparts representing urban areas. However, rural areas are more scarcely populated.

Figure 3 shows the distribution of the population represented by each MP in different electoral districts, and the relationship between the share of households involved in agriculture and the number of constituents within a district. Panel A shows that in 202 electoral district areas, the number of citizens (or potential voters) represented by each MP is not proportionally distributed all over the country. Moreover, panel B demonstrates that fewer citizens are represented by each MP in more rural areas where agriculture represents a major economic sector (in areas with larger shares of households involved in agriculture). The significantly negative slope of the univariate regression can be interpreted as evidence of an uneven distribution of MPs and an over-representation of rural areas in the Iranian parliament.[14] Consequently, rural elites may be able to exert pressure on their respective MPs and, as a result, promote their interests in the parliament. Thus, the agricultural sector's resistance against the 2010 reforms may have been amplified by the biased distribution of seats in the Iranian parliament and gives an advantage to this sector in comparison to other industries.

3.3 Reform Process of 2010–2011

Although the general discussion about cutting subsidies was focused on by the media, the energy subsidies for groundwater pumping was largely disregarded by the public. Insufficiently informed farmers were not made aware of how the compensation scheme to be implemented. This lack of knowledge created a large amount of unease among farmers and, subsequently, among MPs who represent respective communities. With the political representation system biased towards rural areas, this distress may have been transferred to the national level.

Electricity tariffs increased dramatically, but not for the agricultural sector. During 2008–2009, energy tariffs in the agricultural sector for normal hours[15] and normal tariffs[16] were 13.3 rials/kWh[17] (in contrast, the off-peak[18] hour normal tariff was 3.3 rials/kWh and the peak hours[19] normal tariff was 33.3 rials/kWh). The normal hours/normal tariff for agricultural use was 5.5 times less than for domestic use[20] (74.4 rials/kWh) and 10 times less than for industry use (134.51 rials/kWh) at that time (NKEPD Co. 2011). In December of 2010, the subsidy reforms were implemented. Electricity tariffs for groundwater pumping increased more than tenfold to 140 rials/kWh for normal tariffs as outlined in the subsidy reform (Behzad 2010, Deputy of Minister of Energy, MOE, announcement on 21.12.2010). In just one month, the pumping tariff was reduced to 120 rials/kWh under the subsidy reform (Nikbakht 2011, Deputy Minister of Agriculture, announcement on 26.01.2011). As the irrigation season approached, the pressure on the government and the MOE increased, and the government gave up their original plan. Finally, in April

[13]Source: https://www.parliran.ir/?siteid=1&fkeyid=&siteid=1&pageid=189, last access 23.05.2016.

[14]The standard error of the intercept in the regression panel B of Fig. 3 is 0.014 and for the slope is 0.06, which is highly significant.

[15]In summer the normal hours are 7:00–19:00, and in winter between 6:00–18:00.

[16]There were two categories of tariffs before the reform. Each user had a quota limit up to certain level of consumption. The cheaper tariff, which covered the energy consumption less than quota level, was called the "normal tariff". Consumption above the quota was more expensive and was called "free tariff".

[17]kWh: Kilowatt hours.

[18]In summer the off-peak hours are 23:00–7:00, and in winter between 22:00–6:00.

[19]In summer the peak hours are 19:00–23:00, and in winter between 18:00–22:00.

[20]As the domestic consumption has a stepwise tariff structure, we have calculated the average electricity price based on a 300 kWh monthly household energy consumption (Source: https://barghnews.com/fa/news/8333, last accessed 25.5.2016).

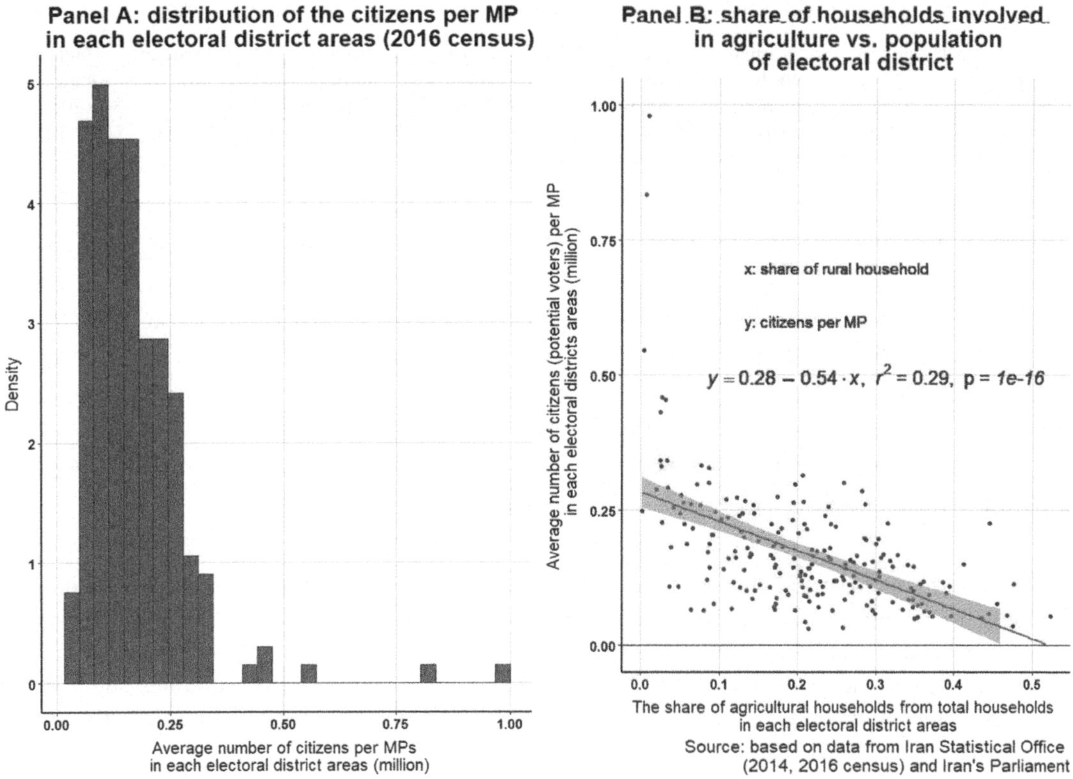

Fig. 3 The distribution of citizens (potential voters) per MP across electoral districts and its relationship with the share of households involved in agriculture

2011, the Minister of Agriculture announced the reduction of the irrigation pumping tariff to 80 rials/kWh for the normal tariff (Khalilian 2011). Furthermore, if farmers only pumped water in non-peak hours, they were charged just 40 rials/kWh, five times less than the newly reformed urban and industrial electricity tariff. Table 7 shows the electricity tariffs for normal, off-peak, and peak hours for different types of uses before and after the implementation of the subsidy reform. We see that price increases in agriculture did not reach the levels of domestic and industrial use despite the initial plan of the government to equalize electricity prices among consumers.

It is evident from this four-month process (end of 2010 up to spring 2011) that the real effects of the price/subsidy reform have been largely eradicated by the amendments to the original plan. In contrast to agriculture, the plan was successfully implemented for domestic and industrial users. Considering Iran's high inflation rate and the fact that the nominal tariff of electricity in agriculture has changed slightly, we can say that the policy was unsuccessful for energy consumption for irrigation. Additionally, the real price/tariff of electricity actually reduced as the tariff increase after 2011 was much less than the inflation rate in the country during the same period (see Table 2 for annual inflation rates). The electrification of the groundwater pumps has still been going on and, as a result, the number of wells and the energy consumption in agriculture has increased.

3.4 The Subsidy Reform Effects on Rural Life

Iranian economic studies have anticipated hard times for the country's agricultural sector. However, of the few studies that were conducted

Table 7 Electricity tariff dynamics before and after the reforms for different consumers (rials/kWh)

	Usage	Normal tariff (2008–2010)	Free tariff (2008–2010)	Revised tariff (End 2010–Spring 2011)	Revised tariff (Spring 2011–2013)	General tariff (2014–2015)	General tariff (2018–2019)
Off-peak hours	Industry	33.68	33.68	170	170	253	303
	Domestic	17.7	46.2	460	460	636	762
	Agriculture	**3.3**	**4.8**	**67.5**	**40**	**55**	**66**
Normal hours	Industry	134.51	134.51	340	340	506	606
	Domestic	74.4	197	460	460	636	762
	Agriculture	**13.3**	**19.4**	**135**	**80**	**110**	**132**
Peak hours	Industry	444	444	680	680	1012	1212
	Domestic	187.5	492.7	460	460	636	762
	Agriculture	**33.3**	**48.6**	**270**	**160**	**220**	**264**

Source North Kerman Electrical Power Distribution Co. and Tavanir

after the reform, major negative effects due to the phasing out of subsidies were not found. For instance, Eskandary et al. (2017) used the input–output matrix approach with data from 2006 to 2014 and showed that agricultural production decreased, but the total value of agricultural products increased by approximately 1.5% as a result of increased agricultural prices. On the other hand, Zare-Chahouki and Sanaei (2017) analyzed available statistics before and after the reform implementation, finding that the consumption of pesticides and fertilizers decreased by 87.5 and 55% during the 2009–2013 period. However, increases in prices for meat, dairy products, and fodder crops have kept agricultural production steady and even encouraged overgrazing. They also found a side effect from the increased fuel prices: rural residents substituted gas with wood and, as a result, deforestation increased. In another study from the Urumiyeh region in northwestern Iran, the effect of the subsidy reform was analyzed from a socio-economic point of view (Ahmadi et al. 2016). The authors conducted a structured survey among rural residents and used a structural equation model to analyze their welfare before and after the subsidy reform. Considering the lump-sum payments for each household stipulated under the compensation scheme of the price reform, as well as other socio-economic factors such as consumption patterns, social conditions, and saving patterns, the model shows that the socio-economic conditions of the rural households were improved. The findings suggest that the price/subsidy reform did not affect rural life as negatively as was expected. It should be mentioned that no study up until now has shown that water extraction changed due to energy price increases. In sum, it appears that, with minor increases in agricultural energy prices and major increases in the prices of food products, there was not a major reduction in agricultural production in Iran. In the future, as more data may become available, it will be possible to analyze the effects of subsidy cuts in more detail.

4 A Path for the Future

During the years after the reform, the announcements of the parliament's agricultural commission and energy commission showed how pressure from farmers was transferred to the parliament via MPs.[21] It is clear that before any compensatory scheme for farmers to be implemented, changing the energy pricing policy for

[21]A list of the announcements of the MPs or Parliament Commissions regarding any further changes to groundwater pumping energy prices since 2011 can be provided upon the request.

agriculture will meet heavy resistance and cause unease among farmers, which is likely to be transferred to MPs and professional parliamentary commissions. This may essentially halt any reforms attempting to further phase out subsidies. Without a deep analysis of the political economy of energy subsidies for irrigation groundwater pumping, the reform is unlikely to work because the resistance from subsidy recipients will stop further progress in this area. The last major political discussion about regulating the subsidization of irrigation water at the national level in Iran occurred in 2015–16, just before the most recent parliamentary election in February of 2016. In 2015, 96 MPs came up with an initiative to fully subsidize irrigation water for agriculture and set the price of water at zero.[22] Fortunately, the plan was temporarily forgotten as the MPs engaged in the parliamentary election campaign in the winter of 2016.

It is clear that the self-sufficiency programs and the phasing out of indirect subsidies are contradictory and cannot be implemented at the same time. Additionally, a master plan for supporting all affected rural communities and smallholders is needed. It must be mentioned that the good governance of water resources is not limited to only the elimination of water and energy related subsidies, but it also refers to multilevel systems of social, economic, and administrative instruments for water resources management and delivery of water services (Rogers and Hall 2003). Definitely, analyzing all dimensions of water governance is beyond the scope of this paper. However, using economic tools for managing water resources is one of the major issues of water governance. As is presented in this paper, economic tools cannot be implemented in a simple way as long as many people heavily depend on the provision of cheap water for their livelihoods. Different solutions are possible for tackling the political economy of the energy-water nexus. For instance, in order to manage groundwater resources in India, energy rationing was suggested as one solution (Shah et al. 2008). Following this approach, free energy is provided for wells for just a few hours per day. Simultaneously, no under-priced electricity will be provided anymore. The acceptability and the consequences of this policy in Iran should be studied in more detail; no studies have examined solutions in this direction up until now. Moreover, there are some attempts to introduce community-based groundwater resources management by the Iranian Water Policy Research Institute (IWPRI), although the successful application of this approach in the field on a noticeable scale has not been reported yet.[23] These are two examples of managing groundwater resources without using direct economic tools. Both approaches require laws and regulations to be altered and responsible institutions should keep this on their agendas. Nevertheless, the elimination of energy subsidies and their possible consequences is a policy option that cannot be postponed, and it requires careful consideration of the incentives of all involved stakeholders.

Following the political economy of the water-energy-food nexus, it would seem prudent and practical to focus on finding alternative income sources for rural communities that would be dramatically affected by substantial changes in groundwater abstraction costs. Eco-tourism, supporting small rural industries (e.g. carpet production, etc.), and tourism based on regional history and cultures are some potential ways for creating alternative livelihoods. New innovations and the latest developments in modern technologies can also help create jobs in rural Iranian areas. For instance, the expansion of solar energy through a decentralized energy generation system can be one of these solutions. All of these possibilities should be analyzed by separate studies and their feasibility should be researched. Since 2011, the government of Iran has started to provide facilities for the expansion of photovoltaic energy in Iran in order to use its massive solar energy potentials. This entered into a new phase in 2015 after a new system of feed in tariffs

[22]Source: https://www3.irna.ir/fa/News/81947715/, last accessed 23.5.2016.

[23]IWPRI (2015): http://iwpri.ir/home/single/127 (last accessed 11.7.2018).

(FiT) was introduced. The idea was to buy energy at higher prices from small to large scale energy producers (IEA 2017). Sadeghi et al. (2017) showed that the expansion of renewable energy generation in Iran has the potential to generate a substantial number of direct and indirect jobs. However, the potential of this approach has not really been explored as a rural development tool for creating alternative sustainable livelihood for rural regions. Thus, von Heyking and Jaghdani (2017) have compared the solar energy generation on a small scale in Iran and Germany by considering the latest tariffs available and macro-economic factors such as interest rates and discount rates. This study shows that, in spite of higher tariffs in Iran, higher interest and discount rates can make such a small investment without governmental help less profitable. Furthermore, the expansion of renewable energy on a small scale will be much more profitable in macroeconomic conditions that are more stable than today's conditions. The possibility of using solar energy for rural development and its advantages or disadvantages requires a separate study that is beyond the scope of this paper. It is nevertheless noteworthy that there is growing international interest in the interconnections between solar energy and groundwater pumping. The Consultative Group for International Agricultural Research (CGIAR) has considered solar pumps for shallow and abundant aquifers as one of the ten innovations for future rural development (Dinesh et al. 2017). From a first glance, these innovations seem irrelevant for Iran because its wells are mainly deep in the depleted aquifers and they need more energy for pumping. However, the idea of selling energy locally can be considered a possible alternative livelihood strategy in the rural areas of Iran. It must be mentioned, however, that very cheap solar energy can lead to further depletion of groundwater resources when the official electricity tariffs are high. Within such a future scenario, generated electricity may be used for pumping water. Therefore, the nexus between the cost of solar energy generation, the official FiT of renewable energy, official electricity tariffs, and water extraction is a separate study that should be considered in the future as solar energy becomes available at cheaper prices.

The solution to the Iranian WEF nexus should be studied and tackled from different dimensions of nexus governance. In this study, we have discussed the political economy aspects of this relationship and showed that phasing out energy subsidies for irrigation pumping cannot be done easily when millions of people are dependent on that. However, creating alternative livelihood strategies for rural populations can be a starting point in applying economic tools to groundwater management.

5 Conclusions

In this chapter, we have analyzed historical trends of groundwater consumption in Iran's agricultural sector and presented an overview of the water management institutions which encourage intensive groundwater use in Iran. The current usage appears to be far from sustainable and it will threaten both food production in different rural communities and freshwater availability for basic needs. Therefore, in the near future, SDG2 and SDG6 are unlikely to be accomplished by supporting agricultural production with the help of irrigation water from groundwater resources. Cheap energy due to heavy subsidies will have devastating effects on freshwater resources and SDG2 and SDG6 for Iran. Although rising energy prices can represent an economically acceptable approach for managing groundwater consumption, it cannot be easily implemented. The outcomes of the subsidy reforms have shown that, even though reform was successful in cutting energy subsidies in non-agricultural sectors, the program did not achieve the desired results in the agricultural sector. Although energy price management can enable the sustainable management of groundwater resources, an alternative has to be presented to farmers to reduce their opposition and potential political distress. Results suggest that a reform process that stipulates subsidies retrenchment can be accelerated if the central government acknowledges and deals with dominating agricultural interests directly. Understanding the incentives and capabilities of rural elites may help address their concerns on a national level in a more balanced fashion. Liberalization processes in agriculture should ideally be guided by a strategic document that would outline basic principles and a vision for Iranian agricultural development. It should, therefore, be inspired by nexus thinking, integrating sectors and considering resource use in an integrated way. This document should be developed in a transparent fashion with participation from all stakeholders involved - another important aspect of nexus thinking. As we have seen in this paper, the parliamentary resistance has dismantled any reform in the agricultural sector. No plan was presented to compensate the affected stakeholders and farmers throughout the reform process. This weak planning stopped any real further reductions in energy prices and groundwater consumption. It would be useful to focus more on considering alternative income resources for rural livelihoods before tackling the subsidy for electricity. In conclusion, the political economy is an extremely important aspect of the governance of the WEF nexus in Iran at the macro and micro levels. The promotion of rural industries using less water could be an alternative that should be considered in the next reform. Apart from the reform policy on any water use management policy, the promotion of solar energy production in rural areas and the sale of energy to the government is a potential alternative for rural areas that can be considered.

Acknowledgements We appreciate the useful comments from Prof. Bernhard Brümmer and two anonymous reviewers on this manuscript. The authors received no financial support for the research, authorship, and/or publication of this article. This research was conducted during the presence of the two authours at the University of Göttingen and Leibniz Institute of Agricultural Development in Transition Economies (IAMO) in Germany.

References

Acemoglu D, Robinson JA (2001) A theory of political transitions. Am Econ Rev 938–963

Ahmadi A, Mojaradi G, Badsar M (2016) Study of socio-economic impact of Targeted Subsidies Law on rural households' quality of life in Urmia Township. Agric Ext Educ Res (in Persian) 9(3):43–54. http://jaeer.srbiau.ac.ir/article_10378.html

Ahmadi Yazdi A (2018) Free electricity for low energy consuming farmers. Donya-e-Eqtesad Newspaper (in Persian). https://www.donya-e-eqtesad.com/fa/tiny/news-3386100, Newspaper No: 4323, News No: 3386100

Akbari A (2018) 350 thousand illegal wells. Mehrnews (in Persian), Newscode: 4263304. https://www.mehrnews.com/news/4263304/. Accessed 21 June 2018

Alijani F, Salarpur M, Sabuhi M (2012) Application of CGE model on the elimination of production subsidy in agriculture. J Agric Econ Dev (in Persian) 26 (3):218–227. https://jead.um.ac.ir/article/view/17778

Bagherzadeh A (2017) Review of electricity consumption in Iranian agriculture (originally in Persian). Agricultural Planning, Economic and Rural Development Research Institute (APERDRI), Tehran, Iran. http://assc.ir/upload/upload/iranassc_masrafbarqkeshavarzi.pdf https://assc.ir/upload/upload/iranassc_masrafbarqkeshavarzi.pdf

Balali H, Khalilian S, Ahmadian M, Pelet-Kaleh ST (2008) Analysis of effects of energy subsidies adjustment on groundwater balance and exploitation. Agric Res (in Persian) 8(3):85–106

Balali H, Montashloo M (2014) The effects of free energy price on soil and water resources in agriculture; mathematical programing approach. Appl Econ Stud Iran (in Persian) 3(10):201–218

Bates RH (1983) Essays on the political economy of rural Africa, vol 38. University of California Press

Behzad M (2010) Support for agriculture and industry by electricity tariffs announcement (in Persian). TAVANIR. https://news.tavanir.org.ir/press/print.php?id=20361

Chitchian H (2014) Agriculture uses 92 percent of water but produces 1/3 of possible production. IRNA News Agency (originally in Persian). https://www.irna.ir/news/81254583/

Chitchian H (2015a) 20 hours free electricity for pumping. Tasnim News Agency (in Persian), newscode: 1013880. https://nasimonline.ir/Content/Detail/1013880/152. Accessed 28 Feb 2016

Chitchian H (2015b) Farmers pay 110 Rials for electricity instead of 800 Rials. Tasnim News Agency (in Persian), newscode: 667112. https://www.tasnimnews.com/fa/news/1393/12/08/667112. Accessed 28 Feb 2016

Coates D, Connor R, Leclerc L, Rast W, Schumann K, Webber M (2012) Water demand: what drives consumption? In: WWAP (World Water Assessment Programme) (ed) The United Nations world water development report 4: managing water under uncertainty and risk, vol 1, pp 44–76. United Nations Educational, Scientific and Cultural Organisation (UNESCO), Paris, France de Fraiture C, Perry C (2007) Why is agricultural water demand unresponsive at low price ranges? In: Molle F, Berkoff J (eds) Irrigation water pricing: the gap between theory and practice, 1st edn. CABI, pp 94–107. https://horizon.documentation.ird.fr/exl-doc/pleins_textes/divers16-09/010045237.pdf

Dinar A (2000) Political economy of water pricing reforms. In: Dinar A (ed) The political economy of water pricing reforms, 1st edn. Oxford University Press, New York, N.Y., pp 1–25. https://documents.worldbank.org/curated/en/2000/04/437127/political-economy-water-pricing-reforms

Dinesh D, Campbell B, Bonilla-Findji O, Richards M (2017) 10 best bet innovations for adaptation in agriculture: a supplement to the UNFCCC NAP Technical Guidelines | CCAFS: CGIAR research program on Climate Change, Agriculture and Food Security (CCAFS Working Paper No. 2015). Wageningen, The Netherlands. https://ccafs.cgiar.org/publications/10-best-bet-innovations-adaptation-agriculture-supplement-unfccc-nap-technical#.WthvYpcuDcs

Dixit A, Londregan J (1996) The determinants of success of special interests in redistributive politics. J Polit 58 (4):1132–1155

Eskandary, M., Mohammadi, H., Mirzaei, H., Kahkha, A. (2017). Effects of energy subsidy reform on economic variables in agricultural sector. J Agric Econ Res (in Persian) 9(33):125–144. http://jae.miau.ac.ir/article_2184.html

Falahatian H (2015) Production of electric power by private sector. Government news agency (originally in Persian). http://dolat.ir/detail/272480 . Accessed 20 May 2016

Garrett G (1998) Global markets and national politics: collision course or virtuous circle? Int Organ 52 (4):787–824

Hajrasuliha M (2015) 30 BCM reduction in Iran renewable water resources. Fars News Agency (originally in Persian), Sept 2015. https://www.farsnews.ir/news/13940705000135 . Accessed 28 Feb 2016

Heidari K, Parmeh Z (2011) An estimate of the impact of bread and energy carriers price correction on household expenditure. Rahburd (originally in Persian) 11 (57):181–195. http://rahbord.csr.ir/Article/13940916153352500483

ICSU (2017) A guide to SDG interactions: from science to implementation. Paris, France. https://doi.org/10.24948/2017.01

IEA (2017) Renewable energy development fund. https://www.iea.org/policiesandmeasures/renewableenergy/?country=Iran. Accessed 28 Apr 2017

Jomehpour M (2009) Qanat irrigation systems as important and ingenious agricultural heritage: case study of the qanats of Kashan, Iran. Int J Environ Stud 66(3):297–315. https://doi.org/10.1080/00207230902752629

Kalantari I (2015) We extract million of years reservoirs. Shargh Newspaper (originally in Persian), 30 July 2015. https://www.magiran.com/npview.asp?ID=3199980

Khalilian S (2011) Good news of agricultural ministry. Javanonline (originally in Persian), Apr 2011. https://www.javanonline.ir/vdcenp8evjh8e7i.b9bj.html. Accessed 8 Nov 2011

Kurian M, Ardakanian R, Gonçalves Veiga L, Meyer K (2016) Institutions and the nexus approach. In: Resources, services and risks. How can data observatories bridge the science-policy divide in environmental governance? Springer, Cham, pp 5–30. https://doi.org/10.1007/978-3-319-28706-5_2

Madani K (2014) Water management in Iran: what is causing the looming crisis? J Environ Stud Sci 4 (4):315–328. https://doi.org/10.1007/s13412-014-0182-z

Mahsuli E (2014) Average export price of electricity is 8–11 cents. ILNA News Agency (originally in Persian), newscode: 223485, Nov 2011. https://www.ilna.news/fa/tiny/news-223485 . Accessed 28 Feb 2016

Majlis of Iran (1999) Electo pumps for wells. Official Newspaper Islamic Republic of Iran (originally in Persian), Tehran, Iran. https://rc.majlis.ir/fa/law/show/93203. Accessed 21 Dec 2018

Majlis of Iran (2004a) Financing casualties of drought and coldness. Official Newspaper Islamic Republic of Iran (originally in Persian), Tehran, Iran. https://rc.majlis.ir/fa/law/show/94180. Accessed 21 Dec 2018

Majlis of Iran (2004b) Iran fourth year development plan (March 2006–March 2011). Official Newspaper Islamic Republic of Iran (originally in Persian), Tehran. https://rc.majlis.ir/fa/law/show/94202. Accessed 21 Dec 2018

Majlis of Iran (2009) Price policy reform for elimination of subsidies. Official Newspaper Islamic Republic of Iran (originally in Persian), Tehran. https://rc.majlis.ir/fa/law/show/789036. Accessed 21 Dec 2018

Majlis of Iran (2010) Recognition the illegal wells. Official Newspaper Islamic Republic of Iran (originally in Persian), Tehran. https://rc.majlis.ir/fa/law/show/782294 Accessed 21 Dec 2018

Majlis of Iran (2015) Free irrigation water. Islamic Parliament Research Center of the Islamic Republic of Iran (originally in Persian), Tehran. https://rc.majlis.ir/fa/legal_draft/state/952698 Accessed 21 Dec 2018

Mazaheri M, Khiayabani N (2017) Analyzing the effects of reducing energy subsidies on income distribution in Iran by a computable general equilibrium model. J Appl Econ Stud Iran (originally in Persian) 6 (21):19–41. https://doi.org/10.22084/AES.2017.1796

Meidani R (2015a) 300 illegal wells. Tasnim News Agency (in Persian), newscode: 901138. https://www.tasnimnews.com/fa/news/1394/08/07/901138. Accessed 28 Feb 2016

Meidani R (2015b) 200 illegal wells. IRNA News Agency (in Persian), newscode: 81710526 (4805189), Aug 2015. https://www.irna.ir/fa/News/81710526/. Accessed 28 Feb 2016

Moazedi A, Taravat M, Jahromi HN, Madani K, Rashedi A, Rahimian S (2011) Energy-water meter: a novel solution for groundwater monitoring and management. In: World environmental and water

resources congress 2011. American Society of Civil Engineers, Reston, VA, pp 962–969. https://doi.org/10.1061/41173(414)99
Modaberi S (2015) Water crisis in 350 aquifers. IRNA News Agency (in Persian), news code: 81897497 (5074848). https://www.irna.ir/fa/News/81897497/. Accessed 10 May 2016
Morady MA, Amidpour M, Hurijafari H, Mehrazma I (2014) Agriculture sector energy demand analyzing and modeling. Iran Energy J (in Persian) 17(2). https://necjournals.ir/article-1-507-fa.html
Musavi SN, Farajzadeh Z, Taheri F (2013) The welfare effect of energy subsidy reduction in Iranian agriculture sector. J Agric Econ Dev (in Persian) 26(4):298–306. https://www.sid.ir/fa/journal/ViewPaper.aspx?id=190998
Nikbakht MA (2011) Agricultural electricity tariffs announcement. Rajanews (in Persian). https://www.rajanews.com/PrintFriendly.asp?id=77342. Accessed 15 Mar 2011
NKEPD Co. (2011) North Kerman Electrical Power Distribution Company electricity tariffs (document in Persian). https://www.nked.co.ir/index.php?option=com_content&view=article&id=13&Itemid=17&lang=fa. Accessed 4 Nov 2011
Nozaripur A (2015) Iran will be Somalia till 2025 by reduction in water resources. IRNA News Agency (in Persian), news code: 81500135(4499154). https://www.irna.ir/fa/News/81500135/. Accessed 28 Feb 2016
OECD (2012) Meeting the water reform challenge. OECD, Paris, France. https://doi.org/10.1787/9789264170001-en
Pfeiffer L, Lin C-YC (2014) The effects of energy prices on agricultural groundwater extraction from the High Plains Aquifer. Am J Agr Econ. https://doi.org/10.1093/ajae/aau020
Purihosseini A (2015) Privatisation of electric power since 2005. BarghNews (in Persian). Bargh News Agency (in Persian). https://barghnews.com/fa/news/10386. Accessed 20 May 2016
Richey AS, Thomas BF, Lo M-H, Famiglietti JS, Swenson S, Rodell M (2015) Uncertainty in global groundwater storage estimates in a Total Groundwater Stress framework. Water Resour Res 51(7):5198–5216. https://doi.org/10.1002/2015WR017351
Ringler C, Willenbockel D, Perez N, Rosegrant M, Zhu T, Matthews N (2016) Global linkages among energy, food and water: an economic assessment. J Environ Stud Sci 6(1):161–171. https://doi.org/10.1007/s13412-016-0386-5
Rogers P, Hall AW (2003) Effective water governance (TEC background paper No. Issue 7). Stockholm, Sweden
Sadeghi Z, Shamsodin pour F, Mirzae HR (2017) The feasibility of creating jobs in renewable energy using input-output approach: a case study of solar and wind energy. Iran J Econ (in Persian) 5(0):145–175. https://doi.org/10.22054/JIEE.2017.7307
Scott CA, Shah T (2004) Groundwater overdraft reduction through agricultural energy policy: insights from India and Mexico. Int J Water Resour Dev 20(2):149–164. https://doi.org/10.1080/0790062042000206156
Scott CA, Kurian M, Wescoat JL Jr (2015) The water-energy-food nexus: enhancing adaptive capacity to complex global challenges. In: Kurian M, Ardakanian R (eds) Governing the nexus. Springer International Publishing, pp 15–38. https://doi.org/10.1007/978-3-319-05747-7_2
Shah T, Bhatt S, Shah RK, Talati J (2008) Groundwater governance through electricity supply management: assessing an innovative intervention in Gujarat, western India. Agric Water Manag 95(11):1233–1242. https://ideas.repec.org/a/eee/agiwat/v95y2008i11p1233-1242.html
Shah T, Giordano M, Mukherji A (2012) Political economy of the energy-groundwater nexus in India: exploring issues and assessing policy options. Hydrogeol J 20(5):995–1006. https://doi.org/10.1007/s10040-011-0816-0
Shah T, Scott CA, Kishore A, Sharma A (2004) Energy-irrigation nexus in South Asia: improving groundwater conservation and power sector viability (Research Report No. 70). Colombo, Sri Lanka
Shahraki J, Hosseini SM, Khazai S (2016) Analysis of the effects of the irrigation water subsidy reform on the Iranian agricultural sector. Agric Econ Res (in Persian) 8(32):61–77. https://www.sid.ir/fa/journal/ViewPaper.aspx?ID=278873
Statistical Center of Iran (2014) Detailed results of the general agricultural census in Iran (originally in Persian). Tehran, Iran. https://www.amar.org.ir/Portals/0/keshavarzi93/results/agri93-99.pdf
Tehran Chamber of Commerce (2017) Agriculture economic growth. Eghtesadonline (in Persian). https://www.eghtesadonline.com/n/14BW . Accessed 5 July 2018
The World Bank (2008) The political economy of policy reform. Washington DC, USA. https://siteresources.worldbank.org/EXTSOCIALDEV/Resources/The_Political_Economy_of_Policy_Reform_Issues_and_Implications_for_Policy_Dialogue_and_Development_Operations.pdf
Theesfeld I (2010) Institutional challenges for national groundwater governance: policies and issues. Ground Water 48(1):131–142. https://doi.org/10.1111/j.1745-6584.2009.00624.x
Thomson H (2014) Food and power: authoritarian regime durability and agricultural policy. Faculty of Graduate School of the University of Minnesota.
UNDP (2018) Sustainable development goals. https://www.undp.org/content/undp/en/home/sustainable-development-goals.html. Accessed 20 June 2018
UNU-FLORES (nd) The nexus approach to environmental resources management. https://flores.unu.edu/en/research/nexus. Accessed 2 July 2018
von Heyking C-A, Jaghdani TJ (2017) Expansion of photovoltaic technology (PV) as a solution for water energy nexus in rural areas of Iran: comparative case

study between Germany and Iran (No. 1709). DARE Discussion Papers. Georg-August University of Göttingen, Department of Agricultural Economics and Rural Development (DARE), Göttingen, Germany. https://www.econstor.eu/handle/10419/168369

Zare-Chahouki M, Sanaei A (2017) Analysis of price policy reform consequences on the natural resources in Iran. Strateg Res J Agric Sci Nat Resour (in Persian) 2(2):107–122. https://www.magiran.com/view.asp?Type=pdf&ID=1739012&l=en

Zilberman D, Sproul T, Rajagopal D, Sexton S, Hellegers P (2008) Rising energy prices and the economics of water in agriculture. Water Policy 10:11

Political Economy of Energy Subsidies for Groundwater Irrigation in Mendoza, Argentina

Félix Sebastián Riera and Bernhard Brümmer

Abstract

Natural resources' policies in Latin America are rarely long-term, consistent and power independent. The Argentinian province of Mendoza achieves both characteristics, groundwater management showed flaws with ensuring quality and availability. Energy subsidies for agricultural irrigation have relied too much on a permanent policy, and are subject to political maneuvering whenever their stability is at risk. Following a tripod framework to review the institutional settings of the water-energy nexus, we review the policy effectiveness and analyze policy outcomes in light of the nexus. Although the majority of policy tools have become demand-oriented during the last 20 years, they do not provide consistent economic incentives for agricultural producers when considering environmental degradation of groundwater resources. The economic tools designed for the agricultural sector aimed to sustain profit levels by influencing the market conditions rather than improving farm-level efficiency. Under water scarcity periods and subsidized energy, we observed changes in grapevine prices and groundwater table levels that suggest a relationship with the exploitation of the Carrizal aquifer. To explore the resource in a sustainable manner and further save energy and money, subsidy effectiveness should be reconsidered. Policy-makers could gradually shift the annual budget from energy subsidy into optimization programs for water and energy, which would upgrade resource management and stimulate sustainable development.

Keywords

Water-energy nexus · Resources policies · Water scarcity · Groundwater depletion · Grapevine production

1 Introduction

In the area of agriculture and resource economics, challenges for optimization are continuously updated. The increasing demand to provide food with limited resources emphasizes the need for more efficient production under changing environmental conditions (FAO/IWMI 2018). The reinvigorated interest in the water–

F. S. Riera (✉) · B. Brümmer
Department of Agricultural Economics and Rural Development, Georg-August-Universität Göttingen (DARE-GWDG), Platz der Göttinger Sieben 5, 37073 Göttingen, Germany
e-mail: sebary@gmail.com

F. S. Riera
Department of Agricultural Economics and Policy, Facultad de Ciencias Agrarias, Universidad Nacional de Cuyo (FCA-UNCuyo), Almirante Brown 500 Chacras de Coria, M5528AHB Mendoza, Argentina

S. Hülsmann and M. Jampani (eds.), *A Nexus Approach for Sustainable Development*,
https://doi.org/10.1007/978-3-030-57530-4_9

food–energy nexus highlights an increased public concern for responsible and efficient use of natural resources (Allan et al. 2015); however, there is a need to provide effective policy implementation tools that comprehensively take into account the water-energy linkages (Dai et al. 2018). In the arid province of Mendoza, Argentina, groundwater irrigation is vital for agricultural activities. The political will to improve the profitability of small producers has distorted economic incentives and led to the creation of power asymmetries among stakeholders and decision makers. Political and economic analysis is conducted in this paper to reveal the reform arena of public policies that link water and energy in the agricultural sector.

Local governments are responsible for designing solid policies that contribute to the responsible use of natural resources and simultaneously coincide with public preferences. Institutional settings, lack of information, policy implementation time-frames, and political influence may obstruct the optimization path of social welfare (Foster et al. 2016; Dinar 2000; Shah et al. 2012). When it comes to water demand, institutional settings are fundamental to empower stakeholders and set economic incentives. As a multi-purpose resource, water is demanded as both a production input and a resource for direct consumption. Globally, the agricultural sector employs near 70% of the total water supply (Dagnino and Ward 2012). Groundwater use for agriculture became desirable because of the ability to face production challenges, especially in scarcity periods. Overexploitation and poor management of groundwater may lead to irreversible quality degradation (Garduño and Foster 2010). Energy and water policies jointly determine institutional settings and power spaces of the stakeholders (Azpiazu et al. 2014). Abuse of economic tools to maintain political power could jeopardize the sustainability of the resource, while simultaneously providing inadequate economic incentives for water users (Badiani et al. 2012).

This paper seeks to describe the economic incentives and behavior of local stakeholders towards the exploitation of groundwater resources. The analysis focuses on the economic and political framework of the water-energy nexus. The main research questions addressed are as follows: (1) How are the agricultural irrigation practices influenced by the water and energy policies, leading to a threat to the Carrizal aquifer as a complementary irrigation source? (2) Do local stakeholders have incentives to consider the environmental effects of groundwater exploitation for production purposes? The main hypothesis is that energy subsidies for agriculture irrigation drive farmers' behavior towards the exploitation of the aquifer. An integrative review of energy policies from other perspectives, such as political economy, could enhance a better understanding of the implications in the context of water management and the nexus approach.

2 Materials and Methods

2.1 Study Area

The province of Mendoza is located in a semi-arid region in the central-west of Argentina (Fig. 1) and it covers an area of 150,839 km^2. It is characterized by its mountainous area, formed by the Andes mountain range that runs from north to south. The average rainfall is 220 mm per year. It has a dry, continental climate characterized by the level of summer rainfall regime (Morello et al. 2012). During 2013, the provincial GDP reached USD 2.08 billion. The main economic sectors are commerce, services, and the manufacturing industry, which jointly represent 60% of the province's GDP (DEIE 2014). During the last decade, agricultural activities contributed USD 132 million to the regional GDP, representing 7% of the total GDP (Medawar et al. 2011). The total irrigated land in Mendoza reached 267,889 ha in 2017, which represents 85% of the arable land in the province and 25% of the national irrigated area (Calcagno et al. 2000; FAO/PROSAP 2015).

The geographical and environmental characteristics make the province strongly dependent on water resources for economic development.

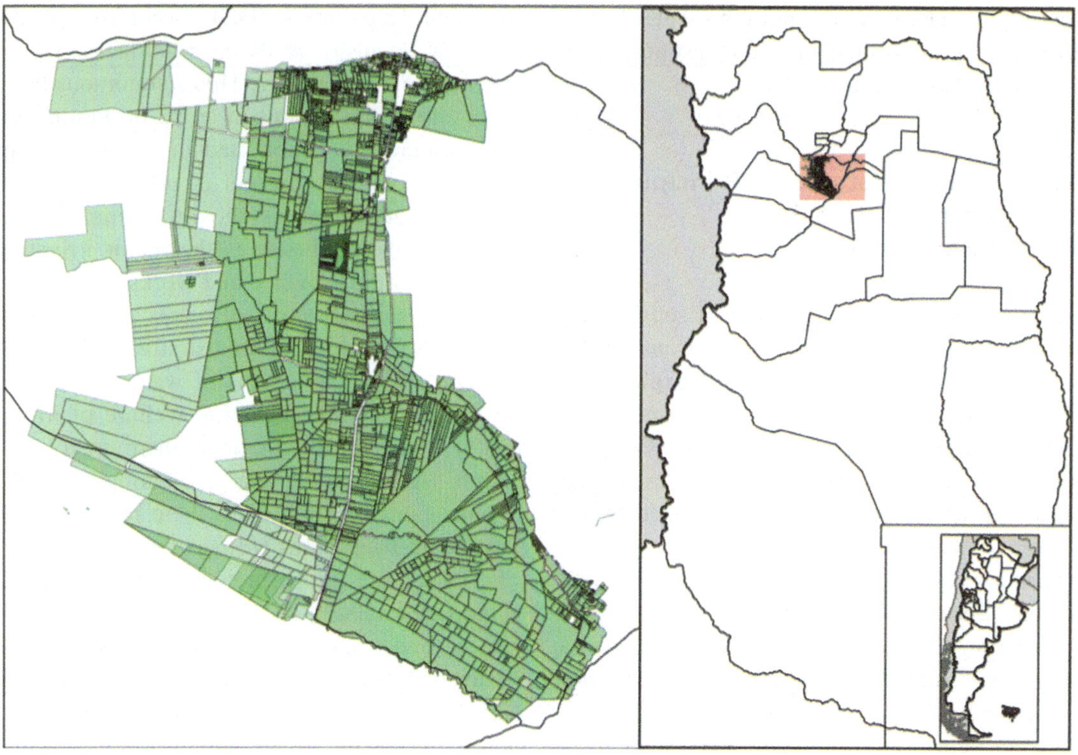

Fig. 1 Land ownership in the study area, location of the Carrizal aquifer and Mendoza Province in Argentina; *source* modified based on IGN (2018)

The location in the Andes Mountains allows primary access to surface water in summer through the largest irrigation system in the country, which was originally designed by the original towns (DGI 2015). Over the Carrizal aquifer, the coexistence of agricultural activities and oil refineries has evolved as a delicate matter for water users and management institutions. The extension of the aquifer is nearly 1,000 km^2, where most of the land is dedicated to agriculture (Fig. 1), in particular grapevine production for wine production. The fragmentation of land rights is represented by the green color in Fig. 1.

2.2 Water Resources

With the melting of snowpack in the high peaks in the spring and summer, water is provided by five rivers to the region. Precipitation as rain provides little input into rivers and occurs mostly in the summertime with high intensities (Maccari 2004). More than 80% of the water supply is employed in agriculture. In particular, the agroindustry demands nearly 13.51 hm^3 of water to produce processed fruits, vegetables, and beverages (Duek et al. 2013).

In terms of irrigation practices and technology adoption, the current policies have led to the formation of two main groups. The first is producers with low technological capacity, which consider the irrigation system a valid approach to accessing quality water at a reasonable cost. Therefore, they welcome discussions to improve the management of the resources as long as the irrigation system remains unchanged. This group is characterized by the traditional practices and use of less modern systems generally associated with higher age groups. The remaining group is made up of market-oriented vineyards, for which technology adoption does not represent a barrier in terms of innovation or the investment costs (Maffioli et al. 2011).

2.3 General Water Law, Principles and Institutional Organization

The main policies and institutional settings that shape the political economy of the water-energy nexus for agricultural producers are explained in this section. Mendoza belongs to the most arid area in Argentina, and the use of water is relevant for every economic activity. The historical relevance of water regulation is represented by the General Water Law (*Ley General de Aguas*), which was issued prior to the provincial constitution (Bermejo 1884). This legislation is the foundation for economic development in the region and further influences the policy design of economic and agricultural tools.

Considered as the foremost legislation on water management, the General Water Law regulates the use, distribution rules, payments and quality (DGI 2016b; Silanes 2013). Water was declared to be an asset of the public domain and three main water-use principles are represented in the General Water Law: inheritance, non-prejudice clause, and specificity. The inheritance principle determines the permanent attachment of water rights with the land property, which avoids the potential to divide and commercialize rights individually. The non-prejudice clause looks after the common welfare of water users since it considers the effects of certain actions or new activities on individuals. Finally, the specificity principle ensures the nullity of contracts that use water for purposes other than those agreed upon (Pinto et al. 2006). The water authority is entitled to provide temporary use of water through the permit instrument known as precarious right, which can be revoked at any time, even without just cause and without the right to prior compensation. Drilling permits constitute the allowance to extract groundwater resources (Reta 2002).

In 1916, the provincial constitution rectified the former water law and constituted the General Irrigation Department (DGI, according to the name in Spanish) as the institutional body to execute the police power. This autonomous body makes its own decisions in terms of administration, resource allocation, and investments. Representation of stakeholders is promoted within the irrigation system as shown in Fig. 2. Consequently, the DGI acts as the major irrigation body within the province and works similarly as a parallel state in terms of water management. The Honorable Administrative Tribunal (HTA) controls the actions of water bodies and privates towards the common welfare of irrigators. Throughout the province, there are 17 Irrigation Associations that conglomerate 135 organizations of water users, here referred to as Watershed Inspections (*Inspecciones de Cauce*). The former are public, non-state-owned organizations that manage and distribute water through the secondary networks represented by the Watershed Inspections. The latter is "ministry legis" by Law 5302 and Law 6405. Their purpose is to engage in the administration and distribution of the water, and the maintenance of secondary networks and derivatives. Officials within the department are elected democratically and they have their own budget (Maccari 2004; Pinto et al. 2006). As a relevant agglomeration of water users, the watershed inspections are composed by producers and enterprises (mostly from agroindustry) that may be consulted by members of the parliament to express their needs in the provincial Congress.

Ideally, the organization of water management should not be static and should respond to the ever-evolving interests of farmers determined by agricultural demand (Jofré 2010). Minor changes in the irrigation system directly result in strategic behavior and the design of complimentary water-use tools. Therefore, any potential changes should be announced in a clear and transparent manner (Erice 2013).

2.3.1 Current Conditions of Irrigation Efficiency

Different definitions of efficiency exist in the field of water management. In general terms, irrigation efficiency is measured as the ratio of

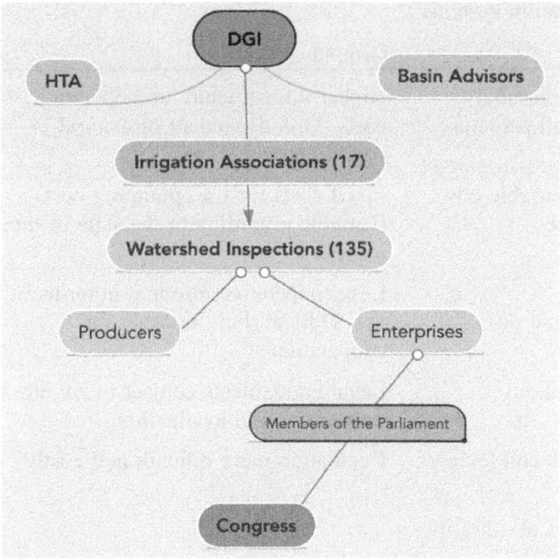

Fig. 2 Scheme of interactions among water management institutions in Mendoza; *source* based on DGI (2015), Maccari (2004), Severino (2005), OEI-DGI (2006)

the water volume beneficially used with respect to the received volume (Morábito 2005). In general, a global indicator of irrigation systems measures the efficiency with a combination of effectiveness ratios that qualify the water management performance. Every stage of the irrigation system is important in the overall efficiency, which depends on the coating state of the channels, distribution rules, in farm use, and other factors. The Carrizal aquifer is located in the northern basin of Mendoza province, where the irrigation system efficiency varies from 28 to 40%. In other words, of the 100 L of water available in the system, the farmer receives between 28 and 40 L (Bos and Chambouleyron 1999; Jofré and Duek 2012; Morábito et al. 2012). On average, with the methods currently practiced, irrigation efficiency is low at the parcel level and the DGI estimation ranges between 30 to 50% (DGI 2016a). At the provincial level, distribution efficiency is between 70 and 90%, depending on the condition of the channels (Morábito et al. 2007; OEI-DGI 2006). The main causes of the low efficiency of irrigation are as follows:

i. Reduced percentage of canal lining at the provincial level;
ii. High infiltration because of the light soils and the phenomenon of clear waters;
iii. Lack of irrigation planning to deliver water according to the actual cropping needs;
iv. Inadequate distribution systems that deliver a large supply of water in a short period of time, leading to losses and waste;
v. Incomplete maintenance of the majority of the irrigation and drainage network.

In short, the technological level of irrigation at the provincial level could be markedly improved if changes in irrigation methods are introduced, such as scheduled rotations and irrigation according to a crop plan. In addition, infrastructure improvements in irrigation and drainage could be made supported by capacity building on water management practices (FAO/PROSAP 2015).

2.3.2 Surface and Groundwater Irrigation

In several regions of Mendoza, surface irrigation overlaps with groundwater irrigation. Between 1960 and 1980, the local and national governments promoted the expansion of the agricultural frontier into more arid areas, at the expense of increasing the exploitation of underground water resources (OEI-DGI 2006). The incentives included tax exemptions and subsidized credit lines for farm technology and pumping equipment.

As shown in Table 1, the irrigation alternatives differ not only in the origin of the resource but also in the physical and institutional management aspects. The conjunctive use of both resources carried out in a responsible manner could lead to improvements in groundwater quality and better use of the existing systems.

Table 1 Comparison of surface water and groundwater irrigation systems

Aspect	Surface water	Groundwater
Physical access	Depends on the natural conditions (seasonal patterns, rainfall, etc.) but also on infrastructure for delivery	Higher infrastructure and operation costs. Less dependent on natural conditions
Abstraction costs	Fix costs normally subsidized and variable costs according to farm characteristics and management	Fixed costs for use, pumping costs (variable according to the state of the source)
Distribution and equity of the public domain	Directly and visible to users More dependency on management and co-operation	Less cooperative resource in terms of use. Difficult but desirable for cooperation
Legal access and entitlements	Managed under specific water allocation, generally with legal entitlements	Legal entitlements subject to zoning restrictions and availability
Asymmetry of information	Availability and quality easy to check and review	Regulation more difficult and costly

Source Own elaboration based on OECD (2015), Theesfeld et al. (2010)

2.3.3 Characteristics of the Carrizal Aquifer

The Carrizal aquifer represents a sub-basin and is the main recharge area of the northern basin (Table 2). Within this area, the development of oil and petrochemical industries has exploited the natural resources, increasing pressure on the environment (Altamirano et al. 2005). Overexploitation of groundwater resources leads to quality degradation, which can be divided into local and diffuse pollution (Margat and van der Gun 2013; Lohn et al. 2000; Oikos 2004). Saline intrusion is a typical effect of excessive and inefficient irrigation that leads to water contamination. Under water scarcity periods, groundwater demand increases, thus excessive pumping undermines the natural harmony of the aquifer as the extraction rate exceeds the replenishment rate (Kupper et al. 2002; Morábito 2005).

The hydrology of the northern basin changed after the construction of the Potrerillos Dam during the early 2000s. Downstream the dam

Table 2 Basin characteristics

Northern basin	Area or volume
Storage capacity	30,000 hm^3
Underground extension	22,800 km^2
Renewable resource	700 hm^3/year
Carrizal aquifer	
Groundwater abstraction	66.7 hm^3/year
Area above the aquifer (Luján Sur)	20,000 ha
Agricultural land served	5,000 ha
Grape for wine production	3,250 ha
Vegetables	1,300 ha
Olives and pastures	450 ha
Irrigation means	
Surface only	1,330 ha
Groundwater only	1,330 ha
Conjoint use	1,330 ha

Source Own based on Foster and Garduño (2005), Hernández et al. (2012), IDR (2016), OEI-DGI (2006)

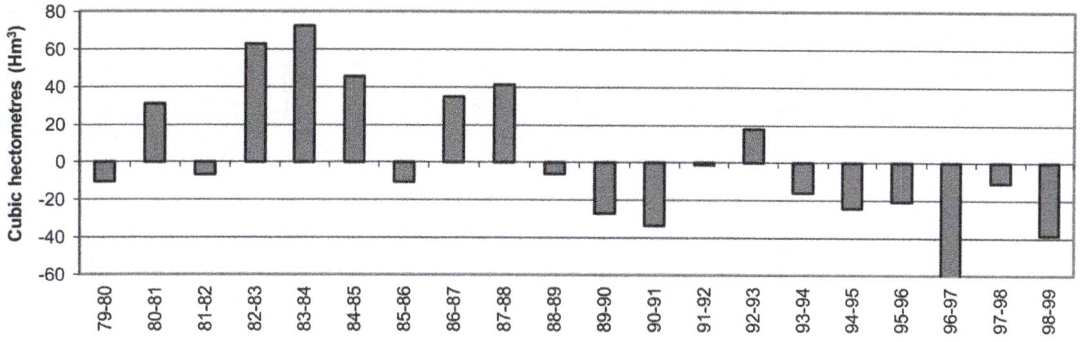

Fig. 3 Annual changes in the storage of the aquifer (1979–1999); *data source* based on Hernández et al. (2012)

construction the river carries less sediment, therefore the distributed waters filtrate easily on the ground; this phenomenon is known as "clear waters". Until 1999, an accurate estimation of the groundwater abstraction in the Carrizal aquifer was 66.7 km^3 per year (Hernández et al. 2012). According to Hernández et al. (2012), between 1979 and 1999, the average pumped water in the Carrizal aquifer was 61.235 hm^3 as shown in Fig. 3. Although the information about the current storage of the aquifer (after the year 2000) is currently classified because of the increasing public concern over groundwater salinity increase in the past (Erice 2013), the water authority commented on the positive resilience of the aquifer supported by the zoning restriction.

3 Energy and Subsidy Information

In Mendoza, the energy production increased at lower rates than the total demand in the period 2003–2013. The province does not perform satisfactorily on energy self-sufficiency, as imported energy represents nearly 20% of total consumption (EPRE 2013). Since 2008, water institutions have managed the resource under a water scarcity scenario, which means that the snowfall during winter does not fulfill the expected demand for irrigation during spring and summer. The lower surface water supplied translates into increased energy demand for pumping groundwater.

Increasing demand for subsidized energy from 2004 to 2014 grew 53% in a decade (Fig. 4). Since 2015, the adjustment of macroeconomic variables and reorientation of expenditure has led to a formal devaluation of national currency and lower subsidy share in energy prices (DEIE 2014; EPRE 2018). Despite the national trend of diminishing energy subsidies, agricultural producers still benefit the most from the provincial subsidy scheme that continued support for small and medium farmers. The provincial budget grew in real terms from USD 1.22 million in 2015 to USD 3.6 million in 2018. The average cost per subsidized kWh grew from USD 2.5 in 2015 to USD 3.74 in 2018.

Promoting agricultural irrigation by subsidizing energy prices is a policy tool that seeks to leverage small agricultural producers that are not capable of improving their production efficiency mainly due to scale (farm size) or a previous year of economic losses. The subsidy is available upon request by farmers, but not available for consumers. A tariff adjustment was issued in 2008 but properties smaller than 50 ha were exempt from the tariff increment. In addition, farmers that do not receive surface water may qualify as well. Since the subsidy is attached to a property (agricultural parcel) and not to a specific person, strategic behavior by stakeholders could lower the efficiency of the energy policy.

Regulated by Law 6498, the irrigation tariff establishes compensation from the provincial state to the energy distributors. Moreover, the law determines tariff segments according to the time slot that energy is consumed (EPRE 2018). The time slot for the high-demand period has

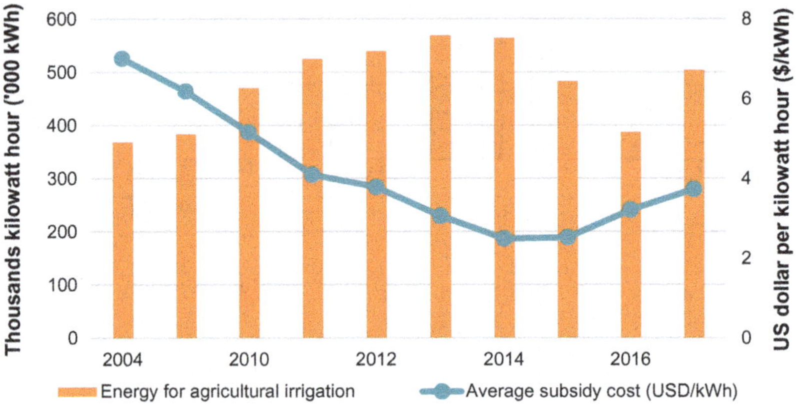

Fig. 4 Energy consumed for agricultural irrigation in Mendoza (2004–2017); *source* based on collected data and DEIE (2014), EPRE (2017, 2018)

Table 3 Energy subsidy scheme for irrigation according to pumping equipment (2012 data)

Pumping equipment power	Lower voltage		Medium voltage	
	High-demand (%)	Low-demand (%)	High-demand (%)	Low-demand (%)
<10 kW	57.4	79.0	63.2	79.0
10 < kW < 300	57.3	79.0	63.1	79.0
>300 kW	50.5	69.6	55.6	69.6

Source Own based on EPRE (2017)

changed continuously, establishing one or two periods of higher pricing during the day. In fact, these changes seek to segment the demand for targeted pricing as exposed in Table 3. However, as pointed out by Severino (2005), these time slots do not correspond with the national energy market that provides local distributors.

There is a strong relationship between subsidy and energy tariffs. During 2012, energy subsidies reached up to 57 and 79% of the total energy tariffs for groundwater pumping in high-demand periods and low-demand periods, respectively. On a yearly basis, the provincial energy regulator (EPRE) seeks to improve policy targeting by visiting beneficiaries randomly and checking their subsidy qualification. This action has contributed to improving the policy targeting by decreasing the list of beneficiaries by 10%. Normally, the subsidized power for agriculture irrigation is near 4 MWh per year, from which 20% is estimated as loss from inefficiency due to improper pumping equipment (Severino 2016).

Although the real cost of providing energy has changed over time, the subsidy rate has remained untouched. In 2015, the new national administration announced the lowering the subsidy share of energy tariffs. Nevertheless, agricultural beneficiaries continued to receive the subsidy provided by the province, where the total amount of subsidy budget increased to USD 1.89 million, representing a 55% increase. Attempts to withdraw the energy subsidies for agricultural irrigation have not been successful in the past.

Currency devaluation started in 2002 and followed with adjustments in the valuation of the Argentinean peso over a decade. This implied local adjustments in the energy prices despite government support as shown in Fig. 5, where prices for (A) high-demand and (B) low-demand periods are exposed. Energy consumption for pumping groundwater increased substantially from 2009 until 2011 mainly driven by the mentioned water scarcity period.

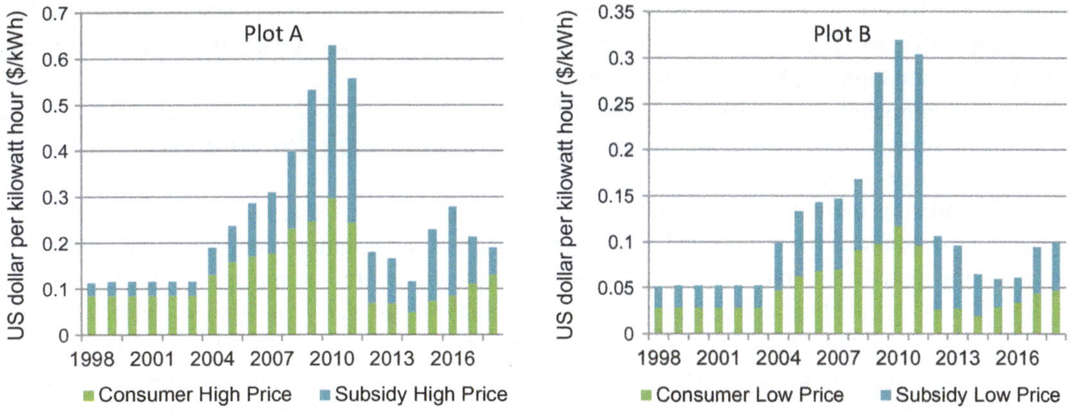

Fig. 5 Quarterly energy tariffs for irrigation. Prices for high- and low-demand periods; *source* based on collected data and DEIE (2014), EPRE (2016, 2018)

4 Agricultural and Economic Tools

Although the agricultural contribution to the GDP of Mendoza remains below 10%, the sector becomes economically relevant when the multiplicative effects are considered. Since our focus is on grape production, we review the fiscal policies and economic tools employed in different sectors to improve the economic performance of vineyards over the last 25 years.

Prior to the economic crisis of 2001, wine-related associations and institutions drafted a restructuring plan to improve the strategic opportunities of the industry. This plan included technology adoption, variety improvement, and development of new markets (COVIAR/OVA 2018; Azpiazu and Basualdo 2001). Foreign direct investment assisted in this process since vineyards and wineries had not fully utilized the present infrastructure and institutions but were prepared in terms of quality analysis frameworks and procedures. The new focus on targeted markets struggled to effectively increase the market share. Argentinean wine is welcomed in countries with high purchasing power, but it is relatively expensive compared to other producers. This is because of the high production costs and lack of proper infrastructure that increases the cost of logistics and relatively higher importation levies. Trade logistics represent 17% of the production and commercialization costs of Argentinean wine (COVIAR/OVA 2018).

Regarding the price paid to producers, small and medium vineyards have historically faced an uneven situation for bargaining power possibly related to market failures as information asymmetry and moral hazard. In grapevine production, the former implies that different decisions would be made with uneven information; the latter considers the non-compliance with specific tasks of an agreement. Moreover, the increase in export-oriented wineries has led to higher mistrust in the free-market, as they require some quality standards achieved by specific vineyard management practices. A considerable share of small and medium producers found alternatives to cope with the moral hazard of the specific grapevine management as they organized themselves into collectives to increase their bargaining power through the shield of cooperatives or signed a contract agreement with the grapevine buyer. Agreements for partial or total production improved the relationship between grapevine producers and winemakers.

Governments and wine institutions also intervene to regulate market prices by regulating supply in grape derivatives as wine and must. Jointly, regional provincial governments agreed on the share of production that will be transformed into a must to regulate the wine supply and further sustain prices. More recently, the abundance of base wine in stock and good harvest years led to lower market prices of grapevines, a situation that had to be solved with government intervention. In 2016, the national government spent USD 11.2 million to improve the market price through reductions in wine stock. Considering the external factors in grape production, institutions have deployed programs for assessing potential climate risks and minimizing effects of contingencies: for example, through programs like hail storm protection and rural insurance (Gibbons et al. 2016). Institutions seek to help farmers to improve their agronomic management practices through smarter use of agrochemicals. Uncertainty in the local currency jeopardized the competitive gains of micro-devaluations since input providers adjusted faster to real values in USD.

When it comes to production factors, grapevine remains a labor-intensive crop in Mendoza, with a variable share of intermediate inputs according to location, grape quality, and training system. On average, the total production cost is USD 3,824 per hectare, composed of labor (72.1%), machinery (11.4%), agrochemicals (9%), and physical inputs (7.4%) (COVIAR/OVA 2018). Regarding labor policies, institutions lagged behind in terms of formality and flexibility of the labor force. In addition to the higher employment costs for firms and small vineyards, the deployment of alternative labor standards for seasonal and permanent workers did not affect the level of employment in the sector. On the contrary, these measures actually led to a lower share of permanent labor with higher benefits. Seasonal workers can count on a regulated salary that is constantly updated by institutions. These measures improve resource allocation and flexibility but also incentivize farm workers to seek short-term employment (Van den Bosch 2008). Altogether, the review of the agricultural policies shows an active public sector in the regulation of all production factors. However, in the framework of the political economy, we will further discuss the general environment of these regulations along with the effectiveness and suitability of the economic measures.

5 Framework for Analysis

For those farmers that have a well, groundwater becomes a common pool resource with low-excludability for pumping it (OECD 2015), which means that stakeholders with a drilling permit can pump water at the marginal cost whenever they need it and they cannot be excluded from using the aquifer's resource. Natural conditions of underground resources have historically raised concerns about characteristics such as boundaries of the reserve, the hydrogeological uncertainties, irreversibility of mismanagement, and information asymmetries (Booker et al. 2012; NRC 1997; Theesfeld 2010). Following the complexity of the political economy of underground resources, the institutional tripod framework is used in this study to assess the tripartite institutional performance by decoupling the roles of organizations and stakeholders at different levels (OECD 2015; Meinzen-Dick 2007). This methodology helps to understand the underlying power structures, decision making stages, and incentives of participants in the political process.

Meinzen-Dick (2007) introduced the framework acknowledging that there is no single solution for all water problems in policy analysis. An objective manner of analysis is to decompose the policy instruments into regulatory, economic, and voluntary. The regulatory instruments frame the command and control of water policies, i.e., the ownership of permits, pollution standards, and abstraction. In most cases, water permits are attached to agricultural land and are non-tradable.

OECD (2015) and Theesfeld (2010) agree on the importance of analyzing institutions involved, power structures, and the independence of decision makers to comprehend the political process of water policy. The economic instruments reflect the financial incentives that may drive the decision of the stakeholders; this could be directly influenced by groundwater fees related to infrastructure, location, and services (Zilberman et al. 2008). Furthermore, the joint analysis of physical conditions and institutional settings that consider asymmetric information are critical factors for the design and implementation of policies (Dinar 2000). A systematic review of planning and policy instruments is essential to achieve a comprehensive governance structure of public institutions (Theesfeld et al. 2010). Energy policies that subsidize groundwater withdrawals are commonly referred to as ill-conceived policies (Bailis 2011). Since the marginal cost of acquiring water for irrigation decreases, it is possible that economic agents continue or start employing water inefficiently.

As expected, diminishing subsidies translate into higher tariffs for producers, who initially constrained their energy consumption for agricultural irrigation. Later, the prevailing scenario of water scarcity and the recovery of grapevine prices led to a subsequent increase in groundwater use. However, there remains no good estimate of the demand function of groundwater for agriculture in the study area. Regarding other areas in the northern basin in Mendoza, the price elasticity is −0.57 for producers that only use groundwater and −1.28 for users with access to both types of irrigation systems (Barbazza 2005). This means that a 1% increase in groundwater price will diminish consumption by 0.57% for farmers that rely completely on groundwater, whilst a 1% increase in water abstraction will diminish consumption by 1.28% for conjoint users.

Other studies conducted in India analyzed the effect of a 10% reduction in energy subsidies and found a reduction of the pumped water between 4.4 and 6.7% (Badiani and Jessoe 2011; OECD 2015; Shah et al. 2012). For Mexico, Sun et al. (2016) showed that doubling the cost of pumping would only reduce demand by 6%. However, the total withdrawal of the energy subsidy would decrease pumping by 15% in the short run and settle at 19% in the long term (OECD 2015). Often, energy subsidies for irrigation efficiency are interpreted as a double-edged sword in groundwater management (OECD 2015). Acquiring higher efficiency standards on irrigation is more beneficial for farmers but could deteriorate soil quality or aquifer recharge. Some additional measures should foster the cultivation of less water demanding crops to avoid the negative effects of the measure. Coady et al. (2015) propose that efficient pricing from energy producers to suppliers should equal the cost of production. In addition, Pigouvian taxation is a tool to correct externalities that are not covered by other political measures. Moreover, Sun et al. (2016) showed that the effectiveness of electricity price-based policies is certainly a reason to consider the Pigouvian tax for groundwater as a common pool resource.

6 Results and Discussion

The review of economic tools and policies utilized in the agricultural sector presented above depicts the active participation of the state. One perspective could call this state-dependence the result of the institutions and organizations that have not done enough to ensure the independence of the farming sector as well as not putting in place self-regulatory processes for water institutions. Another view is that proper regulation has been developed as a result of the state intervention but the implementation is failing because of weak monitoring.

The Carrizal aquifer provides irrigation to a prestigious wine region and has been the subject of numerous political and environmental conflicts with respect to water management, resource quality, and groundwater exploitation reviewed in detail in Table 5. Initially, the zoning restriction aimed to control the water quality pollution and monitor the oil pollution in the aquifer. Assisted by the National Institute of Water (INA), the DGI seeks to improve the understanding of the aquifer characteristics, promoting research with pumping trials, water table monitoring, and quality control. The budget for these expenses is now obtained from groundwater users through the annual fee, which has increased considerably. Soon after the zoning restriction was issued, the judicial sector enforced a review of the water quality in surrounding areas. The INA performed water quality analysis in the aquifer, confirming that the alarming quality decrement of irrigation water characterized by high salinity levels.

In response to political pressure and considering the economic weight of oil refinement for the provincial GDP, the provincial government called for new water quality analysis and determined the validity of oil exploitations (Severino 2005). At the same time, fiscal oilfields (YPF, according to the Spanish name) faced political and institutional consequences and had to reimburse the damage to most affected farmers. In 2004, DGI ordered YPF damage control actions and they have worked collaboratively on groundwater quality and water table level monitoring. Simultaneously, attempts to review for adjustments with regard to energy tariffs for agricultural irrigation in the province were neutralized by local legislators. Supported by the private sector, their argument focused on the difficult scenario for farmers, in particular for those solely dependent on groundwater irrigation. Nothing was mentioned about responsible exploitation or sustainable use of the resource.

On average, the energy subsidy for irrigation increased by over 20% in the last 20 years, reaching 55.6% in 2016 (Fig. 6). As subsidized energy became the normal scenario, farmers began to make their decisions based on other variables such as output and input prices, with no considerations of the potential diffuse pollution effects of the groundwater overdraft. The recomposed grapevine price (60% increase) seemed to have an effect on pumped groundwater in 2016 as indicated by the fact that the water table reached the lowest in two decades (−90 m), as farmers would have a greater income to cope with energy costs even though the government had considerably adjusted the energy tariffs.

Although there is an agreement on quality monitoring across water institutions relating to the increased levels of salinization and resource depletion of the aquifer over the years (Foster and Garduño 2005; OEI-DGI 2006; Reta 2005), water quality remains affected by industrial activity and agricultural practices, in particular with regards to phosphorus levels (Lavie et al. 2010). Currently, the annual consumption of subsidized energy for agricultural irrigation is at 53%, 9.5% higher than that of the last decade in terms of the total energy demanded. In the past, water authorities have created conditions for improving resource management to diminish pollution in the long term (Jofré et al. 2012). However, to achieve earlier results, stakeholders need to be stimulated to act collectively in resource exploitation through economic tools that internalize trade-off decisions between productivity and environmental effects

Fig. 6 Evolution of grapevine prices, average energy subsidy (AvSubsidy), and depth of the groundwater table; *source* based on collected data and COVIAR/OVA (2018), Hernández et al. (2012), OEI-DGI (2006)

Table 4 Current policy tools on water management

Orientation	Instrumoent	Regulatory approaches	Economic instruments	Collective management approaches
Demand side approaches	*Extensive margin (wells)*	Drilling permit requirement	–	Association of groundwater users
	Intensive margin (use)	*Direct*: Flowmeter	*Direct*: Higher annual fee Energy subsidies	–
		Indirect: empowerment of water institutions	*Indirect*: assistance to improve infrastructure	*Indirect*: determination of turn scheme
Supply side approaches	*Additional supply for storage*	–	–	Construction of reservoirs
	Additional supply for use	Surface water supply: Turn scheme	Financing infrastructure	Collective management plans

Source Own based on DGI (2018), Erice (2013), OECD (2015) and Theesfeld et al. (2010)

(Ostrom 1990, 2014). In terms of water management, current policies were clustered in three main instrumental approaches and classified according to the orientation (Table 4). While several policy instruments have been designed for both orientations (demand and supply), some regulations have not been fully implemented or actions need a longer time for proper assessment.

Participation and disputes related to water–energy policies have had a rich history over the last 20 years. Several external effects drive the excessive pumping of groundwater; the water scarcity period since 2009 implies a lower volume of surface water to deal with higher temperatures and uncertainty in rainfall patterns. Regarding the economic sphere, low profitability thwarted the incentives for improving irrigation efficiency at the parcel level.

In other words, the reform arena was not suitable for the relaxation of the subsidy scheme. This situation led to the depleted water table levels in the aquifer and increments in the depth of water extracted (Álvarez et al. 2011; Foster and Garduño 2005; Puebla et al. 2005). During 2017, the agricultural sector utilized 505,000 kWh for irrigation purposes. Historically, the energy utilized for agricultural irrigation has been between 8 and 10% of the total provincial consumption, with the exception of 2016, when the joint effect of subsidy cutbacks and currency devaluation constrained the demand to 390,000 kWh. From the 300 MW of installed energy capacity for agricultural irrigation, the Ministry of Energy estimates that 15% is inefficiently used. This represents USD 6 million of government expenditure (EPRE 2018).

Undoubtedly, the policy planning has been undermined by several economic and environmental facts during the last 20 years. The review of the political treatment of pollution accusations and the attempts to modify the agricultural irrigation subsidies have revealed the weaknesses of decision makers. At the first sign of modifying the status quo of acquired subsidies for water abstraction, lobbyists and watershed inspectors are quick to respond to protect their interests. State subsidies for energy irrigation are assigned to a compensation fund that ameliorates the bi-monthly costs during periods of high demand. This financial aid compensates for fixed costs of infrastructure and lowers the tariff prices for the high- and low-demand periods. Considering the state budget and the energy provided, the average cost of the subsidized energy is USD 3.74 per kWh, the highest in the last five years.

However, the awareness of water availability and quality problems by specific institutions (DGI and INA) shows some indication of willingness to improve management and quality. Moreover, if policy tools continue to stimulate technology adoption and the government maintains this orientation, water-use efficiency will increase in the agricultural sector contributing to the achievement of specific targets of the Sustainable Development Goals (SDGs). In particular, with respect to the change in water-use efficiency (goal 6.4.1), protection of water-related ecosystems (goal 6.6), and improvement of energy intensity (goal 7.3.1).

7 Conclusions and Outlook

This review of the institutional settings and the political disputes about water resource quality and management reveals the public sensitivity towards pollution in common pool resources such as the Carrizal aquifer. This is particularly the case when quality degradation is not diffuse but local and the guilty can easily be identified. In order to gain political credibility, public institutions need to show responsibility to deal with groundwater issues by explaining and analyzing the risks and benefits (Foster and Garduño 2012). A credible threat of losing drilling permits could create enough incentives to improve groundwater management (Livingston and Garrido 2004). This scenario would motivate water and energy institutions to boost efficiency, reaching higher standards on integrated water resource management.

In this paper, the beneficiaries of the irrigation policies are the agricultural producers. Findings indicate joint implications of water, agricultural, and energy policies for groundwater availability. The DGI remains the highest authority in irrigation water management in the province. The tripod analysis revealed an unbalanced set of policy tools. There are more policy measures oriented to the demand side and relevant participation of collective action involved in the management within the framework. More policies on the supply side could be more difficult and possibly expensive to design but they would have a bigger effect on the water management system. Some suggestions on the supply side include fostering wastewater reuse, solar pumping systems, and better storage facilities. As stated by Abler and Shortle (1991), these political changes will be viable if they positively affect the institutions' budget, gain confidence from large stakeholders in the political sphere, and optimize the administrative and enforcement costs. This calls for avoiding policy objectives that could backfire in the wrong direction. However, it is expected that lowering energy subsidies for agricultural irrigation will correct the economic incentives and thus diminish groundwater use. In the past, no clear and consistent policies were made to improve the targeting of beneficiaries towards paying the full costs of energy tariffs.

Under these conditions, the stakeholders may perceive that business as usual is acceptable as there are no changes in policies and incentives. If the recent modification of electricity tariffs imposed by the national government comes along with better targeting of subsidy beneficiaries, the marginal cost of water abstraction would increase, which may improve the irrigation practices and intercept overexploitation of the aquifer. The provision of a subsidy for groundwater abstraction may improve the living standard of less profitable farmers but it is not the appropriate approach to improve their livelihood long term. Conversely, when policies are not complemented with instructive and participatory approaches that improve water management, farmers will continue to rely on their traditional irrigation practices with no changes in the marginal productivity of water and increasing production costs in the medium term, as more energy would be used for irrigation. From the environmental perspective, advancing towards a more responsible taxation scheme in developing countries implies the consistent political will to withdraw inefficient subsidies supported by technical evidence. This research contributes to the analysis from the political economy perspective by unveiling the motivation of different groups of stakeholders. In this line, more research is desirable as it would not only contribute to policy-design but will ease implementation in developing countries.

Acknowledgements This research was jointly funded by the German Service of Academic Exchange (DAAD) and the Georg-August-Universität Göttingen. The coordination with the General Direction of Irrigation (DGI) and the Statistics Bureau of the province (DEIE) allowed detailed planning for fieldwork execution between November 2016 and January 2017. The Ente Provincial Regulador Electrico (EPRE) contributed with the subsidized energy data for estimating groundwater volumes used for agricultural irrigation. The authors also thank Dr. Tinoush Jamali Jaghdani for the constructive review in the early stages of this research. This manuscript has improved considerably thanks to the insights of the two anonymous reviewers.

References

Abler DG, Shortle JS (1991) The political economy of water quality protection from agricultural chemicals. Northeastern J Agric Resour Econ 20:53–60

Allan T, Keulertz M, Woertz E (2015) The water–food–energy nexus: an introduction to nexus concepts and some conceptual and operational problems. Int J Water Resour Dev 31:301–311. http://www.tandfonline.com/doi/full/10.1080/07900627.2015.1029118

Altamirano JC, Lana NB, D'Angelo JA, Ciocco N (2005) Polybrominated diphenyl ethers in Mendoza river basin and el Carrizal reservoir: levels, distribution and correlation with physico-chemical variables 1–4

Álvarez A, D'Elía M, Paris M, Fasciolo GE, Barbazza C (2011) Evaluación de la contaminación de acuíferos producida por actividades de saneamiento y re-uso de efluentes en el norte de la provincia de Mendoza. Rev de la Fac de Cs Agr 43(1):19–39. http://bdigital.uncu.edu.ar/2720

Azpiazu D, Basualdo E (2001) El complejo vitivinícola argentino en los noventa: potencialidades y restricciones. CEPAL

Azpiazu D, Bonofiglio N, Nahón C (2014) Agua y Energía. Mapa de situación y problemáticas regulatorias de los servicios públicos en el interior del país. No. 18, FLACSO—Fac. Lat de Cs Sociales. http://biblioteca.clacso.edu.ar/ar/libros/argentina/flacso/tesis/dt18.pdf

Badiani R, Jessoe KK (2011) The impact of electricity subsidies on groundwater extraction and agricultural production. http://are.berkeley.edu/documents/seminar/JessoeDraft.pdf

Badiani R, Jessoe KK, Plant S (2012) Development and the environment: the implications of agricultural electricity subsidies in India. J Environ Dev 21 (2):244–262. http://jed.sagepub.com/cgi/doi/10.1177/1070496512442507

Bailis R (2011) Energy and poverty: the perspective of poor countries. In: Galarraga I, González-Eguino M, Markandya A (eds) Handbook of sustainable energy. EE Publishing, pp 505–538. http://www.elgaronline.com/view/9781849801157.00035.xml

Barbazza CV (2005) Funciones de demanda de agua subterránea para el Este mendocino. In CONAGUA, Mendoza, Argentina

Bermejo M (1884) Ley de aguas. http://www.agua.gob.ar/dgi/ley-aguas

Booker JF, Howitt RE, Michelsen AM, Young RA (2012) Economics and the modeling of water resources and policies. Nat Res Model 25(1):168–218. http://doi.wiley.com/10.1111/j.1939-7445.2011.00105.x

Bos MG, Chambouleyron JL (1999) Parámetros de desempeño de la agricultura de riego de Mendoza, Argentina. http://hdl.handle.net/10568/36302

Calcagno AN, Mendiburo N, Gaviño Novillo M (2000) Informe sobre la gestión del agua en la República Argentina. World Water Vision

Coady D, Parry I, Sears L, Shang B (2015) How large are global energy subsidies? Fiscal Aff:30. https://www.imf.org/external/pubs/ft/wp/2015/wp15105.pdf

COVIAR/OVA (2018) Impacto de la Vitivinicultura en la Economía Argentina. Corporación Vitivinícola Argentina; Observatorio Vitivinícola Argentino; COVIAR. https://goo.gl/A9AUDQ

Dagnino M, Ward FA (2012) Economics of agricultural water conservation: empirical analysis and policy implications. Int J Water Resour Dev 28(4):577–600

Dai J, Wu S, Han G, Weinberg J, Xie X, Wu X, Song X, Jia B, Xue W, Yang Q (2018) Water-energy nexus: a review of methods and tools for macro-assessment. App Energy 210:393–408. https://linkinghub.elsevier.com/retrieve/pii/S0306261917312242

DEIE (2014) Anuario económico https://goo.gl/D4yV4a. Accessed 1 Jan 2015

DGI (2016a) Balance Hídrico. Río Mendoza. Mendoza, Argentina

DGI (2016b) Departamento General de Irrigación. Resolución 641:6148–6149

DGI (2018) Servidor de mapas—IDE Irrigación. http://ide.irrigacion.gov.ar. Accessed 14 May 2018

DGI (2015) Aqualibro, 2nd edn. Irrigación Edita, Mendoza

Dinar A (ed) (2000) The political economy of water pricing reforms, 1st edn. Oxford University Press, New York. http://documents.worldbank.org/curated/en/2000/04/437127/political-economy-water-pricing-reforms

Duek AE, Fasciolo GE, Quiles ME, Zoia O (2013) Uso del agua en la industria alimenticia de Mendoza. In Congreso nacional del agua xxiv (conagua), San Juan, Argentina

EPRE (2013) Informe de la demanda Mendoza y Cuyo. Ente Provincial Regulador Eléctrico (EPRE). https://goo.gl/1Bx8gg. Accessed 13 Jan 2016

EPRE (2016) Cuadros tarifarios. http://www.epremendoza.gov.ar/subcategoria.php?subcategoria{_}id=59. Accessed 13 Jan 2016

EPRE (2017) Informe de la demanda Mendoza y Cuyo. Ente Provincial Regulador Eléctrico (EPRE). https://goo.gl/vxwMF2. Accessed 20 March 2019

EPRE (2018) Informe de la demanda Mendoza y Cuyo. Ente Provincial Regulador Eléctrico (EPRE). https://goo.gl/H6ZXr8. Accessed 20 March 2019

Erice MV (2013) Necesidad de transparencia y legalidad en la gestión y uso de las aguas subterráneas (El reciente caso de la Provincia de Mendoza). Derecho y Ciencias Sociales Octubre (9):191–202

FAO/IWMI (2018) More people, more food, worse water? A global review of water pollution from agriculture. In: Mateo-Sagasta J, Marjani Zadeh S, Turral H (eds) FAO; IWMI, Rome. http://www.fao.org/documents/card/en/c/CA0146EN. Accessed 20 March 2019

FAO/PROSAP (2015) Estudio del potencial de ampliación del riego en Argentina. Desarrollo institucional para la inversión Org de las NU para la Agr y la Aliment (FAO); Programa de Servicios Agrícolas Provinciales (PROSAP)

Foster S, Garduño H (2005) Gestión sustentable del agua subterránea. Argentina: Enfoque de gestión integrada para la conservación del agua subterránea en los acuíferos de Mendoza. World Bank. www.worldbank.org/gwmate. Accessed 02 Dec 2016

Foster S, Garduño H (2012) Groundwater-resource governance: are governments and stakeholders responding to the challenge? Hydrol J 317–320

Foster ET, Dinar A, Rapoport A (2016) Comparing alternative modifications of energy subsidies: the case of groundwater extraction 1–32

Garduño H, Foster S (2010) Sustainable groundwater irrigation: approaches to reconciling demand with resources. World Bank. www.worldbank.org/gwmate. Accessed 13 Jan 2016

Gibbons MA, Maffioli A, Rossi MA (2016) Money for Wine? Complementarities in the provision of private and public goods to wine producers

Hernández J, Martinis N, Fornero L (2012) Modelación hidrológica de la cuenca norte de Mendoza. No. 146, Instituto Nacional del Agua (INA)

IDR (2016) Ecoatlas. http://www.ecoatlas.org.ar/index.html. Accessed 26 Jan 2016

IGN (2018) Capas SIG. http://www.ign.gob.ar/NuestrasActividades/InformacionGeoespacial/CapasSIG. Accessed 23 July 2018

Jofré JL (2010) Efectos de las innovaciones productivas en la agricultura sobre la materialidad institucional del régimen hídrico. El caso mendocino entre 1976–2010. Doctoral. Universidad Nacional de Cuyo

Jofré JL, Duek A (2012) Criterios de política hídrica para el ordenamiento territorial. In I encuentro de investigadores en formación en recursos hídricos. Buenos Aires, Argentina: Instituto Nacional del Agua (INA)

Kupper E, Querner E, Morábito JA, Menenti M (2002) Using the SIMGRO regional hydrological model to evaluate salinity control measures in an irrigation area. Agric Water Manage 56(1):1–15. http://linkinghub.elsevier.com/retrieve/pii/S0378377401001950. Accessed 23 July 2018

Lavie E, Morábito JA, Salatino SE, Bermejillo A, Filippini MF (2010) Contaminación por fosfatos en el oasis bajo riego del río Mendoza. Rev De La Fac De Cs Agr 42(1):169–184

Livingston ML, Garrido A (2004) Entering the policy debate: an economic evaluation of groundwater policy in flux. Water Resour Res 40(12). https://agupubs.onlinelibrary.wiley.com/doi/abs/10.1029/2003WR002737. Accessed 13 Jan 2016

Lohn P, Guimaraes R, Bucich N (2000) Evaluación hidroquímica y de la contaminación químico-biológica de la cuenca el Carrizal—zona norte—provincia de Mendoza. Republica Argentina. In XI congresso brasileiro de Águas subterrâneas. Sao Paulo, Brazil, pp 1–23. http://aguassubterraneas.abas.org/asubterraneas/article/view/23923/15985. Accessed 13 Jan 2016

Maccari LC (2004) Proyecto de fortalecimiento institucional. Provincia de Mendoza. PROSAP. https://goo.gl/QJgDwK. Accessed 20 Jan 2016

Maffioli A, Ubfal D, Baré GV, Cerdán-Infantes P (2011) Extension services, product quality and yields: the case of grapes in Argentina. Agric Econ 42(6):727–734. http://doi.wiley.com/10.1111/j.1574-0862.2011.00560.x. Accessed 10 Jan 2016

Margat J, van der Gun J (2013) Groundwater around the world, 1st edn. CRC Press, Boca Raton, FL. http://www.crcpress.com

Medawar A, Perlbach de Maradona I, Pasteris E, Garcia M, Carretero ME, Calderón M (2011) El producto geográfico bruto de la Provincia de Mendoza en los años 2010–2011. DEIE, FCE, UNCuyo

Meinzen-Dick R, (2007) Beyond panaceas in water institutions. In: Proceedings of the national academy of sciences 104(39):15200–15205. http://www.pnas.org/content/104/39/15200.full

Morábito JA (2005) Desempeño del riego por superficie en el área de riego del río Mendoza Eficiencia actual y potencial. Parámetros de riego y recomendaciones para un mejor aprovechamiento agrícola en un marco sustentable. Master Thesis. UNCuyo. Available at: http://bdigital.uncu.edu.ar/objetos{_}digitales/4137/morabito.pdf Accessed December 20, 2015

Morábito JA, Mirábile CM, Salatino SE (2007) Eficiencia de riego superficial, actual y potencial en el área de regadío del río Mendoza Argentina. Ingeniería Del Agua 14(3):199–213

Morábito JA, Salatino SE, Schilardi C (2012) El desempeño del uso agrícola del agua en los oasis de los ríos Mendoza y Tunuyán a través de nuevos indicadores. In VI jornadas de actualización en riego y fertirriego. Prácticas para incrementar la productividad y asegurar la sostenibilidad del uso del agua y del suelo. Mendoza, Argentina

Morello J, Matteucci SD, Rodríguez AF, Silva ME, de Haro JC (2012) Ecorregiones y complejos ecosistémicos argentinos 1st ed. Buenos Aires: Orientación Gráfica Editora. https://books.google.de/books?id=q-RrngEACAAJ. Accessed 20 July 2015

NRC (1997) Valuing ground water. Economic concepts and approaches. National Academy Press, Washington, D.C.

OECD (2015) Drying wells, rising stakes. OECD Publishing, Paris. http://www.oecd-ilibrary.org/agriculture-and-food/drying-wells-rising-stakes{_} 9789264238701-en. Accessed 13 Jan 2016

OEI-DGI (2006) Integración de información, para el diagnóstico y gestión de la calidad del recurso hídrico en cuencas de la provincia de Mendoza. Argentina, OEI-DGI

Oikos (2004) Informe ambiental Oikos. Oikos red ambiental. http://www.oikosredambiental.org/documentos/infoamb2004.pdf. Accessed 20 Dec 2015

Ostrom E (1990) Governing the commons: the evolution of institutions for collective actions, 1st edn. Cambridge University Press, Cambridge

Ostrom E (2014) Institutions and sustainability of ecological systems. In: Galiani S, Sened I (eds) Institutions, property rights, and economic growth. Cambridge University Press, Cambridge, p 339

Pinto M, Rogero GE, Andino MM (2006) Ley de aguas de 1884. Comentada y concordada, 1st edn. Irrigación Edita, Mendoza, Argentina. https://goo.gl/DR8p6s. Accessed 20 Jul 2015

Puebla P, Llop A, Bertranou A, Zoia O, Falótico N, Fasciolo GE, Comellas EA, Reta J (2005) Gestión integral de los recursos hídricos (GIRH). El caso del agua subterránea. http://arxiv.org/abs/arXiv:1011.1669v3. Accessed 19 Jul 2015

Reta J (2002) Argentina (Provincia de Mendoza). In: Garduño H (ed) Administración de Derechos de Agua. FAO, Rome, pp 243–274

Reta J (2005) La contaminación de las aguas subterráneas: el caso de los acuíferos de Ugarteche-Carrizal. In Scoones and Sosa, eds. Conflictos socio-ambientales y políticas públicas en la provincia de Mendoza. OIKOS red ambiental, Mendoza, Argentina, pp 340–357

Severino S (2005) Tarifa eléctrica del riego agrícola en la provincia de Mendoza: Análisis de alternativas de su determinación In XX congreso nacional del agua y iii simposio de recursos hídricos del cono sur. Mendoza, Argentina

Severino S (2016) Entrevista con representativo de EMESA y EPRE.

Shah T, Giordano M, Mukherji A (2012) Political economy of the energy-groundwater nexus in India: exploring issues and assessing policy options. Hydrogeol J 20(5):995–1006

Silanes R (2013) Manuel Bermejo y la ley de aguas, 1st edn. DGI, Mendoza, Argentina

Sun S, Sesmero JP, Schoengold K (2016) The role of common pool problems in irrigation inefficiency: a case study in groundwater pumping in Mexico. Agric Econ 47(1):117–127. http://doi.wiley.com/10.1111/agec.12214

Theesfeld I (2010) Institutional challenges for national groundwater governance: policies and issues. Groundwater 48(1):131–142. http://doi.wiley.com/10.1111/j.1745-6584.2009.00624.x. Accessed 23 Jan 2016

Theesfeld I, Schleyer C, Aznar O (2010) The procedure for institutional compatibility assessment: ex-ante policy assessment from an institutional perspective. J Inst Econ 6(03):377–399

Van den Bosch ME (2008) Un modelo de desarrollo sustentable en las áreas bajo riego de los distritos Ugarteche y El Carrizal. Departamento de Luján de Cuyo. Provincia de Mendoza: Un aporte para el ordenamiento territorial rural. Master thesis. UNCuyo. http://www.bdigital.uncu.edu.ar/4631. Accessed 21 Jul 2015

Zilberman D, Sproul T, Rajagopal D, Sexton S, Hellegers P (2008) Rising energy prices and the economics of water in agriculture. Water Policy 10 (SUPPL. 1):11–21

Rural Resources (including Forestry) in the Local Development of Low Carbon Economy: A Case Study of Poland

Paweł Wiśniewski

Abstract

Due to the significant share in total greenhouse gas (GHG) emissions and sensitivity to climate change, agriculture and rural areas should be an important area of activity in the local development of a low carbon economy. However, this is rarely the case in Poland and other European countries. The paper evaluates the role and importance of rural resources (including forestry) for the local development of a low carbon economy. Based on a SWOT analysis the strengths and weaknesses, as well as the opportunities and threats in the two key elements such as maintenance or increase of the ability to absorb CO_2 and reduction of GHG emissions from agricultural and forest lands are shown. Specific attention was paid to the potential of agriculture and rural areas in terms of the possible use of their resources in order to increase the C sequestration in biomass and soil and reduce GHG emissions, as well as the use of agricultural, agroforestry and agro-food processing with biomass for renewable energy development, including the production of biogas and biofuels. The information obtained from the SWOT analysis shows that this potential is based on considerable resources of agricultural and forest land in Poland. The goal of local governments and agricultural producers should be to identify and eliminate identified weaknesses and potential risks while implementing the relevant principles of a low carbon economy and low carbon development directions of rural areas.

Keywords

Low carbon development · Local low carbon economy plans · Rural resources · Rural areas · Sustainable agriculture · SWOT analysis · Poland

1 Introduction

One of the most important economic and environmental challenges of our time, in the light of climate change, is the transition to an environmentally sustainable economy that allows reducing greenhouse gas (GHG) emissions. In pursuing the aims of the EU climate policy, as well as striving to meet new challenges, Poland (like other countries) must be prepared for the necessity to move to a low carbon economy. This has been recognized in the draft of the National Programme for Development of Low Carbon Economy, adopted by the Ministry of Economy (2015). Unfortunately, work on the final adoption of this programme was halted in 2016 and its

P. Wiśniewski (✉)
Department of Landscape Research and Environmental Management, University of Gdansk, Bażyńskiego 4, 80-309 Gdańsk, Poland
e-mail: pawel.wisniewski@ug.edu.pl

S. Hülsmann and M. Jampani (eds.), *A Nexus Approach for Sustainable Development*,
https://doi.org/10.1007/978-3-030-57530-4_10

status is currently not quite clear. The concept of low carbon development has its roots in the UNFCCC adopted in Rio in 1992 (UNEP 2008). The development of a low carbon economy is one of the priorities adopted by the European Parliament and the Council of the European Union, the 7th General Union Environment Action Programme to 2020 (European Commission 2014). This requires the integration of all aspects of low carbon technologies and practices, efficient energy solutions, clean and renewable energy as well as environmentally friendly technological innovation. It also coincides with the objectives and priorities of the Europe 2020 Strategy for smart, sustainable and inclusive growth (European Commission 2010).

In order to effectively transform the Polish economy, appropriate actions should be planned not only on a national and regional scale but also at the local level. To this end, local low carbon economy plans have been created. These are important strategic documents which are to determine the vision of communes (according to Local Administrative Units—LAU level 2) development towards a low carbon economy and to increase the chances of success for local authorities in applying for EU funds in the 2014–2020 financial perspective. They are equivalent to the Sustainable Energy Action Plans (SEAP)—key documents developed by the signatories of the Covenant of Mayors for Climate and Energy, an association of more than 6 thousand local governments from Europe and beyond. The tasks, which are included in these plans, should focus on low carbon and resource-efficient activities aimed at improving energy efficiency and use of renewable energy sources in all sectors of the economy with the participation of producers, consumers of energy, residents, local authorities, and institutions. Meanwhile, in the currently prepared low carbon economy plans, particular attention is being paid to the energy, construction and transport sectors while agriculture and rural areas are being treated marginally (Wiśniewski and Kistowski 2016, 2017a, b). This trend is observed not only in Poland but also in other European and world countries (Bao 2012; Tinsley 2014; Qu et al. 2016).

It should be noted that agriculture has been identified as a major contributor to atmospheric greenhouse gases on a global scale with about 14% of global GHG emissions coming from this sector (IPCC 2013). According to Richards et al. (2015), agriculture contributes an average of 30% of countries' total GHG emissions. This is higher than the IPCC global estimate of agriculture's contribution to emissions because of the large number of countries where agricultural emissions are low but relatively important in national greenhouse gas budgets. According to these authors, in 42 countries, agriculture contributes more than half of GHG emissions. In 91 countries agriculture contributes $\geq 20\%$ of greenhouse gas emissions. The two regions with the highest average contribution of agricultural emissions are West and East Africa. However, agricultural emissions are also of high importance in Southern Asia and South America—regions that contribute substantially to global agricultural emissions. According to the data presented in Poland's National Inventory Report 2017 (The National Centre for Emissions Management 2017), the total share of GHG emissions from agriculture in Poland is 7.7%. It is slightly lower than in many other European countries (in the EU, an average of 10.1% of greenhouse gases come from agricultural sources), which results from the fact that the Polish economy is based on coal and the role of energy industries in total emissions. It should be stressed, however, that agriculture in Poland is the source of 29.8% of national methane emissions and 78.0% of nitrous oxide emissions. According to the author's earlier research on the assessment of GHG emissions from agriculture at the local level in Poland (Wiśniewski and Kistowski 2018), the share of this sector in total GHG emissions in some Polish communes amounts to as much 20–50%. There is, therefore, an urgent need to draw attention to the role and importance of agriculture and rural areas in the development of low carbon economy at the local level, and to identify activities in these areas, aiming at the reduction of greenhouse gas emissions and improvement of the ability to absorb CO_2, which can be broadly applied in the planning of low carbon development of communes.

Rural areas and related agricultural activities should be an important element in the local development of a low carbon economy and one of the key areas of activity in the currently prepared low carbon economy plans. This is due to the share of agriculture in the total GHG emission, as well as the high potential of rural areas (including forestry) in terms of use their resources to increase carbon sequestration, reduce GHG emissions and climate change mitigation, and also use of agricultural activity to the development of renewable energy, including the production of biogas and biofuels (Kundzewicz and Kozyra 2011; Colomb et al. 2013; Pandey and Agrawal 2014; Nayak et al. 2015; Wiśniewski and Kistowski 2016, 2018; Żukowska et al. 2016; Feliciano et al. 2017; Peter et al. 2017; Vetter et al. 2017). The inclusion of agricultural activities to shaping a low carbon economy is closely linked to the Nexus Approach to the sustainable management of environmental resources and the related Sustainable Development Goals (SDGs) (UN 2015; Liu et al. 2018). Integrated soil and land-use management, addressing carbon sequestration, improving soil fertility and climate change adaptation (by increasing its carbon content, which improves water retention), including also the energy sector (biofuel), should be seen as water-soil-waste-energy nexus, thus a resources-oriented nexus (Hettiarachchi and Ardakanian 2016; Bleischwitz et al. 2018). Several SDGs (e.g., Goals 2—end hunger, 12—sustainable consumption and production, 13—combat climate change, and 15—sustainable use of terrestrial ecosystems and sustainably manage forests) are intrinsically linked to the planning and management for the sustainable use of rural resources and agricultural production space. With regard to rural areas, the concept of sustainable development is particularly important because of the need for taking into account the priority role of the natural environment in the implementation of the production function and the strategic objectives of the development of these areas (Hediger et al. 1998; Sobczyk 2014; Ammirato et al. 2017; Haider et al. 2018). This concept includes the actions which aim at improving the business environment and rural life, without prejudice to the required resources, which should include values of environment, landscape, traditions and cultural heritage. This approach brings together the laws of nature and the economy by following economic activities in the chosen direction of development, adapted to the existing natural conditions, as well as the needs and will of local communities and environmental standards. The realization of the vision of a low carbon economy in rural areas should, therefore, be an essential element of their sustainable development.

The aim of the study is to draw attention to the role of agricultural activity and rural resources (including forestry) in shaping the low carbon economy at the local level in Poland. Actions in these areas have been indicated, enabling reduction of GHG emissions and improvement of CO_2 absorption capacity, which may facilitate the implementation of effective low carbon policy at a commune's level.

2 Material and Methods

Poland, located in Central Europe (Fig. 1), is characterised by a high share of agricultural land in total surface area (approx. 170,000 km^2, 54% of the country's area). Concentration of agricultural land is observed on areas with fertile soils and in the less industrialised central and eastern parts of the country. The lowest shares of agricultural land are observed in areas featuring natural conditions disadvantageous for farming, the highly forested north-western Poland, and the areas characterised by high degree of industrialisation and urbanisation. Arable land takes the most important share in the structure of agricultural land (more than 75%). The zonal soils, dominated by Cambisols and Podzols occupy around 74% of the total area of Poland. The mid-zonal soils, which include alluvial soils, half-bog soils and black earth, Calcaric Leptosols, Gleyic Phaeozems and Gleyic Podzols, occupy roughly 25% of the territory of Poland. The out-of-zone soils are represented in Poland mainly by the Chernozems, having developed on loess. The arable lands in Poland are mainly covered by

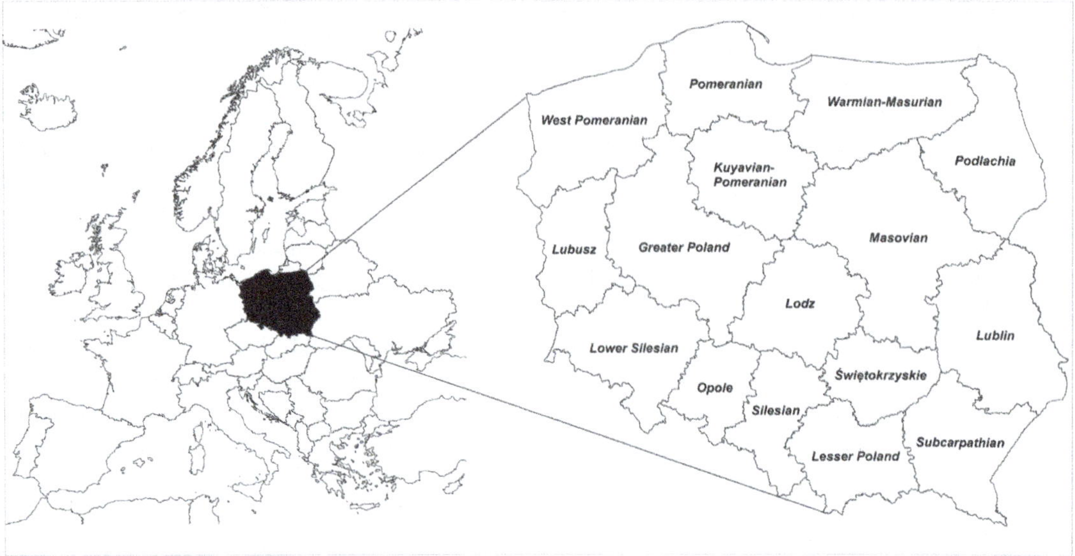

Fig. 1 Location of Poland in Europe and the Polish voivodships (according to Nomenclature of Territorial Units for Statistics—NUTS level 2). *Source* Own elaboration

soils classified as medium or poor quality (Bański 2010; Kabała et al. 2016). The rate of carbon accumulation is reduced by the large proportion of acid soils (nationally, the share of soils with very high or high acidification is over 60%), as these have a low retention capacity and low humus content. The structure of crops on arable lands is dominated by cereals, whose share amounts to approx. 72%. The shares of pulse, industrial and root crops are decidedly smaller. In Poland systems of large-scale management with crop monocultures and simplified crop rotation (especially on the grounds of the former State Agricultural Farm) are still in use, often employing inappropriate cultivation technologies. This method of management triggers erosion processes that contribute to the reduction of soil organic matter and limited soil carbon sequestration.

In order to diagnose the situation of agriculture and rural areas in terms of low carbon economy in Poland, as well as the possible use of their resources towards low carbon development of local governments, a SWOT (Strengths and Weaknesses, Opportunities and Threats) analysis was made. SWOT analysis is frequently used in environmental management as a diagnostic method to identify key factors influencing the success or failure of a specific project. The standard application of SWOT analysis is based on a template, which provides the necessary heuristics to examine future prospects. This investigation is structured in terms of potential that may promote, or barriers that may hinder, the achievement of the specified goals. The result of the SWOT analysis offers insights concerning the trajectory of the strategic planning categorized in "strengths" that should be supported (i.e., inner potential), "opportunities" that have to be sought (i.e., environmental prospects), "weaknesses" that must be overcome (i.e., inner barriers), and "threats" that ought to be alleviated (i.e., environmental hindrances) (Lozano and Vallés 2007; Hovardas 2015). The SWOT analysis carried out in this paper includes a comprehensive literature review, in-depth data analysis (e.g., public statistics data and Agricultural Market Agency data), as well as data and results from earlier studies.

In order to properly diagnose the situation of agriculture and rural areas in terms of low carbon economy in Poland, the main strengths, weaknesses, opportunities and threats were identified in two key elements such as maintenance or increase of the ability to absorb CO_2 and reduction of GHG emissions from agricultural and forest lands and soils. On the basis of the results, proposals on the principles of operation of a low carbon economy and the main directions of low carbon development based on local resources of rural areas (including forestry) in Poland are made.

3 Results and Discussion

3.1 Rural Resources in the Local Development of Low Carbon Economy—SWOT analysis

A. *Strengths*

The strengths of rural areas and their resources in terms of the development of a low carbon economy in Poland are as follow:

- Large area of agricultural land and the ability to use their potential for the development of energy crops. A particular high share of arable land (over 75% of total agricultural land) can be observed in Voivodships (according to Nomenclature of Territorial Units for Statistics—NUTS level 2): Kuyavian-Pomeranian, Lodz, Opole, Pomeranian, Greater Poland and West Pomeranian. Subcarpathian, Podlachia and Warmian-Masurian Voivodeships are, in addition to a high share of arable land, characterized by a relatively high share (over 29%) of permanent grassland (meadows and pastures) (Table 1).
- Large area of forests, especially in Masovian, Warmian-Masurian, Greater Poland and West Pomeranian Voivodships (Table 1). The afforestation rate in Poland is at an average of 29.5% and varies from 21.4% in Lodz Voivodship to 38.2% in Subcarpathian Voivodship and 49.3% in Lubusz Voivodship (The State Forests Information Centre 2017). In approx. 20% of communes nation-wide, (mainly in Lubusz, Subcarpathian, Pomeranian, West Pomeranian, Warmian-Masurian and Podlachia Voivodships) afforestation reaches 40% (Polna 2005). Mixed and deciduous forests, which absorb more CO_2 than the coniferous ones, are of the greatest importance.
- The functioning of agricultural schools and the Agricultural Advisory Centres. There are currently approx. 200 agricultural schools in Poland, including 45 run and supervised by the Ministry of Agriculture and Rural Development. The farm advisory system is formed by the Agricultural Advisory Centre in Brwinów, along with 3 branches and 16 provincial Agricultural Advisory Centers, as well as institutions and non-state actors. Educational and consulting activities of these units should contribute to the implementation of appropriate agricultural practices that help to reduce emissions from agriculture and soils by farmers, including, among others: mulching, conservative plowing or no-till farming, keeping the soil covered with vegetation or leaving the plant residue on it, as well as the adjustment of the supply of nitrogen to the needs of the plants and the adaptation of production systems to maximize the use of animal manure in crop production.

B. *Weaknesses*

Considering weakness, first and foremost, the following should be mentioned:

- A large share of the poor and the poorest arable soils—temporarily or permanently dry, with low productivity due to poor organic matter—that limits the ability to absorb CO_2. Nation-wide, the share of poor and the poorest soils (V and VI quality class) amounts to 34%. A particularly high percentage (over 45%) of the poor and the poorest arable soils occur in Lodz, Masovian, and Podlachia Voivodships (Table 2). The quality of Polish soil is among the lowest in Europe, and the potential production of 1 ha of soil in Poland corresponds to the potential of approx. 0.6 hectares of arable land of all EU countries (Skłodowski and Bielska 2009).
- A large share of acid soils, low retention capacity, and low humus content, which tends

Table 1 Land use in Poland by voivodships in 2015

Voivodship	Agricultural land (thousand ha)	Of which (%)					Forests, trees, and bushes (thousand ha)	Of which (%)	
		Arable land	Orchards	Meadows and permanent pastures	Agricultural land built	Areas under ditches and ponds		Forests	Trees and bushes
Greater Poland	1935.7	78.3	0.8	14.0	5.7	1.2	797.1	98.7	1.3
Kuyavian-Pomeranian	1169.2	84.9	1.0	11.2	2.0	0.9	438.9	97.7	2.3
Lesser Poland	922.7	71.2	3.1	20.7	4.3	0.7	464.0	95.0	5.0
Lodz	1284.4	77.5	2.5	15.7	3.3	1.0	402.3	97.3	2.7
Lower Silesian	1185.3	73.4	0.5	21.6	2.5	2.0	625.8	97.7	2.3
Lublin	1757.4	74.8	1.9	18.5	3.6	1.2	608.6	95.9	4.1
Lubusz	565.1	71.2	0.5	24.1	2.5	1.7	718.8	98.9	1.1
Masovian	2385.1	69.9	3.9	21.9	3.4	0.9	881.0	94.1	5.9
Opole	600.1	81.8	0.5	14.2	2.1	1.4	263.0	98.5	1.5
Podlachia	1214.1	63.4	0.4	32.9	2.6	0.7	643.8	97.9	2.1
Pomeranian	920.2	76.2	0.5	19.8	2.2	1.3	690.5	98.8	1.2
Silesian	628.3	72.5	1.0	21.9	2.9	1.7	412.6	97.7	2.3
Subcarpathian	937.2	64.4	1.1	29.2	4.2	1.1	728.7	94.0	6.0
Świętokrzyskie	751.6	72.3	4.3	18.5	3.9	1.0	345.9	96.9	3.1
Warmian-Masurian	1306.7	67.2	0.2	29.6	2.0	1.0	796.1	96.9	3.1
West Pomeranian	1119.2	77.0	0.4	20.1	1.6	0.9	857.3	97.4	2.6

Source Own elaboration based on the Central Statistical Office (2015b) data

Table 2 The share of the poor and the poorest arable soils, acid and very acid soils in agricultural land, values of ecological stability of land surface and potential of biogas production in Poland by voivodships

Voivodship	The share of the poor and the poorest arable soils (%)	The share of acid and very acid soils in agricultural land (%)	Values of ecological stability of land surface[a]	Number of agricultural biogas plants	The annual efficiency of the agricultural biogas installations (million m^3/year)	Total installed electric capacity from agricultural biogas plants (MW_e)
Greater Poland	41.8	36.0	0.65	8	36.4	9.9
Kuyavian-Pomeranian	24.6	28.0	0.58	6	29.7	8.0
Lesser Poland	30.4	51.0	0.89	2	5.0	1.2
Lodz	46.2	59.0	0.57	4	21.9	5.1
Lower Silesian	21.7	31.0	0.89	9	34.1	9.5
Lublin	23.0	44.0	0.68	7	36.1	9.9
Lubusz	42.8	40.0	1.81	4	11.1	2.8
Masovian	45.1	55.0	0.79	4	20.0	4.8
Opole	21.8	19.0	0.66	1	8.0	2.0
Podlachia	47.1	55.0	1.22	9	30.6	7.9
Pomeranian	34.6	46.0	1.13	9	44.9	12.2
Silesian	35.9	42.0	0.92	3	6.8	2.1
Subcarpathian	27.5	61.0	1.41	3	10.1	2.5
Świętokrzyskie	35.6	39.0	0.79	1	2.5	0.8
Warmian-Masurian	25.5	41.0	1.30	10	37.9	9.5
West Pomeranian	27.1	38.0	1.21	13	50.8	12.7

Source Own elaboration based on Skłodowski and Bielska (2009), Central Statistical Office (2015a) data, Harasim (2015), Agricultural Market Agency (2016) data
[a]Values of ecological stability of land surface in Poland: 0.5–1.0—low stability indicator; 1.01–1.5—medium stability indicator; 1.51–2.0—high stability indicator

to reduce the rate of carbon accumulation. A particularly large share (over 50%) of soils with very high or high acidification characterized by agricultural land in Masovian, Lesser Poland, Lodz, Podlachia and Subcarpathian Voivodships (Table 2).

- Inappropriate variety choice in crop rotation and the rare use of intercrop. In the years 2000–2015, Poland has abandoned the cultivation of perennial crops—grasses or their mixtures with legume plants (Fabaceae)—leaving a large amount of biomass in the form of plant residues and to improve the balance of nitrogen in the soil (Gaweł 2011; Kozłowski et al. 2011). Cultivation of legume mixtures with grasses and grasses represent only 0.6% of the country (Harasim 2015). Their lowest share (0.2%) in the structure of land use in 2014 was characterized by Lower Silesian, Opole, Subcarpathian and Silesian Voivodships, while the highest (from 1.1 to 1.3%)—Podlachia, Warmian-Masurian, and West Pomeranian Voivodships.
- The low level of ecological stability of land surface in Poland calculated by Harasim (2015). Lubusz Voivodship with its extensive forest areas is standing out for its high level of stability. The lowest level of ecological

stability of land surface characterized Lodz and Kuyavian-Pomeranian Voivodships (Table 2). The environmental status of the land surface is shaped mainly by such land use forms as arable lands, forests and permanent grasslands. Land use pattern in Poland is undergoing two major kinds of change: lands sown to annual crops, lands under forest and urban and built-up areas are increasing whereas grasslands, lands under perennial crops and idle lands are decreasing in area (Harasim 2015).

- A system of large-scale management (farms over 15 ha) with monocultures of plants and simplified crop rotation (especially on the lands of former State Agricultural Farm in northern, central and eastern Poland) and often improperly conducted cultivation technology. This method of management triggers erosion processes that contribute to the reduction of soil organic matter and to low soil carbon sequestration (Wiśniewski and Kistowski 2017a).
- Despite the fact that Poland has large resources of the agricultural substrate, there are very few biogas power plants based on agricultural biogas. At the end of 2016, according to the Agricultural Market Agency (2016), only 83 business entities were involved in the production of agricultural biogas in Poland, producing electricity only in cogeneration system in 93 plants with an annual capacity to produce agricultural biogas at the level of 385.9 million m^3 and total installed electric capacity 100.9 MW_e. There is slight progress in the development of biogas plants in Poland. However, compared with other European countries this segment is still relatively small —Poland places 14 amongst the EU member countries in spite of the wide availability of feedstock for agricultural biogas generation (Powałka et al. 2013; Szymańska and Lewandowska 2015). Most agricultural biogas plants are located in West Pomeranian Voivodship, where has 13 plants with a total installed capacity of 12.7 MW_e. For comparison, in Opole and Świętokrzyskie Voivodships located on one of this type of installation with a capacity of 2.0 and 0.8 MW_e (Table 2).
- The dominance of coniferous forests. The changes in species composition and structure of plant communities, transformations of oak-hornbeam habitats and soil related to intense deforestation followed by reforestation of land used in the meantime for agricultural purposes are a consequence of the fact that irrespective of the production potential of habitats, the species preferred for reforestation were those of low climatic and trophic requirements and characterised by fast growing. In the mesotrophic and oligrophic habitats in Poland, the species best meeting the above requirements is *Pinus sylvestris L.*, which for the last two centuries has been predestined to be a dominant species in Polish forests. The monoculture domination of pine for the last two centuries has resulted in simplification and impoverishment of species composition of tree stands leading to the present different forms of secondary communities (Matuszkiewicz et al. 2013; Łaska 2014). Secondary forest communities after afforestation represent often phytocenoses of tree-stands incompatible with the habitat conditions, which reduces the importance of their ecological and protective functions (Łaska 2014; Stefańska–Krzaczek and Pech 2014; Wiśniewski and Kistowski 2015).

C. *Opportunities*

Opportunities to maintain or increase the ability to absorb CO_2 and reduce emissions from agricultural land and soil based on rural resources in Poland should be sought primarily in:

- The growing importance of intercropping, short-rotation plantations, and areas with nitrogen-fixing crops, with a high potential of humus reproductive matter to maintain and increase soil organic matter (legumes, grasses, Fabaceae and their mixtures). Their cultivation has attracted attention both within agri-environment measures, which provide direct payments to EU farmers who subscribe, on a voluntary basis, to environmental commitments related to the preservation of the environment and maintaining the countryside, and for its environmental beneficiary effects.

- The increasing interest in agricultural biogas plants, enabling policy for manure management and other agricultural waste, and the development of crops for the feedstock. In the register of producers of agricultural biogas, according to Agricultural Market Agency (2016), in the period from January 2011 to November 2016, there were 79 new business entities and 85 new installations.
- Implemented concepts of determining a network of ecological corridors and patches. In most of the proposed Polish ecological corridor networks—like in other European countries (e.g., The Pan European Ecological Network, Trans-European Wildlife Network, networks in France, Netherlands, Germany, Austria and the Czech Republic)—forest corridors with predominant cover of forest habitats, of a continuous nature, play a key role. Also important are forest corridors with relatively small patches of a discontinuous nature ("stepping stones") forming a series of neighboring islands of forest ecosystems, field afforestation and roadside tree corridors (complementary to other ecological routes), and beyond-valley strips of pastures with small bushes and trees (Czochański and Wiśniewski 2018).

D. *Threats*

Amongst the potential threats we should, among others, mention:

- The systematic reduction of the agricultural area. From the 1950s, the area of arable land began to decrease and this process is currently underway (Bański 2010). According to the Central Statistical Office of Poland data, in the years 2004–2014, their area decreased by 10.8% (1.8 million ha). A particularly large decline in the agricultural area during this period (about 23–28%) was recorded in the Lesser Poland, Subcarpathian, Silesian, Świętokrzyskie and West Pomeranian Voivodships. According to Mickiewicz et al. (2013), in the years 2002–2010, every year there was a decline by 42,000 ha up to 62,000 ha of arable land. The constant and systematic decrease in this area is related to the urbanization and industrialization of the country, housing development, road expansion, and afforestation.
- Increased risk of agricultural drought. In recent decades an increase in the frequency and severity of summer droughts is reported to be an emerging issue globally (Kundzewicz 2008; Somorowska 2016). Since the end of the sixties of the last century in Poland, one can observe a decrease of rainfall of about 70 mm, so much that some regions of the country started to suffer from insufficient amounts of water. Droughts occurred in the country more and more frequently (Dąbrowska–Zielińska et al. 2011). Analysis carried out by Somorowska (2016) shows a broad variability in the spatial extent of droughts on the territory of Poland. Drought events of differing severity occur in both the winter and summer halves of the year. The percentage of the drought-affected area has witnessed an increase. A relatively large area of Poland exhibits significant drying trends, extending from the south-west towards the center part of the country and, in some cases, to the north-east. This especially pertains to months in the summer half of the year, as well as to the growing season considered as a whole. The drying trends during the growing season extend to approximately 21–27% of the total land area. However, drying signals occur over a much larger percentage of the total area, which amounts to over 50%. Lack of sufficient rainfall during the plant growth season results in the occurrence of agricultural drought, which links various characteristics of meteorological drought to agricultural impacts, focusing on precipitation shortages, differences between actual and potential evapotranspiration, soil water deficits, etc. (Łabędzki and Bąk 2014).

Particularly at risk are the very light and light sandy soils with low capacity to retain water in the soil profile. Limited access to water can be a significant barrier for both food production and the development of green technologies. Energy crops are characterized by the diverse requirements of water and reaction to their satiation in the soil environment (Ostrowski et al. 2009).

- The severity of stress factors affecting forest environment and crops, particularly weather anomalies and extreme phenomena (such as heatwaves, prolonged droughts, floods, storms and heavy showers), which occur in Poland more and more frequently (Stuczyński et al. 2000; Kundzewicz and Matczak 2012; Jaworski and Hilszczański 2013; Bojar et al. 2014). These can contribute, among others, to greater severity of disease and the gradation of plant pests, difficulties in the timely and accurate execution of agrotechnical practices, direct destruction of plants or the crop in their ripening phase, acceleration of the processes of soil erosion, significant limitations of yielding, as well as infectious diseases and forest fires.
- Increased use of physiologically acidic nitrogen fertilizers, which—especially in case of insufficient doses of organic and natural fertilizers containing calcium—may contribute to a further increase in soil acidification.
- The tendency for specialization of farms towards the separation of crop production from animal production, which might result in the exclusion of the use of manure on farms with no livestock with, at the same time, a lack of carbon sequestering practices.

3.2 The Main Directions of Development of Low Carbon Economy at the Local Level in Poland, Based on Rural Resources

Having in view the conditions resulting from the SWOT analysis, one should, foremost, aim to exploit the full potential of agricultural areas and forestry to achieve the desired reduction of GHG emissions and increase carbon sequestration in biomass and soil. This could involve adding organic matter to the soil while reducing its losses, optimization of storage systems, transport and distribution of animal manure on the fields and their proper management, extensive use of agricultural and agro-food processing for the purpose of development of renewable energy (including the production of biogas and biofuels), and to significantly improve energy efficiency and increase the share of renewable energy in plant production and husbandry (Wiśniewski and Kistowski 2018). However, as emphasized by Bennetzen et al. (2016), agricultural GHG emissions can only be reduced to a certain level and a simultaneous focus on other parts of the food system is necessary to increase food security whilst reducing emissions. Amundson and Biardeau (2018) also pay attention to the political and economic barriers to the implementation of soil carbon sequestration on a global scale. Schlesinger and Amundson (2018) suggest caution in ascribing large, potential climate change mitigation to enhanced soil management. They find that all increments to soil organic matter are laudable, however the most promising techniques, including applications of biochar and enhanced silicate weathering, collectively are not likely to balance more than 5% of annual emissions of CO_2 from fossil fuel combustion.

Implementation of the above-mentioned objectives requires the implementation of the relevant principles of the functioning of a low carbon economy and low carbon development directions of rural areas, including in particular:

- The increased use of manure and organic fertilizers by farmers (e.g., compost and green manure plowing) and management of non-used organic waste from the agricultural activity for the purpose of energy production. As Case et al. (2017) point out, processing of organic waste can improve its nutrient availability and content, and thereby increases the agricultural value of the waste when used as fertilisers, while contributing to a more biobased, 'circular' economy. Thus, for example, Danish government policies aim to increase in manure processing (e.g. increasing anaerobic digestion for bioenergy recovery).

- The increase of carbon sequestration in the soil, in particular by preventing erosion and maintaining the correct structure and high content of nutrients in the soil. Since the Common Agricultural Policies (CAP) reform in 2003, many efforts have been made at the European level to promote more environmentally friendly agriculture. In order to oblige farmers to manage their land sustainably, the GAEC (Good Agricultural and Environmental Conditions) were introduced as part of the Cross Compliance mechanism (mechanism that links direct payments to compliance by farmers with basic standards concerning the environment, food safety, animal and plant health and animal welfare, as well as the requirement of maintaining land in good agricultural and environmental condition). Among the standards indicated, the protection of soils against erosion and the maintenance of soil organic matter and soil structure were two pillars to protect and enhance the soil quality and functions (Borrelli et al. 2016). The results of research on different agricultural management practices in central Belgium carried out by Nadeu et al. (2015), showed that soil management strategies targeting C sequestration will be most effective when accompanied by measures that reduce soil erosion given that erosion loss can balance potential C uptake, particularly in sloping areas.
- Measures aimed at bringing the organic matter to the soil while reducing its losses, in particular, to increase biomass production, by the use of organic fertilizers, introduction of grassland and the application of appropriate agricultural practices such as mulching, conservative plowing or no plowing, maintenance of vegetation soil cover or leaving the plant residues.
- The use of intercropping, under-sown crops with a positive rate of reproduction of soil organic matter, plowing and use of crop residues remaining on the field as green manure, as well as composting and using the compost to fertilize with high-value products of animal husbandry. Intercropping is an agricultural practice in which two or more crop species or varieties are grown together in the same field. It is widely practiced by smallholder farmers in developing countries because it enhances the amount of biomass and crop yield that is harvested from a piece of land, and a recent study showed that intercropping increases potential for carbon sequestration and soil organic carbon stocks (Cong et al. 2014, 2015).
- The use of land set-aside or fallow land and uncultivated land for growing energy crops. Energy crops are basic material in the bioenergy industry, and they can also mitigate carbon emissions and have environmental benefits when planted on marginal lands (Wang et al. 2017). Marginal lands are candidates for growing dedicated energy crops such as perennial warm-season grasses (WSGs) and short-rotation woody crops (SRWCs). Experimental data on biomass yields and other ecosystem services from the different marginal lands are, however, few. These few studies have indicated that growing dedicated energy crops on marginal sites can, in general, provide a number of ecosystem services including biomass production, soil water and wind erosion control, soil C sequestration, absorption or retention of pollutants or metals, stabilization or reclamation of minesoils, and improvement in soil properties, among others (Davis et al. 2010; Gelfand et al. 2011, 2016). Research carried out by Dauber et al. (2010) and Meehan et al. (2010) indicate that perennial bioenergy crops (e.g., mixed grasses and forbs) generally have higher biodiversity than annual row crops. Some perennial grasses pose an invasive risk if cultivated outside their native areas, and could potentially have strong negative impacts on biodiversity if they escape plantations and spread over natural areas. Great care must be taken when cultivating these species for biomass to prevent unintended spread. Control

measures may include establishing wide, regularly mown buffer strips along the edges of cultivated fields and regular monitoring (Low and Booth 2007; Searle et al. 2016). The use of marginal land and part of small agricultural parcels located in areas of better soil agricultural suitability is often unprofitable for economic reasons. Currently, in Poland, more than 2.7 million ha of agricultural land is not declared as area for agricultural activity by the farmers. This assessment includes 2.03 million ha of unutilised areas of effective production (parcels >0.3 ha), which constitutes 14.2% of the overall agricultural area. A significant proportion of the unutilised agricultural land constitutes medium and high productivity soils. This situation is particularly visible in Małopolskie, Podkarpackie, Świętokrzyskie, Śląskie, and part of Mazowieckie voivodeships (Pudełko et al. 2018).

- The development of a network of small agricultural biogas plants (constructed by farmers). Bioenergy is a largely CO_2-neutral energy source and it is permanently renewable, as it is produced out of biomass, which is actually living storage of solar energy through photosynthesis. In recent years public opinion regarding large-scale biogas plants became more negative, its economic feasibility dropped dramatically and its implementation is decreasing sharply, among others in Germany, Belgium and the Netherlands. An interesting alternative is the production of renewable energy from small-scale anaerobic digestion, which mainly uses manure and agricultural residues of the farm. In order to promote such a solution, the BioEnergy Farm 2 project was created. This is a project co-financed by the European Union, the aim of which is to provide practical guidance to farmers to produce biogas on farm based scale in small biogas plants and at the same time contributing to farm income and delivering environmental friendly energy (Paterson et al. 2015).
- Liming the acidic and highly acidic soil in order to reduce acidification of agricultural soils. Liming to recommended soil pH values increases productivity, benefits soil structure, improves degraded soils and, when used with other appropriate management practices, can benefit grassland biodiversity. It also reduces some greenhouse gas emissions, among others nitrous oxide (N_2O) (Goulding 2016). As indicated by Kreišmane et al. (2016), if raising the pH of soil, the amount of N_2O emissions in the result of nitrification decreases. Therefore, it is important to perform also maintenance liming if applying nitrogen fertilisers. Oertel et al. (2016) state, however, that there are no significant correlations between NO and N_2O emissions and pH-value. Therefore, emission measurements under field conditions are necessary.
- Increasing green areas and forest cover as well as the introduction of tree planting (especially in areas with poor sandy soils susceptible to erosion) and the creation of forest ecological corridors. Forest management, combined with global environmental changes, increases the capacity of carbon uptake of ordinary managed forests (Dybala et al. 2019). Forest plantations, particularly the ones subject to short-rotation forestry systems, potentially have high soil carbon accumulation, especially in agricultural lands. Moreover, the trees outside forests may represent a sensible carbon stock. These aspects have important implications after the recent recognition of the Land Use, Land Use Change and Forestry (LULUCF) sector in the EU target within the 2030 Climate and Energy Policy Framework (Calfapietra et al. 2015; Nabuurs et al. 2017). The LULUCF regulation is the third pillar of the EU's 2030 climate and energy framework. Together with the Effort Sharing Regulation and the revised ETS directive, it creates a binding legal framework for the EU's efforts to reduce overall greenhouse gas emissions by at least 40% by 2030, compared to 1990 levels (European Commission 2018). The current annual mitigation effect of EU forests via contributions to the forest sink, material substitution and energy substitution is estimated at 13% of total current EU emissions (Nabuurs et al. 2017). Agroforestry systems may provide diverse ecosystem services and

economic benefits that conventional agriculture cannot, e.g. potentially mitigating greenhouse gas emissions by enhancing nutrient cycling, since tree roots can capture nutrients not taken up by crops (Franzluebbers et al. 2017). As Boberek (2015) points out, applicable legislation in Poland not only does not support but even impedes, the implementation of tree planting on farms, including shelterbelts and modern agroforestry systems. This is contrary to the dominant trend in European policies towards sustainable intensification of agriculture that is stressing the importance of climate change-resilient cropping systems for farmers linking climate-change mitigation and adaptation. Boberek (2015) also indicates that agroforestry definition and agroforestry regional policy need to be clarified in the light of a transition of the systems present in Poland that has taken place in recent times. If these issues are addressed to decision-makers, a decline in biodiversity-supporting ecosystem services might be averted, and the design of modern agroforestry systems in Poland promoted. In order to increase the forests' potential to reduce GHG emissions and contribute to climate change mitigation, it is necessary to strengthen the resilience of forests through proper care of newly established forest plantations, the introduction of admixture and biocenotic species in afforestation, and preventing the fragmentation of forest complexes. In areas with re-planted pine monocultures, the systematic reconstruction of forests in order to improve water conditions of soil, soil-protection and soil-forming functions should be pursued (Wiśniewski and Kistowski 2015, 2018).

- Promotion of good practices and low carbon technologies in the agro-food sector. The aim should be to increase farmers' interest in improving the techniques of animal nutrition, appropriate offset rations and adding preparations of nitrogen-binding compounds to the fodders, as well as the use of manure slabs and slurry tanks, improvement of systems to maintain livestock and lowering of methane emissions from stored manure and slurry, which is the result of lowering the storage temperature of manure through recovery and accumulation of heat or construction of installations for the recovery of biogas from slurry fermentation. Concerns over the negative environmental impact from livestock farming across Europe continue to make their mark resulting in new legislation and large research programs. However, despite a huge amount of published material and many available techniques, doubts over the success of national and European initiatives remain. Uptake of the more cost-effective and environmentally-friendly farming methods (such as dietary control, building design and good manure management) is already widespread but unlikely to be enough in itself to ensure that current environmental targets are fully met. Evaluation of the existing and new best available techniques (BAT) is a key to a successful abatement of pollution from the sector and this in turn relies heavily on good measurement strategies. Consideration of the global effect of proposed techniques in the context of the whole farm will be essential for the development of a valid strategy (Loyon et al. 2016).

4 Conclusions

1. Agriculture and rural areas (including forestry)—in contrast to current trends—should be an important element in the local creation of a low carbon economy and an important area of activity in the local plans of a low carbon economy, currently being developed. This is due to the share of agriculture in the total GHG emission, as well as the high potential of these areas in terms of use their resources to increase carbon sequestration, reduce GHG emissions and climate change mitigation, and also use of agricultural activity to the development of renewable energy, including the production of biogas and biofuels.
2. The information obtained from the SWOT analysis shows that this potential is based on

considerable resources of agricultural land and forest land. However, to exploit this potential, many weaknesses must be overcome. These include, among others, a large share of the poor and the poorest arable soils, large share of acid soils, high risk of erosion, low level of ecological stability of the land surface, as well a large share of phytocenoses of tree-stands with pine dominance, which reduces the importance of their ecological and protective functions.

3. Opportunities that have to be sought include, among others, the growing importance of intercropping, short-rotation plantations, and areas with nitrogen-fixing crops, which has attracted attention both within agri-environment measures, which provide direct payments to EU farmers who subscribe, on a voluntary basis, to environmental commitments related to the preservation of the environment and maintaining the countryside and for its environmental beneficiary effects. There is also growing interest in small-scale agricultural biogas plants, which mainly use manure and agricultural residues of the farm.
4. Numerous threats that ought to be alleviated include, among others, constant and systematic decrease in the agricultural area, which is related to the urbanization and industrialization of the country, housing development, road expansion, and afforestation. In recent decades an increase in the frequency and severity of summer droughts is reported. Limited access to water and weather anomalies and extreme phenomena can be a significant barrier for both food production and the development of green technologies.
5. The goal of local governments and agricultural producers should be to identify and eliminate weaknesses and potential risks hindering the maintenance or increase of the ability to absorb CO_2 and reduction of GHG emissions from agricultural lands and soils while implementing the vision of a low carbon economy in rural areas. This requires the implementation of the relevant principles of the functioning of a low carbon economy and low carbon development directions of rural areas, including in particular:
   - the increased use of manure and organic fertilizers by farmers and management of non-used organic waste from the agricultural activity for the purpose of energy production;
   - the increase of carbon sequestration in the soil, in particular by preventing erosion and maintaining the correct structure and high content of nutrients in the soil;
   - introduction of grassland and the application of appropriate agricultural practices such as mulching and intercropping;
   - use of uncultivated and marginal lands for growing energy crops;
   - liming the acidic and highly acidic soil;
   - increasing green areas and forest cover, as well as use agroforestry systems, which may provide diverse ecosystem services and economic benefits that conventional agriculture cannot;
   - introducing good practices and low carbon technologies in the agro-food sector.
6. Policies targeting CO_2 sequestration and reduction of GHG emissions in rural areas should be viewed in the context of integrated resources management, nexus-oriented governance and as tools for climate change adaptation and for achieving several resource-related SDGs.

References

Agricultural Market Agency (2016) Register of producers of agricultural biogas (as at November 4, 2016). Agricultural Market Agency, Warsaw

Ammirato S, Felicetti AM, Della Gala M, Frega N, Palmiro Volpentesta A (2017) Sustainable development for rural areas: a survey on the agritourism rural networks. In: Camarinha–Matos LM, Afsarmanesh H, Fornasiero Rosanna (eds) Collaboration in a data-rich world. PRO-VE 2017. IFIP advances in information and communication technology, vol 506. Springer, Cham, pp 564–574. https://doi.org/10.1007/978-3-319-65151-4_50

Amundson R, Biardeau L (2018) Opinion: soil carbon sequestration is an elusive climate mitigation tool. Proc Natl Acad Sci 115(46):11652–11656. https://doi.org/10.1073/pnas.1815901115

Bański J (2010) Atlas of polish agriculture. Institute of Geography and Spatial Organization Polish Academy of Sciences, Warsaw (in Polish)

Bao W (2012) Low carbon economy and sustainable energy development in rural areas of Sichuan Province, China. Adv Mater Res 476–478:824–827. https://doi.org/10.4028/www.scientific.net/AMR.476-478.824

Bennetzen EH, Smith P, Porter JR (2016) Decoupling of greenhouse gas emissions from global agricultural production: 1970–2050. Glob Change Biol 22:763–781. https://doi.org/10.1111/gcb.13120

Blanco-Canqui H (2016) Growing dedicated energy crops on marginal lands and ecosystem services. Soil Sci Soc Am J 80:845–858. https://doi.org/10.2136/sssaj2016.03.0080

Bleischwitz R, Spataru C, VanDeveer SD, Obersteiner M, van der Voet E, Johnson C, Andrews-Speed P, Boersma T, Hoff H, van Vuuren DP (2018) Resource nexus perspectives towards the United Nations sustainable development goals. Nat Sustain 1:737. https://doi.org/10.1038/s41893-018-0173-2

Boberek R (2015) Agroforestry systems in Poland a preliminary identification. Pap Glob Change 22:37–51. https://doi.org/10.1515/igbp-2015-0014

Bojar W, Knopik L, Żarski J, Sławiński C, Baranowski P, Żarski W (2014) Impact of extreme climate changes on the predicted crops in Poland. Acta Agrophysica 21 (4):415–431

Borrelli P, Paustian K, Panagos P, Jones A, Schütt B, Lugato E (2016) Effect of good agricultural and environmental conditions on erosion and soil organic carbon balance: a national case study. Land Use Policy 50:408–421. https://doi.org/10.1016/j.landusepol.2015.09.033

Calfapietra C, Barbati A, Perugini A, Ferrari B, Guidolotti G, Quatrini A, Corona P (2015) Carbon mitigation potential of different forest ecosystems under climate change and various managements in Italy. Ecosyst Health Sustain 1(8):1–9. https://doi.org/10.1890/EHS15-0023

Case SDC, Oelofse M, Hou Y, Oenema O, Stoumann Jensen L (2017) Farmer perceptions and use of organic waste products as fertilisers—a survey study of potential benefits and barriers. Agric Syst 151:84–95. https://doi.org/10.1016/j.agsy.2016.11.012

Central Statistical Office (2015a) Means of production in agriculture in the 2014/2015 farming year. Statistical Publishing Establishment, Warsaw

Central Statistical Office (2015b) Statistical yearbook of agriculture 2015. Statistical Publishing Establishment, Warsaw (in Polish)

Colomb V, Touchemoulin O, Bockel L, Chotte J-L, Martin S, Tinlot M, Bernoux M (2013) Selection of appropriate calculators for landscape-scale greenhouse gas assessment for agriculture and forestry. Environ Res Lett 8:1–10. https://doi.org/10.1088/1748-9326/8/1/015029

Cong W-F, Hoffland E, Li L, Janssen BH, Van Der Werf W (2015) Intercropping affects the rate of decomposition of soil organic matter and root litter. Plant Soil 391:399–411. https://doi.org/10.1007/s11104-015-2433-5

Cong W-F, Hoffland E, Li L, Six J, Sun J-H, Bao X-G, Zhang F-S, Van Der Werf W (2014) Intercropping enhances soil carbon and nitrogen. Glob Change Biol 21:1715–1726. https://doi.org/10.1111/gcb.12738

Czochański JT, Wiśniewski P (2018) River valleys as ecological corridors—structure, function and importance in the conservation of natural resources. Ecol Quest 29 (1):77–87. https://doi.org/10.12775/EQ.2018.006

Dąbrowska-Zielińska K, Ciołkosz A, Malińska A, Bartold M (2011) Monitoring of agricultural drought in Poland using data derived from environmental satellite images. Geoinf Issues 1(3):87–97

Dauber J, Jones MB, Stout JC (2010) The impact of biomass crop cultivation on temperate biodiversity. Glob Change Biol 2:289–309. https://doi.org/10.1111/j.1757-1707.2010.01058.x

Davis SC, Parton WJ, Dohleman FG, Smith CM, Del Grosso S, Kent AD, DeLucia EH (2010) Comparative biogeochemical cycles of bioenergy crops reveal nitrogen-fixation and low greenhouse gas emissions in a *Miscanthus* × *giganteus* agro-ecosystem. Ecosystems 13:144–156. https://doi.org/10.1007/s10021-009-9306-9

Dybala KE, Matzek V, Gardali T, Seavy NE (2019) Carbon sequestration in riparian forests: a global synthesis and meta-analysis. Glob Change Biol 25 (1):57–67. https://doi.org/10.1111/gcb.14475

European Commission (2010) Europe 2020. A strategy for smart, sustainable and inclusive growth, COM (2010) 2020. European Commission, Brussels

European Commission (2014) General Union Environment Action Programme to 2020. Living well, within the limits of our planet. European Commission, Luxembourg

European Commission (2018) Regulation on land use, land use change and forestry in 2030 climate and energy framework adopted. https://ec.europa.eu. Accessed 09 Apr 2018

Feliciano D, Nayak DR, Vetter SH, Hillier J (2017) CCAFS-MOT—a tool for farmers, extension services and policy-advisors to identify mitigation options for agriculture. Agric Syst 154:100–111. https://doi.org/10.1016/j.agsy.2017.03.006

Franzluebbers AJ, Chappell JC, Shi W, Cubbage FW (2017) Greenhouse gas emissions in an agroforestry system of the southeastern USA. Nutr Cycl Agroecosyst 108:85–100. https://doi.org/10.1007/s10705-016-9809-7

Gaweł E (2011) The role of fine-grained legume plants in a farm. Water Environ Rural Areas 3(35):73–91 (in Polish)

Gelfand I, Zenone T, Jasrotia P, Chen J, Hamilton SK, Robertson GP (2011) Carbon debt of Conservation

Reserve Program (CRP) grasslands converted to bioenergy production. Proc Natl Acad Sci USA 108 (33):13864–13869. https://doi.org/10.1073/pnas.1017277108

Goulding KWT (2016) Soil acidification and the importance of liming agricultural soils with particular reference to the United Kingdom. Soil Use Manag 32(3):390–399. https://doi.org/10.1111/sum.12270

Haider LJ, Boonstra WJ, Peterson GD, Schlüter M (2018) Traps and sustainable development in rural areas: a review. World Dev 101:311–321. https://doi.org/10.1016/j.worlddev.2017.05.038

Harasim A (2015) Land use in Poland in terms of environmental stability. Ann Pol Assoc Agric Agribus Econ 1:66–71 (in Polish)

Hediger W, Dorenbos A, Lehmann B (1998) Sustainable development of rural areas—methodological issues. In: ERSA conference papers, August 28–September 1. European Regional Science Association, Vienna

Hettiarachchi H, Ardakanian R (eds) (2016) Environmental resource management and the nexus approach. Springer International Publishing, Cham

Hovardas T (2015) Strengths, Weaknesses, Opportunities and Threats (SWOT) analysis: a template for addressing the social dimension in the study of socio-scientific issues. Aegean J Environ Sci 1:1–12

IPCC (2013) Climate change 2013. The physical science basis. In: Contribution of working group I to the fifth assessment report of the intergovernmental panel on climate change. Intergovernmental Panel on Climate Change, Cambridge

Jaworski T, Hilszczański J (2013) The effect of temperature and humidity changes on insects development their impact on forest ecosystems in the context of expected climate change. For Res Pap 74(4):345–355. https://doi.org/10.2478/frp-2013-0033

Kabała C, Świtoniak M, Charzyński P (2016) Correlation between the polish soil classification (2011) and international soil classification system world reference base for soil resources (2015). Soil Sci Annu 67 (2):88–100. https://doi.org/10.1515/ssa-2016-0012

Kozłowski S, Swędrzyński A, Zielewicz W (2011) Leguminous plants in the natural environment. Water Environ Rural Areas 4(36):161–181 (in Polish)

Kreišmane D, Naglis-Liepa K, Popluga D, Lēnerts A, Rivža P (2016) Liming effect on nitrogen use efficiency and nitrogen oxide emissions in crop farming. Res Rural Dev 1:30–36

Kundzewicz ZW (2008) Hydrological extremes in the changing world. Folia Geogr Geogr Phys Ser 39:37–52

Kundzewicz ZW, Kozyra J (2011) Reducing impacts of climatic threats to agriculture and rural areas. Pol J Agron 7:68–81 (in Polish)

Kundzewicz ZW, Matczak P (2012) Climate change regional review: Poland. WIREs Clim Change 3:297–311. https://doi.org/10.1002/wcc.175

Liu J, Hull V, Godfray HCJ, Tilman D, Gleick P, Hoff H, Pahl-Wostl C, Xu Z, Chung MG, Sun J, Li S (2018) Nexus approaches to global sustainable development. Nat Sustain 1:466–476. https://doi.org/10.1038/s41893-018-0135-8

Łabędzki L, Bąk B (2014) Meteorological and agricultural drought indices used in drought monitoring in Poland: a review. Meteorol Hydrol Water Manag 2 (2):3–14. https://doi.org/10.26491/mhwm/34265

Łaska G (2014) Ecological consequences of deforestation and afforestation on a post-arable land: changes in the composition and structure of plant communities and transformations of oak-hornbeam habitats and soil. Ecol Quest 20:9–21. https://doi.org/10.12775/EQ.2014.012

Low T, Booth C (2007) The weedy truth about biofuels. Invasive Species Council, Melbourne

Loyon L, Burton CH, Misselbrook T, Webb J, Philippe F-X, Aguilar M, Doreau M, Hassouna M, Veldkamp T, Dourmand J-Y, Bonmati A, Grimm E, Sommer SG (2016) Best available technology for European livestock farms: availability, effectiveness and uptake. J Environ Manag 166:1–11. https://doi.org/10.1016/j.jenvman.2015.09.046

Lozano M, Vallés J (2007) An analysis of the implementation of an environmental management system in a local public administration. J Environ Manag 82 (4):495–511. https://doi.org/10.1016/j.jenvman.2006.01.013

Matuszkiewicz JM, Kowalska A, Kozłowska A, Roo-Zielińska E, Solon J (2013) Differences in plant-species composition, richness and community structure in ancient and post-agricultural pine forests in central Poland. For Ecol Manag 310:567–576. https://doi.org/10.1016/j.foreco.2013.08.060

Meehan TD, Hurlbert AH, Gratton C (2010) Bird communities in future bioenergy landscapes of the Upper Midwest. Proc Natl Acad Sci USA 107(43):18533–18538. https://doi.org/10.1073/pnas.1008475107

Mickiewicz B, Mickiewicz A, Sobala M (2013) Analysis of behind change of arable land area in period between two agricultural censuses (2002–2010). Optim Econ Stud 4(64):13–24 (in Polish)

Ministry of Economy (2015) National programme for development of low carbon economy (draft). Ministry of Economy, Warsaw

Nabuurs GJ, Delacote P, Ellison D, Hanewinkel M, Hetemäki L, Lindner M, Ollikainen M (2017) By 2050 the mitigation effects of EU forests could nearly double through climate smart forestry. Forests 8:1–14. https://doi.org/10.3390/f8120484

Nadeu E, Gobin A, Fiener P, van Wesemael B, van Oost K (2015) Modelling the impact of agricultural management on soil carbon stocks at the regional scale: the role of lateral fluxes. Glob Change Biol 21:3181–3192. https://doi.org/10.1111/gcb.12889

Nayak D, Saetnan E, Cheng K, Wang W, Koslowski F, Cheng Y-F, Zhu WY, Wang J-K, Liu J-X, Moran D, Yan X, Cardenas L, Newbold J, Pan G, Lu Y, Smith P (2015) Management opportunities to mitigate greenhouse gas emissions from Chinese agriculture. Agr Ecosyst Environ 209:108–124. https://doi.org/10.1016/j.agee.2015.04.035

Oertel C, Matschullat J, Zurba K, Zimmermann F, Erasmi S (2016) Greenhouse gas emissions from soils—a review. Chem Erde 76:327–352. https://doi.org/10.1016/j.chemer.2016.04.002

Ostrowski J, Gutkowska A, Tusiński E (2009) The role of water factor in modeling categorization and evaluation of land usefulness for the cultivation of energetic crops. Water Environ Rural Areas 4(28):187–202 (in Polish)

Pandey D, Agrawal M (2014) Carbon footprint estimation in the agriculture sector. In: Muthu SS (ed) Assessment of carbon footprint in different industrial sectors, vol 1. Springer, Singapore, pp 25–47. https://doi.org/10.1007/978-981-4560-41-2_2

Paterson M, Kayser K, Bonhomme S, Majewski E, Amrozy M, Berruto R, Parola F, Bijnagte JW, Gysen M (2015) Implementation guide for small-scale biogas plants. BioEnergy Farm II Publication, KTBL, Darmstadt

Peter C, Helming K, Nendel C (2017) Do greenhouse gas emission calculations from energy crop cultivation reflect actual agricultural management practices?—a review of carbon footprint calculators. Renew Sustain Energy Rev 67:461–476. https://doi.org/10.1016/j.rser.2016.09.059

Polna M (2005) Changes in the woodiness of Poland in the years 1990–2001. Acta Sci Pol Silvarum Colendarum Ratio et Ind Lignaria 4(1):51–60 (in Polish)

Powałka M, Klepacka A, Skudlarski J, Golisz E (2013) The agricultural biogas sector in Poland as compared to other European Union member-countries. Probl World Agric 13(3):203–212 (in Polish)

Pudełko R, Kozak M, Jędrejek A, Gałczyńska M, Pomianek B (2018) Regionalisation of unutilised agricultural area in Poland. Pol J Soil Sci 1:119–132. https://doi.org/10.17951/pjss/2018.51.1.119

Qu F, Sun R, Guo Z, Yu F (eds) (2016) Ecological economics and harmonious society. Springer, Singapore

Richards M, Wollenberg E, Buglion-Gluck S (2015) Agriculture's contributions to national emissions. CCAFS Info Brief. CGIAR Research Program on Climate Change, Agriculture and Food Security (CCAFS), Copenhagen

Schlesinger WH, Amundson R (2018) Managing for soil carbon sequestration: let's get realistic. Glob Change Biol 25(2):386–389. https://doi.org/10.1111/gcb.14478

Searle S, Petrenko C, Baz E, Malins C (2016) Crops of the biofrontier: in search of opportunities for sustainable energy cropping. The International Council on Clean Transportation, Washington D.C.

Skłodowski P, Bielska A (2009) Properties and fertility of soils in Poland—a basis for the formation of agro-environmental relations. Water Environ Rural Areas 4 (28):203–214 (in Polish)

Sobczyk W (2014) Sustainable development of rural areas. Probl Sustain Dev 1:119–126

Somorowska U (2016) Changes in drought conditions in Poland over the past 60 years evaluated by the standardized precipitation-evapotranspiration index. Acta Geophys 64(6):2530–2549. https://doi.org/10.1515/acgeo-2016-0110

Stefańska-Krzaczek E, Pech P (2014) Vegetation diversity of the Scots pine stands in different forest sites in the Turawa Forest District. For Res Pap 75(1):77–87. https://doi.org/10.2478/frp-2014-0008

Stuczyński T, Demidowicz G, Deputat T, Górski T, Krasowicz S, Kuś J (2000) Adaptation scenarios of agriculture in poland to future climate changes. Environ Monit Assess 61:133–144. https://doi.org/10.1023/A:1006378420994

Szymańska D, Lewandowska A (2015) Biogas power plants in Poland—structure, capacity, and spatial distribution. Sustainability 7:16801–16819. https://doi.org/10.3390/su71215846

The National Centre for Emissions Management (2017) Greenhouse gas inventory for 1988–2015. The National Centre for Emissions Management, Warsaw

The State Forests Information Centre (2017) Forests in Poland 2017. The State Forests Information Centre 2017, Warsaw

Tinsley S (2014) Environmental management in a low carbon economy. Routledge, London, and New York

UN (2015) Transforming our world: the 2030 Agenda for sustainable development A/RES/70/1. United Nations, New York

UNEP (2008) UNEP 2008 annual report. United Nations Environment Programme, Nairobi

Vetter SH, Sapkota TB, Hillier J, Stirling CM, Macdiarmid JI, Aleksandrowicz L, Green R, Joy EJM, Dangour AD, Smith P (2017) Greenhouse gas emissions from agricultural food production to supply Indian diets: implications for climate change mitigation. Agr Ecosyst Environ 237:234–241. https://doi.org/10.1016/j.agee.2016.12.024

Wang J, Gao M, He X, Zhang Q, Leo N, Xu C (2017) Evaluation of potential productivity of woody energy crops on marginal land in China. Chin Geogr Sci 27:963–973. https://doi.org/10.1007/s11769-017-0916-4

Wiśniewski P, Kistowski M (2015) Structure and importance of soil-protecting forests in the areas administered by the RDSF Toruń. Bull Geogr Phys Geogr Ser 8:123–132. https://doi.org/10.1515/bgeo-2015-0010

Wiśniewski P, Kistowski M (2016) Local low carbon economy plans in the context of low carbon rural development. J Ecol Eng 17(4):112–119. https://doi.org/10.12911/22998993/63960

Wiśniewski P, Kistowski M (2017a) Agriculture and rural areas in the local planning of low carbon economy in light of the idea of sustainable development—results from a case study in north-central Poland. Fresenius Environ Bull 26(8):4927–4935

Wiśniewski P, Kistowski M (2017b) Low-carbon development of rural areas in Poland against low-carbon economy planning. UG Publishing House, Gdańsk (in Polish)

Wiśniewski P, Kistowski M (2018) Assessment of greenhouse gas emissions from agricultural sources in order to plan for needs of low carbon economy at local level in Poland. Geogr Tidsskr Dan J Geogr. https://doi.org/10.1080/00167223.2018.1436447

Żukowska G, Myszura M, Baran S, Wesołowska S, Pawłowska M, Dobrowolski L (2016) Agriculture vs. alleviating the climate change. Probl Sustai Dev 2:67–74

Opportunities and Challenges to Adopting Sustainable Watershed Management Interventions: An Overview of Experiences from Ethiopia

Gebreyesus Brhane Tesfahunegn and
Elias T. Ayuk

Abstract

Watershed management techniques in view of the water–soil–waste nexus approach and its link to food security have been widely studied in Ethiopia. However, there has been limited uptake and implementation of the recommended technologies in many watersheds. This paper aims to illustrate the experiences and lessons about watershed management interventions in the Tigray region, northern Ethiopia and identifies opportunities and challenges to the adoption of these practices. Information was collected based on discussions with farmers and development agents, field observations, personal experiences, and literature review. Positive impacts of the nexus-oriented watershed management interventions included 50% increase in income, 56% increase in food security, 62% increase in vegetation coverage, 63% increase in surface and groundwater resources, 62% restoration of soil fertility, sediment accumulation behind bunds (up to 65 t ha^{-1} $year^{-1}$), and 80% reduction in soil loss. In Ethiopia, opportunities for the widespread adoption of watershed management interventions include improvement in socioeconomic conditions, experiences on natural resources restoration, farm diversification and intensification, and understanding the complementarity of crop and livestock systems. On the other hand, adoption of these interventions was challenged by a lack of improved agronomic practices, poor crop and livestock management, unpredictable climate, the large area of land under severe degradation, and high competition between crop and livestock systems. In order to improve the adoption of nexus-oriented watershed management interventions, a conceptual research strategy including local and scientific knowledge and approaches (from participatory problem identification to monitoring, and evaluation strategy) is suggested.

Keywords

Income · Farm diversification · Food security · Land degradation · Natural resources restoration

1 Introduction

Ethiopia has been seriously affected by land degradation in the form of soil erosion, forest degradation, and soil fertility decline. Despite the

G. B. Tesfahunegn (✉)
College of Agriculture, Aksum University, Shire-Campus, P. O. Box 314, Shire, Ethiopia
e-mail: gebre33@gmail.com

G. B. Tesfahunegn · E. T. Ayuk
International House 2nd Floor, United Nations University Institute for Natural Resources in Africa (UNU-INRA), University of Ghana, Legon Campus, Accra, Ghana
e-mail: ayuk@unu.edu

S. Hülsmann and M. Jampani (eds.), *A Nexus Approach for Sustainable Development*,
https://doi.org/10.1007/978-3-030-57530-4_11

recent massive restoration measures implemented on degraded landscapes in the country, the problem remains strongly manifested in northern Ethiopia. Soil erosion rates up to 130 t ha^{-1} $year^{-1}$ from cropland (FAO 1986) and soil nutrient depletion of 60 kg ha^{-1} of combined nitrogen (N) and phosphorus (P) (Stoorvogel and Smaling 1990) have been reported in Ethiopia. For small watersheds in northern Ethiopia, soil loss rates of 30 t ha^{-1} $year^{-1}$ (Tesfahunegn 2011) and 10 t ha^{-1} $year^{-1}$ (Fenta et al. 2016) and combined N and P nutrient export of 9.2 kg ha^{-1} $year^{-1}$ (Tesfahunegn and Vlek 2013) were estimated. The severe erosion could be associated with forest cover, which decreased to the level of less than 3% in the early 1990s, and other characteristics of poor management (Bishaw 2001). In other areas in Ethiopia, the relatively stable and higher percentages of forest coverage and soil and water conservation (SWC) practices were reported to decrease the severity of erosion (Solomon et al. 2018).

To reduce erosion related problems, the government of Ethiopia in collaboration with international donors has been implementing various mechanical and biological SWC measures in the country since the 1970s. The SWC implementation programme from the 1970s to 1990s was top-down (government-led, incentive based (food-for-work)) and has paid more attention to the implementation of structural SWC measures targeting to control soil erosion and improve resources such as vegetation, water availability. In the early 2000s, community-based integrated watershed management was introduced to achieve the broader integrated natural resources management and to improve the livelihoods (Bishaw 2001; Gebregziabher et al. 2016). Watersheds have been recognized as appropriate unit to integrate land resource management, where management is not only limited to improving (and jointly considering in the sense of a nexus approach) land, water, biodiversity, and food security (Karlberg et al. 2014; Johnson and Karlberg 2017), but also concerned with integration for self-reliance and holistic economic development of the community and assures minimum disturbance to the environment (Alemu and Kidane 2014; Karlberg et al. 2015). The community-based integrated watershed management approach in Ethiopia involves the most affected people across all phases of the development activities and demands different disciplines (biophysical, social, and economic sciences) and administrative units such as village, sub-district, and region (Lakew et al. 2005; Kerse 2017).

Considering the contribution of agriculture to the national economy (employment) and food security, the government of Ethiopia has implemented the policy of 'Agriculture Development-Led Industrialization (ADLI)' since 1993. The primary objective of this policy is to support for achieving rapid and sustainable economic growth by improving the productivity of the agriculture sector as a priority sector since the early 1990s (Dube et al. 2019). To achieve the goals of this policy on all lands in the country, a strategy on sustainable watershed resources management including soil, water and biodiversity conservation has been put in place. Watershed management is supposed to be practiced as a means to increase rain-fed and irrigation agricultural production, conserve natural resources and reduce poverty in Ethiopia, a country characterized by low agricultural productivity, severe natural resource degradation, and high levels of poverty (Kerr 2002). Under the current conditions in Ethiopia (low agricultural productivity, severe natural resource degradation, and high levels of poverty), strengthening the participatory approaches could support the successful implementation of integrated watershed management, in the context of achieving integrated water–soil–waste management (nexus approach) at the local level. This nexus approach is particularly important in view of the current sustainable development goals (SDGs) (Hülsmann and Ardakanian 2018), which have targets to improve degraded land resources (SDG 15), eliminate poverty (SDG 1), and achieve zero hunger (SDG 2) by 2030 (United Nations 2015).

Sustainable watershed management interventions as integrated water–energy–food (WEF) management nexus approach can inform sector planning, policy, and technology decisions by identifying potential trade-offs and exploring synergies in the presence of climate change challenges (FAO 2014). The link between the WEF nexus approach, its resources-focused complement, the water–soil–waste nexus approach (Hülsmann and Ardakanian 2018) and SDGs is well recognized as the majority of the goals are directly related to the sustainable use of resources such as land, soil, water, food, and energy. Integrated watershed management as the judicious use of resources such as land, water, biodiversity, and farm inputs to obtain optimum production with minimum disturbance to the environment can support the WEF security nexus (FAO 2014), and thereby the SDGs. The implication is that the nexus approach is a pillar and an enabler for achieving sustainable resource and economic development (FAO 2014; SEI 2014). For the purpose of this study, we aim to highlight resources and in particular food. Therefore, when talking about the nexus approach we refer to the water–soil–waste–food security nexus if not specified otherwise.

The success of watershed projects is determined by improving community livelihoods and the environmental services offered to the people, such as reducing on-site and off-site effects of erosion (Kerr 2002; Verhoeven 2013), which is in line with the goals of the nexus approach. In northern Ethiopia, examples of unsuccessful negative off-site effects or indicators of erosion include siltation of micro-dams and irrigation channels (Behailu 2002; Tamene 2005; Fantaw 2007; Wang et al. 2014), and flooding of downstream agricultural fields (Fantaw 2007). Such off-site impacts could be associated with the inappropriateness of management options and implementation strategy. In general, considerable information has been developed pertaining to watershed management interventions from the view of the nexus approach in northern Ethiopia (e.g., Verhoeven 2013; Karlberg et al. 2014, 2015; Johnson and Karlberg 2017) However, there has been limited uptake and implementation at a larger scale of the recommended approaches, management options, and technologies by local people at the watershed scale. To achieve sustainable development through the implementation of ecologically friendly and locally acceptable technologies, there is a need to address nexus problems within a watershed by developing, testing, selecting, and disseminating relevant technologies to the local community (Karlberg et al. 2014, 2015; Johnson and Karlberg 2017).

Indicators of successful achievement of the implemented watershed management practices of a nexus approach in Ethiopia include reclaimed degraded land, improved soil fertility, water resource development, increased agricultural production, and thereby improved community and environmental wellbeing (Mekonen and Tesfahunegn 2011; Alemu and Kidane 2014; Gebregziabher et al. 2016). However, there are many challenges to implementing and adopting sustainable watershed management practices from the perspective of the nexus approach in northern Ethiopia. Hence, understanding the opportunities and challenges for the adoption of sustainable watershed management interventions that can positively impact the nexus approach based on the existing experiences in Ethiopia is crucial for designing, introducing, and disseminating an alternative research strategy.

This paper aims to assess the impacts of introduced nexus-oriented watershed management practices and understand the opportunities and challenges to the adoption of sustainable watershed management practices in northern Ethiopia. The paper further explores a research strategy that can enhance the adoption of the best watershed management technologies to achieve the nexus approach in Ethiopia, which has been identified as a priority area for the SDGs.

2 Materials and Methods

This paper examines the experience of watershed management practices in view of the nexus approach in the Tigray region, northern Ethiopia (Fig. 1). In this region, elevation varies between 1800 and 2650 m above sea level. The main rainy season extends from June to the first week of September in most parts of the region. The mean annual rainfall is about 650 mm and the mean monthly temperature during the growing season is 20 °C. The main soil types are Cambisols on undulating plains and rolling land, Lithosols on hilly and steep to very steep land, and Vertisols on flat and plateau landforms (Tesfahunegn and Vlek 2013). The vegetation is sparse as a result of centuries of overexploitation and consists of shrubs and small trees with little economic value. The farming system is principally a mixed crop-livestock system, but farmers' livelihood mainly depends on crop cultivation (Tesfahunegn and Vlek 2013).

The watersheds in the Tigray region of northern Ethiopia were selected for this study because of severe degradation in the region due to long-cultivation and high population pressure. To reduce such challenges, various different watershed management interventions have been introduced since the 1990s. Thus, in this study, the impacts, and challenges, and opportunities of the implemented watershed management intervention were assessed from a nexus perspective, with the intention of identifying relevant lessons and experiences to find alternative solutions.

In this study, a range of different approaches was used to collect and organize the information. This included the researchers' personal experiences (field observations, research, and teaching) and literature review (published and unpublished sources). Data on the impacts of watershed management interventions as a nexus approach were organized using the "before" and "after" intervention implementation (Table 1), based on different survey tools (interviews, observations, transect walks, group discussions, and key informants' interviews).

Four small watersheds of similar size (Abraha-Atsbaha, Gerebshelela, Mai-Negus and Medego), were selected to represent the farming system and implementation of watershed management interventions in the highlands of the Tigray region. The selected watersheds were visited several times by the first author to observe

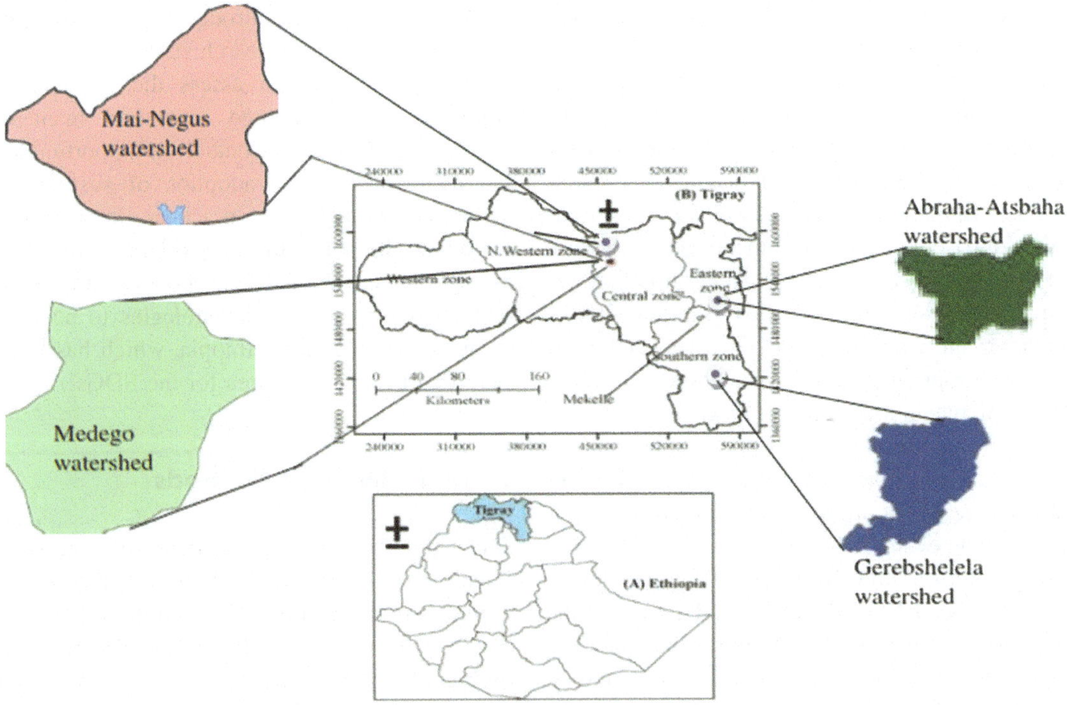

Fig. 1 Map of Ethiopia and Tigray region with the selected watersheds in northern Ethiopia

Table 1 Some performance indicators of watershed management interventions as water-soil-waste management nexus approach from the perspectives of the Ethiopia highlands (*Source* modified after Mekonen and Tesfahunegn 2011)

Performance criterion	Activity after intervention	Indicator used
Watershed management	Inventory of interventions introduced	– Type of interventions (see below) – Visual assessment of watershed management investments (e.g., photos)
Water recharge and storage	– Measurement and visualization of surface and subsurface water levels – Amount of surface water storage	– Change of groundwater recharge rates and number of wells – Change in well depths to access groundwater – Change in number of springs and their duration of water – Capacity of surface water storage – Change in irrigated area – Change in village level water supply
Soil moisture retention	– Determination of time series, intra- and inter-season and year variations in soil moisture	– Change in moisture statuses – Decrease in soil moisture stress – Change in cropping patterns and intensity – Relative change in yields and biomass
Level of soil erosion and soil quality	– Measurement of erosion rate – Sediment deposition rate – Presence, expansion and development of new active erosion features – Determination of soil fertility and nutrients	– Changes in rates of soil loss, sediment deposition and runoff data – Visual assessment and field measurement of rill and gully erosion status (before and after interventions) – Changes in soil fertility and nutrients and soil physical properties using before and after
Productivity of lands	– Assessing change and trend of productivity of lands (arable, grazing, plantation, exclosure) (yield, biomass quantities) – Value added to social, economic and environmental situations	– Relative change in farmland area, biomass and yield (more than, same as or less than the pre-program) – Changes in biomass of other land use types
Vegetation	Determine land use and cover dynamics (before and after interventions)	– Change and trends in vegetation cover – Change in species composition and diversity

the physical environment including the implemented watershed management practices. Formal and informal discussions were arranged to discuss the impacts, challenges, and opportunities for the introduction, implementation, and adoption of watershed management interventions with regard to the nexus approach with 80 farmers and 12 development agents (DAs). The 80 farmers (20 from each watershed) were selected using simple random sampling, whereas all three DAs from each watershed were included in the discussions during the data collection about impacts, opportunities, and challenges of watershed management interventions. The data collected from these discussions were used to triangulate information extracted from the existing literature. The gender proportion of the farmers involved in the discussions was 87% male and 13% female. Of the 12 DAs involved in the discussions, 75% were male and 25% were female. The reference decade for assessing the implemented interventions in most watersheds of the Tigray region by most researchers was the 2000s because this marked the beginning of the implementation of programmes focused on community-based integrated watershed development (Mekonen and Tesfahunegn 2011; Gebregziabher et al. 2016). In addition, published literature, which has used aerial photographs since 1965 and Landsat images from the

1970s to 2012, was reviewed with an assessment of the effects of watershed interventions on land use and land cover dynamics and their impacts on erosion (e.g., Alemayehu et al. 2009; Fenta et al. 2016).

The different watershed management interventions that have been implemented since the 2000s can be categorized in physical and biological (Mekonen and Tesfahunegn 2011; Alemu and Kidane 2014). Physical interventions include: hillside stone terraces, soil bunds, stone bunds and check-dams while biological interventions mean grassed waterways, reforestation, exclosure (protected areas), pits and agronomic practices (compost, hedge cropping). The implemented interventions varied across sites as only physical SWC was dominant in some watersheds, whereas integrated biological and physical SWC were both well established in others. Qualitative and quantitative data were collected and reviewed, then analyzed using descriptive statistics. Outputs were assessed using graphical and statistical indicators (e.g., mean, percentage).

3 Results and Discussion

3.1 Overview of the Impacts of Watershed Interventions in Ethiopia

The implemented SWC measures (physical and biological) as part of watershed management practices in the last three decades were successful in reducing the challenges of land degradation at many sites in the Tigray region (Table 2). In some watersheds, vegetation coverage improved by 85%, the number of new wells increased 7- to 9-fold, and soil loss decreased by 80% compared to the situation before the implementation of the watershed management interventions. The food security of the local community increased by 56% after the watershed management interventions were implemented. An example of such a successful watershed intervention is Abraha-Atsbaha (Gebregziabher et al. 2016). Field observation, discussion with farmers and DAs, and existing reports indicated that the Gerebshelela, Mai-Negus, and Medego watersheds had a less successful response to the integrated watershed interventions implementation. This could be because of the lower appropriateness of the interventions to local conditions, social and cultural factors, and governance issues.

In agreement with results presented here, in the Agula watershed (northern Ethiopia the area under irrigation increased from 7 ha before the intervention to 222.4 ha post-intervention and the area under dense forest also increased from 32.4 to 98 ha (Alemayehu et al. 2009). However, the nature and scale of the impacts of the watershed management interventions vary considerably across the different watersheds in the Tigray region. The factors that contribute to the variability in the extent of success of watershed management interventions are multidimensional, including biophysical (e.g., lithology and upstream–downstream hydrological linkages), and institutional and socioeconomic elements (Gebregziabher et al. 2016). This variability could also be associated with variability in the understanding of the implementation policy (capacity), and commitment of the implementers (e.g., extension staff, researchers, and decision makers) and the level of acceptance of the local people towards the adoption and dissemination of watershed management technologies (World Bank 2008; ATA-MOA 2014).

In addition, field observations and reports have indicated that stone bunds built across the landscape (watershed) retained (deposited) sediment of up to 65 t ha^{-1} year^{-1} (Mekonen and Tesfahunegn 2011), and a sediment depth of 0.20–0.90 m in the bunds on cultivated land and up to 1.5 m in check dams (Fig. 2). Moreover, sheet and rill erosion decreased as much as 68% in the Tigray region (Gebremichael et al. 2005; Mekonen and Tesfahunegn 2011; Fenta et al. 2016). Reduction in soil loss by 80%, increased production area by 20 to 50% as marginal land was rehabilitated and brought back into cultivation, and a threefold increment in crop production was reported for the Abraha-Atsbaha watershed in the Tigray region (Worku and Tripathi 2015; Gebregziabher et al. 2016). Such positive impacts of watershed management interventions in many

Table 2 Indicators of successful achievements of the implemented watershed management interventions in the Tigray region, northern Ethiopia (*Source* Alemayehu et al. 2009; Mekonen and Tesfahunegn 2011, Alemu and Kidane 2014; Gebregziabher et al. 2016; discussion with farmers and DAs in 2017). Note: This information was processed and converted from text format to table by the authors of this paper

Indicator	Intervention-induced change in indicator (%)
Improved farm income	50
Improved food security	56
Reduced risk of crop failure due to moisture stress and climate shocks	30
Improved vegetation restoration and land cover in less successful watershed	40
Improved vegetation restoration and land cover in more successful watershed	85
Average of all the above	*52*

watersheds have substantially contributed to economic development and food security (Fig. 2) compared with before the implementation of the interventions. According to the local farmers' view and aerial photos, before the introduction of interventions, there was no water in the gully and there was barely any fruit and vegetable production in the village (Alemayehu et al. 2009; Mekonen and Tesfahunegn 2011).

The positive impacts of watershed management interventions are interlinked and have effects on the nexus (Fig. 3). For instance, SWC measures (e.g., terraces, trenches, and check dams) reduce soil and nutrient losses and improve water retention, soil carbon sequestration, and agricultural production. Such measures also reduce flood risks and sediment load and thus improve the water availability for irrigation and power generation downstream and thereby agricultural production and overall economic development in both the upstream and downstream parts of a watershed (Personal observation; Alemu and Kidane 2014).

3.2 Opportunities for the Successful Adoption of Watershed Interventions

Previous reports supplemented by information acquired from farmer and DA discussions at the selected watersheds (Abraha-Atsbaha, Gereb-shelela, Mai-Negus, and Medego) have indicated that there are many opportunities for the successful implementation of interventions at watershed level in the Tigray region. The commonly reported opportunities for successful adoption of watershed practices by the farmers can be grouped into: (i) socioeconomic/institutional-enabling environment, (ii) awareness on diversification and intensification, (iii) improved knowledge about the complementarity of mixed crop-livestock systems, and (iv) lessons and experiences acquired from interventions that improved the natural resources of the degraded landscape and thereby enhanced crop production (Gebremichael et al. 2005; Mekonen and Tesfahunegn 2011; Gebregziabher et al. 2016).

3.2.1 Socioeconomic/Institutional Enabling Environment

Ethiopia has pursued the ADLI policy and comprehensive natural resources management strategy for more than two decades. As a result, this has presented an opportunity for farmers and experts to have access to extension services such as training (short and long term), farmer's days, field visits, experience sharing, and application of participatory approaches (Fig. 4a). To support the dissemination and adoption of watershed management interventions to the community in the Tigray region, the number of public universities has expanded from one in the mid-1990s by more than four-fold in 2016, and research

Fig. 2 Field level indicators of impacts of watershed management intervention in Medego watershed, Ethiopia: **a** Terraces filled by soil accumulation; **b** Fruits irrigated using water accumulated behind check-dam; and **c** Vegetables (onion) irrigation. (*Source* Mekonen and Tesfahunegn 2011; Field observation in 2017)

centers from one in the early 2000s by more than ten-fold in 2016. A similar incremental trend for health centers and health extension systems and other sectors such as environmental agent branch offices has been reported in Ethiopia (Dube et al. 2019). Such expansion of public and private institutions has been supported to develop strong links with farmers and has created an opportunity to test, adapt, and disseminate different knowledge, skills, and attitudes with regard to watershed management interventions in the region. The local farmers and DAs have witnessed that farmers' capacity in terms of physical, capital, and mental assets (readiness and willingness) have improved over time. These opportunities also facilitated the adoption of watershed interventions despite the suspicion that there might be some risks (e.g., waterlogging) associated with watershed technologies.

3.2.2 Awareness of Diversification and Intensification

The increased awareness and readiness by farmers to implement diversification and intensification of crops, vegetables, fruits, and livestock (Fig. 4b, c) has been reported as an opportunity for increased uptake of recommended watershed interventions. Diversification and intensification have been practiced under rainfed and irrigated agriculture even though such practices occur most often in areas with irrigation systems. As a result of integrated watershed management interventions, farmer access to water from surface and groundwater sources has increased the irrigated agricultural area. Farmers harvested crops at least twice per year as a result of irrigation practice implementation as compared to a watershed without appropriate interventions (Alemayehu et al.

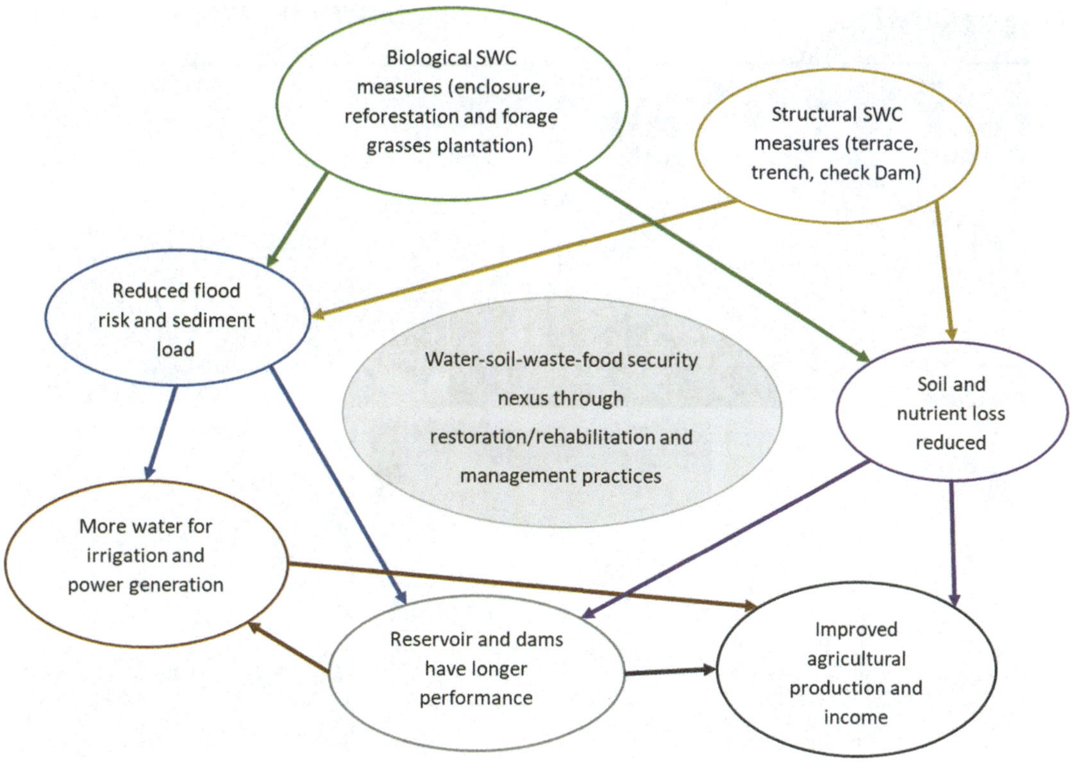

Fig. 3 Multi-dimensional effects of SWC measures on degraded land rehabilitation as related to the Water-soil-waste-food security nexus. (*Source* Modified after Alemu and Kidane 2014)

2009; Mekonen and Tesfahunegn 2011; Gebregziabher et al. 2016; Personal observation).

3.2.3 Improved Knowledge of the Complementarity of Mixed Crop-Livestock Systems

In the investigated watersheds, improvement of farmers' knowledge on the complementarity of the crop-livestock system was reported by the farmers as an opportunity to implement sustainable watershed management interventions. For example, all the farmers involved in the group discussions confirmed that livestock is used as a source of manure (composting), farm power (ploughing), income diversification, and an opportunity for coping with risk. Similarly, crop and livestock complement each other to increase human nutrition. Crop residue is used as a source of livestock feed and for soil fertility improvement, e.g., beans are used as sources of feed and soil fertility, cattle as a source of manure and for ploughing (Fig. 5a, b).

3.2.4 Evidence of the Improvement of Landscape Natural Resources

Long-term watershed interventions implemented on the marginal/degraded landscape, severely degraded gullies, and field borders have improved biomass of trees, bushes, and grasses and water resources availability (field observations e.g., Fig. 5b; Mekonen and Tesfahunegn 2011). According to all the farmers and DAs involved during the group discussions, practical experiences of natural resources improvement on degraded landscape and used for apiculture, and integration of biomass produced from stabilized gully as the source of feed for the dairy farm can be used as a pilot case study to be scaled-up under similar conditions. Surface and groundwater resources have increased substantially and

(a)

(b)

(c)

Fig. 4 **a** Farmers' participation in an experience sharing event about field trials during field tour day, **b** Improvement in crop diversification (rainfed and irrigation) and **c** Improvement in livestock diversification in watersheds in the Tigray region. (*Source* Photos by 1st author, 2012)

farmers have used these resources to provide irrigation. After the improvement of natural resources in the landscape, such as vegetation and water, environmentally friendly interventions, such as apiculture (bee production), are commonly practiced on exclosures and homesteads by farmers and youths to improve their livelihoods. Meaza (2015) and Kerse (2017) reported that income from honey production increased by 300% over three years and incomes from vegetables and spice production have also tripled under watershed management in the Tigray watersheds. The increment in the availability of biomass of grass and trees such as elephant grass, Sesbania (*Sesbania sesban*) and Leucaena (*Leucaena leucocephala*) on the edge of a gully and marginalized land have encouraged farmers to adopt zero grazing while

Fig. 5 **a** Complementarity of the crop-livestock system as feed, soil fertility and ploughing and **b** Degraded landscape natural resources improvement used as the source of feeds for dairy and apiculture in Ethiopia. (*Source* Photos by first author 2014)

practicing fattening (ox, sheep, and goat), and dairy farming (Fig. 5b). In the Abraha-Atsbaha watershed, improvement of fodder availability by 100% has been reported by Gebregziabher et al. (2016). Such successful examples of implemented watershed interventions can present an opportunity for adoption at a larger scale in northern Ethiopia.

3.3 Challenges to the Adoption of Integrated Watershed Interventions in Ethiopia

The discussion with farmers from the four watersheds in the Tigray region and reports from previous studies (e.g., Mekonnen and Fekadu 2015; Meshesha and Tripathi 2015; Gebregziabher et al. 2016) were used to identify the main challenges to the adoption of watershed interventions in the region in the past, present, and future. These were grouped as follows: (i) traditional agronomic and livestock practices and genetically related constraints; (ii) climate-related challenges; (iii) presence of a large degraded land mass (erosion, sedimentation, and soil fertility); and (iv) socio-economic or institutional challenges.

3.3.1 Traditional Practices and Climate-Related Challenges

A major challenge to the adoption of sustainable watershed interventions by farmers is the lack of suitable improved agronomic practices and technologies such as seed, planting equipment, post-harvest and tillage technology, and lack of improved livestock breeds and management practices. Farmers have experienced the rapid deterioration of the potential of the introduced seeds after just two years. Such introduced seeds have been reported by all the farmers in the study watersheds as a low potential yield that fell below that of the local seed after around two years. In addition, the prevalence of pests and diseases, livestock feed shortage, use of crop residue as energy source and animal feed, use of animal dung as an energy source (despite the high cost of inorganic fertilizers), labor shortage when implementing labor intensive SWC, manure transportation to distant fields, and low manure availability were noted as challenges to adoption of watershed management practices in many areas in northern Ethiopia. In line with this notion, previous reports estimated that the burning of dung as fuel instead of using it as a soil conditioner has caused a reduction in grain production by 550,000 t annually (Ethiopia Environmental Protection Authority 2003; Gashie 2005).

Poor rainfed and irrigation water management coupled with overgrazing instead of zero grazing (Fig. 6a), diseases and parasites that cause high animal mortality, lack of sufficiently well-managed pasture land in terms of nutritional quality and quantity (Fig. 6b), and drinking water shortage during the dry season challenged the wider dissemination of technologies to address the nexus approach. Livestock energy wastage (animals walking long distances), the land degradation problem (crusting and erosion) due to animal trampling (Fig. 6b), high chances of disease spread as large numbers of animals use the same water points (Fig. 6d), and unreliable rainfall (frequent drought) were also identified as impediments to the adoption of the best watershed interventions in northern Ethiopia. Uncontrolled and open grazing on communal land is a common practice in Ethiopia, even on sloppy lands and this increases soil erosion relative to the effects of the restoration measures. Moreover, farmers may be reluctant to implement SWC measures on communal land because they are inclined towards short-term benefits such as livestock feed, timber, and fuel wood sources (Mekonnen and Fekadu 2015).

3.3.2 Area of Degraded Land Versus Managed Watershed

The existence of a large area of communal marginal land and the associated problem of sedimentation of water bodies including man-made reservoirs and the slow process of rehabilitation (Fig. 7a–f) challenges the adoption of sustainable watershed interventions in Ethiopia. Degraded marginal land needs huge conservation efforts in terms of labor and materials (Mekonnen and Fekadu 2015). Tackling this challenge demands considerable harmonized efforts (technical, financial, and policy strategies for resource mobilization) targeted to the extent of the problem to be addressed in each watershed. Degraded land can be used as an opportunity to adopt sustainable watershed management practices; however, the degraded land area has strong connectivity with the landscape under intervention, and remediation can demand considerable labor and time. The linkages between landscapes under intervention and those without intervention discourage the hope and vision of local people as they learn that landscape interventions demand considerable labor and time and have no immediate or short-term benefits. Such limitations slow the uptake of watershed management practices and dissemination by farmers in northern Ethiopia.

3.3.3 Socioeconomic and Institutional Challenges

Socioeconomic factors such as the lack of market information, agricultural processing technologies, capital, poor infrastructure, and gaps in the research–extension–farmer linkages have challenged the wider adoption of successful sustainable watershed management interventions. Farmers lack interest in long-term SWC investments and prefer interventions and watershed technologies that offer benefits over a short period of time. While farmers have been implementing SWC practices on the ground through a campaign (local people mobilization), the standard design cannot be implemented directly for every slope, soil type, and orientation of the landscape. In addition, financial constraints for frequent technical supervision, monitoring, evaluation, and discussions of feedback activities are restricted, which partially challenges the quality and sustainability of work (Mekonnen and Fekadu 2015). There is no sufficiently well institutionalized system to regulate the issue of timely maintenance of the implemented SWC structures because the development agents tasked with coordinating the practices focus more on the seedling plantation and constructing new physical structures than on maintaining the existing structures.

In addition, resources such as land, water, agricultural input, and labor competition between crops, livestock, and natural resources in a landscape (e.g., afforestation, exclosure)

Fig. 6 Different practices in the watershed: **a** Overgrazed land; **b** Protected pasture land; **c** Degradation by cattle trampling; **d** Cattle use the same water source in Ethiopia. (*Source* Photos by 1st author 2011)

represent bottlenecks to the successful adoption of watershed management interventions. For example, competition for water by both crop and livestock is crucial as off-season production of crops and forage requires water for irrigation. Similarly, land for crop and forage production, land for grazing and exclosure, and labor to manage the fields and livestock are all highly competitive (Fig. 8a, b).

The majority of the participants in the group discussions (93%) identified key constraints for vegetative SWC measures' establishment at the watershed-scale. These include high seedling transportation cost, poor handling while transporting and planting, high livestock interferences, free grazing, and moisture stress. Local people understand that uncontrolled livestock movement coupled with moisture stress drastically hinders proper growth and survival of planted seedlings. Measures to improve such constraints have already been attempted and include nursery establishment for seedling production in the nearby plantation sites, selection of appropriate multi-purpose trees, introducing area enclosures, timely or early fencing and planting, and practicing SWC. Moreover, previous studies (e.g., MoAR 2010; Mekonnen and Fekadu 2015) confirmed that reducing constraints for the

Fig. 7 Landscape under severe land degradation due to water erosion in Tigray region,: **a** Bare land with shallow soil depth; **b** Root exposure; **c** Rocky landscape; **d** Gully expansion; **e** Connectivity to reservoir sediment deposition; and **f** dam outlet sedimentation (siltation). (*Source* Photos by 1st author 2010)

Fig. 8 Water and land resources competition between crop and livestock farming in Ethiopia: **a** Water and labor for the crop, vegetables and fruits irrigation; and **b** Water and labor for livestock farming (forage plant, fattening, dairy farm). (*Source* Photos by 1st author 2014)

introduction and adoption of vegetative measures can improve the performance of physical SWC measures to reduce runoff and soil loss, improving soil fertility and thereby agricultural productivity.

3.4 Suggested Research Strategy to Reduce the Challenges for Adoption of Interventions in Ethiopia

To address the outlined challenges and increase the chance of successful implementation of identified management options, we propose a generalized research strategy (Fig. 9) to address the nexus approach at the watershed scale. The strategy has four stages, even though it is open for improvement using inputs coming from different angles. These stages are as follows: (i) Stage 1: baseline data collection, (ii) Stage 2: decision-making (selection of suitable interventions through participatory approach), (iii) Stage 3: implementation and monitoring, and (iv) Stage 4: evaluation. The overall goal of the strategy described in Fig. 9 is similar to that of Manschatz et al. (2015) and Benavides et al. (2019) as all three approaches demand baseline information, participation of stakeholders, possible site-specific management solutions, an evaluation stage, and project sustainability.

Stage 1: Baseline data Benavides et al. (2019) reported that establishing baseline information is crucial for the introduction of scientifically sound sustainable interventions. They also stated that collecting baseline data is a key practice in many environmental fields, as it is impossible to carry out "before" and "after" comparisons and that it is difficult to engage stakeholders without baseline data. It is essential to increase the number of monitored watersheds, producing baseline data such as socioeconomic, institutional, farm attributes, and biophysical information. Communication of baseline data with all stakeholders to create a common understanding is also important in further steps. Both quantitative and qualitative data should be collected from primary and secondary data sources. These baseline data will be used to identify the causes and extent of land degradation and its indicators (erosion, sedimentation, soil fertility, vegetation, and agricultural production) using the following two approaches. The first approach is the extraction of the knowledge of local communities (farmers) using participatory tools (transect walks, group discussions, and interviews), and the second approach is the application of scientific (expert) knowledge (soil sampling, modelling). For this reason, this first stage is also known as the *problem identification stage*. To facilitate scaling-up and out of the practicality of local knowledge, a comparison of the locally derived data with the scientific approach is necessary using statistical indicators. An overview of Stage 1 is shown at the top of Fig. 9. The information obtained at Stage 1 will be used to ensure all stakeholders are participants in the project and to provide a basis for comparison after the completion or exit of the project.

Stage 2: Decision-making stage. On the basis of the results in Stage 1, different alternative solutions that improve the degraded condition of a watershed (to address the nexus problem) and thereby the livelihoods of the community will be discussed with all the stakeholders. This will be used as an input for scenario development (identification of the optimal interventions). The scenarios developed will be analyzed and the results interpreted at this stage based on both local knowledge and scientific perspectives. Finally, the best scenario considering both socioeconomic and ecological factors will be selected. At this stage, all stakeholders should agree on the resources available and their mobilization for the implementation of the selected scenario as the best practice at the appropriate site. At this stage, it is necessary to initiate and motivate watershed communities to create awareness and develop a sense of ownership by the farming community. This can be achieved through (i) discussion with communities to reach consensus on key issues and resources in line with the nexus approach, (ii) organizing

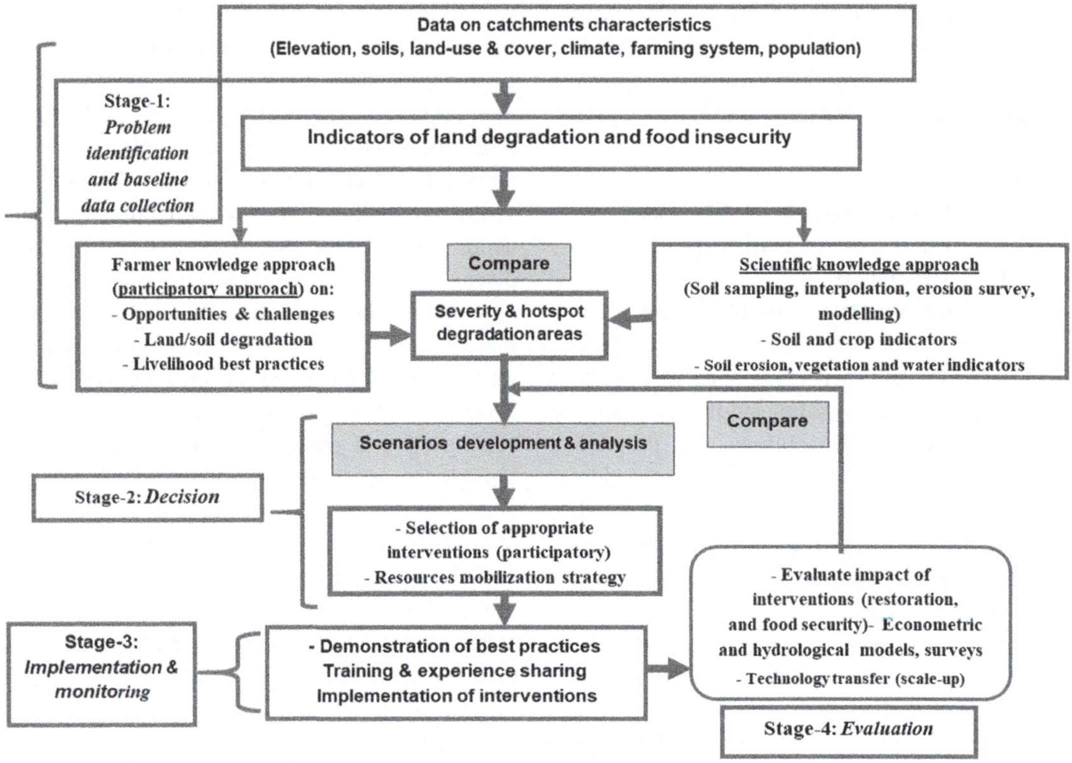

Fig. 9 Conceptual research strategy to support the successful adoption of watershed management interventions to achieve the nexus approach at watershed-scale

communities into working groups to coordinate planning and implementation of possible technologies, and (iii) formulation of community bylaws (regulations) with clear enforcement guidelines starting from implementation to post implementation (Field observations; Mekonnen and Fekadu 2015).

Stage 3: Implementation and monitoring. At this stage, intervention(s) selected as the best scenario in Stage 2 will be implemented. However, a demonstration of the best practices or interventions will be carried out to check if further technical verification is required to learn more about its implementation and the possible benefits. Training, experience sharing, field days, and tours will be used to improve the capacity of the community and extension workers at this stage. Monitoring during implementation includes checking of intervention application according to the scientific specification or recommendation at the desired site and time. Immediate post-implementation monitoring is also included in this stage.

Stage 4: Evaluation. At this stage, evaluation of the impacts of the interventions will be executed. Data collected at this stage will also be compared with Stage 1 (baseline data). Lessons from this stage are important to provide feedback for up-scaling of best practices. Different evaluation approaches and criteria from both local community knowledge and understanding, and from scientific views will be used. Promising interventions will be identified and discussed with the community on how and where to disseminate to address the nexus problem. Qualitative and quantitative indicators and models can be used to evaluate the performance of introduced watershed interventions.

4 Conclusions

Integrated watershed management interventions such as SWC were designed as a strategy in addressing the objectives of the nexus approach (water–soil–waste–food security) through increasing the ground and surface water availability, increasing vegetative biomass, reducing soil erosion and flooding, and improving soil fertility. Such impacts of watershed management implemented for more than 2–3 decades have contributed to the reduction of the water and food insecurity in the Tigray region. This study confirms that watershed management interventions have a key role in improving the land cover of watersheds, decreasing soil erosion, increasing soil moisture, reducing sedimentation and runoff, stabilization of gullies and river banks, rehabilitation of degraded land, and increased recharge in the subsurface water, thereby contributing to poverty alleviation and sustainable livelihoods. However, the results of previous projects have revealed some variability in the extent of success of watershed management interventions, which suggests that there are differences in the motive and capacities of the implementers and community. Even more, reports have revealed that watershed management practices have not, or only to a limited extent been adopted in parts of Ethiopia, which demands further investigation to understand and identify opportunities, root causes, and specific challenges. As part of a solution to improve the chance of success for introduced technologies, the research strategy proposed in this paper suggests that we should combine efforts of researchers (multiple disciplines), decision makers, farmers and extension workers, building on their knowledge, skill, and motivations when designing future approaches. The proposed strategy, in a participatory approach, includes baseline data collection intervention selection, problem identification, designing (planning), development of technology (technology testing), learning, implementation, monitoring and evaluation, and providing feedback. Governance should continue providing technical (access to training, experience sharing, awareness creation forums, information) and financial (credit services, incentives) support to the local community to take actions that sustain natural resource management and thereby improve agricultural production in northern Ethiopia.

Acknowledgements The authors are grateful to UNU-FLORES for the financial support for the first author to present this paper in the Dresden Nexus Conference (17–19 May 2017) in Germany. The support from UNU-INRA and Aksum University (Ethiopia) during fieldwork, development, and organization of this paper is also highly appreciated.

References

Alemayehu F, Taha N, Nyssen J, Girma A, Zenebe A, Behailu M, Deckers J, Poesen J (2009) The impacts of watershed management on land use and land cover dynamics in Eastern Tigray (Ethiopia). Resour Conserv Recycl 53:192–198

Alemu B, Kidane D (2014) The implication of integrated watershed management for rehabilitation of degraded lands: case study of Ethiopian highlands. J Agric Biodivers Res 3(6):78–90

ATA-MOA (Ethiopian Agricultural Transformation Agency (ATA)—Federal Democratic Republic of Ethiopia Ministry of Agriculture (MOA)) (2014) National Strategy for Ethiopia's agricultural extension system- vision, systemic bottlenecks and priority interventions

Behailu M (2002) Assessment and optimisation of traditional irrigation of vertisols in northern Ethiopia: a case study at Gumselasa micro-dam using maize as an indicator crop. Unpublished Ph.D. Thesis, Gent, Belgium

Benavides L, Avellán T, Caucci S, Hahn A, Kirschke S (2019) Müller A (2019) Assessing sustainability of wastewater management systems in a multi-scalar, transdisciplinary manner in Latin America. Water 11 (2):249. https://doi.org/10.3390/w11020249

Bishaw B (2001) Deforestation and land degradation on the Ethiopian highlands: a strategy for physical recovery. Paper presented at International Conference on Contemporary Development Issues in Ethiopia, Kalamazoo, Michigan, U.S.A. https://www.fao.org/

forestry/5856-05a0e08cfda318d93eb26e620f87190b.pdf. Accessed 28 Mar 2017
Dube AK, Fawole WO, Govindasamy R, Ozkan B (2019) Agricultural development led industrialization in Ethiopia: structural break analysis. Int J Agric, For Life Sci 3(1):193–201
Ethiopia Environmental Protection Authority. Ethiopian Environmental Report (2003) Commercial Printing Press. Addis Ababa, Ethiopia
Food and Agriculture Organization of the United Nations, FAO (1986) Ethiopian highland reclamation study. Final report, vol 1–2 (1986). Food and Agriculture Organization of the United Nations, Rome
FAO (2014) Walking the Nexus Talk: assessing the water-energy-food nexus in the context of the sustainable energy for all initiative. FAO environment and natural resources working paper # 58. https://www.fao.org/3/a-i3959e.pdf
Fantaw A (2007) An overview of salt affected soils and their management in Ethiopia. Paper presented in the third international workshop on water management (Water Man) project, 19–21 Sep 2007, Haramaya University, Haramaya, Ethiopia. https://www.wau.boku.ac.at/fileadmin/data/H03000/H81000/H81500/waterman/Workshop_III_Haramaya.pdf. Accessed 27 Jun 2017
Fenta AA, Yasuda H, Shimizu K, Haregeweyn N, Negussie A (2016) Dynamics of soil erosion as influenced by watershed management practices: a Case study of the Agula watershed in the semi-arid highlands of northern Ethiopia". Environ Manage 58 (5):889–905
Gashie W (2005) Factors controlling households energy use: implication for the conservation of the environment. Master Thesis, Addis Ababa University, Ethiopia
Gebregziabher G, Abera DA., Gebresamuel G, Giordano M, Langan S (2016) An assessment of integrated watershed management in Ethiopia. Colombo, Sri Lanka: International Water Management Institute (IWMI). IWMI Working Paper 170, p 28. https://doi.org/10.5337/2016.214. Accessed 23 Feb 2017
Gebremichael D, Nyssen J, Poesen J, Deckers J, Haile M, Govers G, Moeyersons J (2005) Effectiveness of stone bunds on controlling soil erosion on cropland in the Tigray highlands, northern Ethiopia. Soil Use Manage 21(3):287–297
Harry V (2013) The politics of African energy development: Ethiopia's hydro-agricultural state-building strategy and clashing paradigms of water security. Philos Trans Roy Soc A-Math Phys Eng Sci 371 (2002):20120411. https://doi.org/10.1098/rsta.2012.0411
Hülsmann S, Ardakanian R (2018) The nexus approach as tool for achieving SDGs: trends and needs. In: Hülsmann S, Ardakanian R (eds) Managing water, soil and waste resources to achieve sustainable development goals, pp 1–9. https://doi.org/10.1007/978-3-319-75163-4_1
Johnson OW, Karlberg L (2017) Co-exploring the water–energy–food nexus: facilitating dialogue through participatory scenario building. Front Environ Sci 5. https://doi.org/10.3389/fenvs.2017.00024
Karlberg L, Hoff H, Amsalu T, Andersson K, Binnington T, Flores-Lopez F, de Bruin A, Gebrehiwot SG, Gedif B, zur Heide F, Johnson O, Osbeck M, Young C (2014) Applying the nexus—meeting Ethiopia's development goals by addressing links between water, energy and food. SEI Policy Brief. Stockholm Environment Institute, Stockholm
Karlberg L, Hoff H, Amsalu T, Andersson K, Binnington T, Flores-Lopez F, de Bruin A, Gebrehiwot SG, Gedif B, zur Heide F, Johnson O, Osbeck M, Young C, (2015) Tackling complexity: understanding the food–energy–environment nexus in Ethiopia's Lake Tana Sub-Basin. Water Altern-An Interdisc J Water Polit Dev 8(1):710–734
Kerr J (2002) Watershed development, environmental services, and poverty alleviation in India. World Dev 30(8):1387–1400
Kerse BL (2017) A review paper on: the multi-functional implication of integrated watershed management: the new approach to degraded land rehabilitation in Ethiopia. J Biol, Agric Healthc 7(17):21–30
Lakew D, Carucci V, Asrat W, Yitayew A (eds) (2005) Community based participatory watershed development. Ministry of Agriculture and Rural Development, Addis Ababa, Ethiopia
Mannschatz T, Buchroithner MF, Hülsmann S (2015) Visualization of water services in africa: data applications for nexus governance. In: Kurian M, Ardakanian R (eds) Governing the nexus. Springer International Publishing, pp 189–217. https://doi.org/10.1007/978-3-319-05747-79
Meaza H (2015) The role of community based watershed management for climate change adaptation in Adwa, Central Tigray Zone, Ethiopia. Int J Weather, Clim Change Conserv Res 1(1):11–35
Mekonen K, Tesfahunegn GB (2011) Impact assessment of soil and water conservation measures at Medego watershed in Tigray, northern Ethiopia. Maejo Int J Sci Technol 5(03):312–330
Mekonnen GT, Fekadu A (2015) Experiences and challenges of integrated watershed management in Central Zone of Southern Ethiopia. Int J Curr Res 7 (10):20973–20979
Meshesha TW, Tripathi SK (2015) An Evaluation of watershed management practice in Ethiopia: a preliminary review. Int J Environ Sci 4(1):24–30
Ministry of Agriculture and Rural Development (MoAR) (2010) Sustainable land management technologies and approaches in Ethiopia: published by the Sustainable Land Management Project (SLMP), Natural resources management sector, Ministry of Agriculture and Rural Development of the Federal Democratic Republic of Ethiopia
Stockholm Environment Institute (SEI) (2014) Cross-sectoral integration in the sustainable development

goals: a nexus approach. Discussion Brief. https://www.idaea.csic.es/medspring/sites/default/files/Cross-sectoral-integration-in-the-sustainable-development-goals-a-nexus-approach.pdf. Accessed 9 Jul 2018

Solomon N, Hishe H, Annang T, Pabi O, Asante IK, Birhane E (2018) Forest cover change, key drivers and community perception in WujigMahgo Waren Forest of northern Ethiopia. Land 7(32):1–16

Stoorvogel JJ, Smaling EMA (1990) Assessment of soil nutrient depletion in Sub-Saharan Africa, 1983–2000. Report 28. DLO Winand Staring Center for Integrated Land, Soil and Water Research, Wageningen, the Netherlands.

Tamene L (2005) Reservoir siltation in the drylands of northern Ethiopia: causes, source areas and management options. Ph.D. Thesis, University of Bonn, Germany

Tesfahunegn GB (2011) Soil erosion modeling and soil quality evaluation for catchment management strategies in northern Ethiopia. Ecology and Development Series No. 8, Center for Development Research, University of Bonn, Germany, (ISBN 978-3-940124-35-7, ISSN: 1864-0443)

Tesfahunegn GB, Vlek PLG (2013) Assessing sediment-nutrient export rate and soil degradation in mai-negus catchment, Northern Ethiopia. ISRN Soil Science, Article ID 748561. https://doi.org/10.1155/2013/748561

United Nations, (UN) (2015) Transforming our world: the 2030 agenda for sustainable development. Resolution adopted by the General Assembly on 25 (Sept) A/RES/70/1. United Nations General Assembly, New York

Wang X, Dong S, Yang B, Li Y, Su X (2014) The effects of grassland degradation on plant diversity, primary productivity and soil fertility in the alpine region of Asia's headwaters. Environ Monit Assess 186:6903–6917

World Bank (2008) Project Appraisal Document on a Proposed Grant in the Amount of US$20 million and Proposed Grant from the Global Environment Facility Trust Fund in the Amount of US$9 Million to the Federal Democratic Republic of Ethiopia for Sustainable Land Management Project. Environmental & Natural Resources Management Sustainable Development Department AFTSN Africa Region, The Document of the World Bank, Report No: 42927-ET

Worku T, Tripathi SK (2015) Watershed management in highlands of Ethiopia: a review. Open Access Libr J 02:e1481. https://doi.org/10.4236/oalib.1101481

The Potential Contribution of Cultural Ecological Knowledge to Resources Management in a Volcanic River Basin

Vicky Ariyanti, Peter Scholten, and Jurian Edelenbos

Abstract

This study examined the question "What are the main elements of water-related cultural ecological knowledge (CEK) in a volcanic river basin?" The CEK is defined as a body of knowledge, both written and orally transmitted, produced by many generations' observation on the human-nature relationship. This study illustrates that the water-related CEK is interrelated with other resources, the volcano and the lahar (debris flow), in the hydrological cycle. This research applied a qualitative methodology for the case study of the Opak sub-basin at Mt. Merapi volcano, within the unique setting of Yogyakarta, the only ruling kingdom in the Republic of Indonesia. Previous research on traditional ecological knowledge and indigenous technical knowledge provided the conceptual framework. The results provide the main elements of CEK and its formulation patterns. The main elements were historical experiences, legitimatization process, and mechanism of transfer. Although the formulation process stemmed from historical experiences of natural conditions and how humans managed the environment, a legitimatization process is necessary before it is formed as mechanisms of transfer for the future generation. These mechanisms were divided into the following five forms: (1) philosophy and values as the core and more practical forms such as (2) internalization (rituals and myth), (3) resource management practices, (4) artifacts, and (5) five-senses wisdom.

Keywords

Water management · Socio-ecological system · Integrated resources management · Mechanisms of transfer

1 Introduction

In previous decades, water resources were mainly managed using engineering approaches (Pahl-Wostl et al. 2008; Snellen and Schrevel 2004) because engineering offered the rewarding benefit of controlling water, when and where it is desired, and disposing of it after use. However, after some decades or even centuries of practicing intensive engineering, ecology has been shown to suffer substantially (Pahl-Wostl et al. 2008; Savenije and van der Zaag 2008). This issue supported the rise of "integrated water

V. Ariyanti (✉) · P. Scholten · J. Edelenbos
IHS, Institute for Housing and Urban Development Studies, Erasmus University Rotterdam, Rotterdam, The Netherlands
e-mail: ariyanti@ihs.nl

P. Scholten
e-mail: scholten@ihs.nl

J. Edelenbos
e-mail: edelenbos@essb.eur.nl

S. Hülsmann and M. Jampani (eds.), *A Nexus Approach for Sustainable Development*,
https://doi.org/10.1007/978-3-030-57530-4_12

resources management" (IWRM), which was supposed to manage water more holistically (von Hofwegen and Jaspers 1999; Biswas 2004). The IWRM concept addresses all related resources connected in the hydrological cycle (i.e., water, land, and geology). The IWRM thus contributes to the objectives of the sustainable development goals (SDGs) (United Nations 2015). Therefore, this study highlights SDG #6 (ensure availability and sustainable management of water and sanitation for all), which is essential, especially during a disaster; SDG #15 (protect, restore, and promote sustainable use of terrestrial ecosystems, sustainably manage forests, combat desertification, and reverse land degradation and biodiversity loss), which applies also to volcanic river basins; and SDG #11 (make cities and human settlements inclusive, safe, resilient, and sustainable), as the case study is located in a highly populated urban setting.

The IWRM concept uses the river basin as the unit of management (Graefe 2011; Burton 1995). However, when the Indonesian policy adopted the IWRM, it used river basin territory (WS) as the unit of management (Government of Indonesia, Inter-Agency Task Force on Water Sector Policy Reform 1999). The WS may consist of several catchments in one management unit. It is a variation of the river basin concept, used to acknowledge the existence of catchments, and also to recognize the earlier unit of management in water projects, such as dam projects, irrigation areas, and river projects. This change was the result of the reformation of the water sector by the World Bank in 2000 following the Indonesian financial crises in 1997–1999 and was conducted through the Water Sector Adjustment Loan (WATSAL). However, the implementation of the Indonesian IWRM approach for this case study is complicated by the volcanic river basin context. Thus, a definition of a volcanic river basin was developed as a water catchment with the geo-morphological condition (Graefe 2011; Burton 1995) of an active volcano as its origin and its termination at sea. Within this context, the lahar, as a hybrid product of volcanic materials and rainwater (De Bélizal et al. 2013; Pierson and Scott 1985), can be considered one type of interrelation in managing water resources and the volcano in this river basin.

This research investigates the water resource management using a cultural ecological knowledge (CEK) lens focusing on the volcanic river basin setting. This CEK lens is derived from earlier research on traditional ecological knowledge (Berkes et al. 2000) and indigenous technical knowledge (Howes and Chambers 1980; Agrawal 1995). The working definition of CEK can be summarized as the collective body of knowledge and beliefs passed through generations, which explain the human-nature relationships. Since CEK usage is not considered in the current Indonesian water policy, studying its impact on resources management should be relevant to both policymakers and academicians.

There have been various studies addressing the cultural vulnerability of communities in the Mt. Merapi area, Yogyakarta, Indonesia (Donovan et al. 2012), including the perception of local knowledge on volcanic eruption (Dove 2007; Schlehe 1996), historical findings of temples by the rivers buried under centuries of lahar material (Degroot 2009), and geological relevance of the philosophical axis, an imaginary axis connecting the volcano and the sea, which explains the geological activities′ interplay between the volcanic eruptions and tectonic earthquakes (Troll et al. 2015). In addition, research on other volcanic regions in the world has revealed that culture holds an important role in community resiliency (Cashman and Cronin 2008; Chester 2005; Paton et al. 2008). However, there is little research addressing the interrelations among water, lahar, volcano, and local cultural settings. The IWRM approach also seldom explicitly addresses the conjunctions between social conditions, culture, and water resources (Nikolakis et al. 2015; Ross and Pickering 2002).

The first aim of this study was to identify the CEK elements in water resource management in a volcanic river basin, using the case of the Opak sub-basin. Secondly, it aimed to explain the formulation process of CEK, as it is important to know, which knowledge persists through the long process and why. This formulation process is a general approach, which can be used to grasp CEK in other locations. This understanding of CEK could contribute to improving the integrated resource management in this basin.

2 CEK Conceptual Framework

The concept of the socio-ecological system (Folke 2006) is important in CEK definition as it provides an introduction into the complex interrelations among humans and nature, where culture as a product of human knowledge impacts the use of, adaptations to, and perceptions of the environment (Steward 1972). The CEK definition used in this paper is based on earlier studies on similar terms, such as traditional ecological knowledge (Berkes et al. 2000), indigenous knowledge (Agrawal 1995), local knowledge (Geertz 2000), and indigenous technical knowledge (Howes and Chambers 1980). However, these terms mostly refer to oral transmitted knowledge, and there is a chance that written knowledge rooted in a culture may be available. Thus, the term "cultural" is used rather than "traditional." Based on these earlier works, the CEK can be recognized through three main characteristics:

(1) Historical experiences in the generation of knowledge, utilizing indicators in response adaptation, conservation or protection, and managing complex natural dynamics (Berkes et al. 2000). These indicators are used to obtain a more in-depth understanding of empirical experiences through the history of managing environmental resources. Therefore, our research began with asking the experiences of local people in managing water resources in this volcanic river basin.
(2) The mechanisms of transfer in current society (Berkes et al. 2000) were the second part of the assessment, aiming to determine the forms of CEK in the current society. There are several possible forms, such as the institutionalized structures and dynamics of the community and cultural internalization; myths, rituals, worldviews, philosophies, resource management practices, and cultural values; and taboo, regulation, and ethics.
(3) The existence of "five-senses wisdom" in the human-nature relationship, which is divided into substantive and methodological subfeatures based on the human five senses (visual, touch, hearing, taste, and smell) (Agrawal 1995; Howes and Chambers 1980). This approach involves using these indicators to explain natural conditions and how people should react to such conditions.

The relations of the CEK characteristics are hypothetically arranged as shown in Fig. 1. The CEK is based on historical experiences, developed into various mechanisms of transfer. Initially, the CEK conceptual framework was depicted as a linear sequence (Fig. 1), but based on the study, results were conceptualized in a more dynamical and cyclical setting (see Fig. 3).

As for the IWRM approach implemented in this volcanic river basin, the CEK relates to any of the following four dimensions of IWRM (Savenije and Zaag 2008): (1) natural: types of water resources (Graefe 2011; Savenije and Zaag 2008); (2) spatial: characteristics of up-, mid-, and downstream river basin and their interrelations (Molle 2009; von Hofwegen and Jaspers 1999); (3) temporal: seasonal and long-term patterns (Savenije and van der Zaag 2008); and (4) human: actors in water governance multilevel interrelationships (Gupta and Pahl-Wostl 2013; Hooghe and Marks 2001; Jaspers 2003). In addition, within a volcanic river basin, the lahar (volcanic debris flow), a water-related hazard, is also addressed. Problematically, most of the literature on lahars refers to them only as a risk, and not as a resource (Bignami et al. 2012). Therefore, here we consider lahar management along the four dimensions of IWRM as they relate to CEK.

Representing an arguably even more holistic approach than IWRM, the nexus approach comes in various different varieties, for example, water–energy–food, water–soil–waste, water–energy, and water–environment (Lal 2015; Kurian and Ardakanian 2015; Rasul and Sharma 2016), which all aim toward the sustainability and resiliency of human-environmental systems. The approach is in line with the socio-ecological

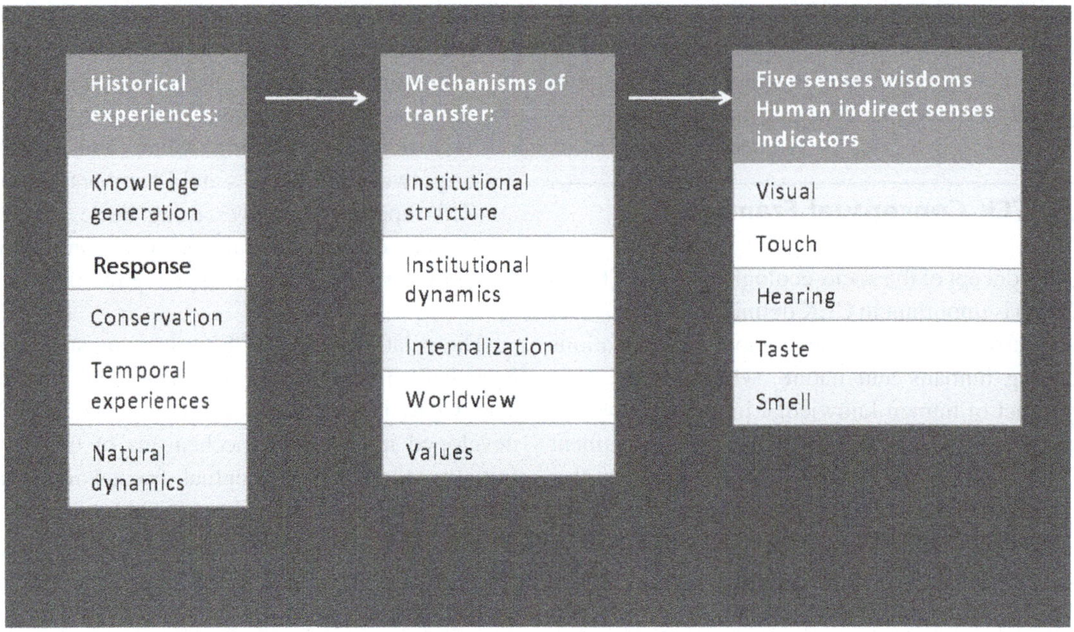

Fig. 1 Initial conceptual frameworks for CEK

system concept (Folke 2006) as the departure point of CEK. Within the current research, the nexus for a volcanic river basin management was defined as the interconnectedness of the water–lahar–volcano, located in each of the dimensions of IWRM and lahar management explained above. For the spatial dimension, the basin connects to the volcano in the upstream part of the river, the urban area in the midstream, and the sea downstream (Zalewski 2013; Manga 1997). The basin also has temporal dimensions of the cycle of underground–surface water, where the volcano acts as a groundwater reservoir (Delcamp et al. 2016) and the cycle of volcano activities (Newhall et al. 2000). The basin also has specific natural dimensions, including the disturbance of the discharge in rivers during and after the eruption (Iles and Hegerl 2015), the rain-triggered lahar flow in the rivers (Rodolfo 1991; Leung et al. 2003), and the temperature and chemical change in water quality during volcanic activities (Ohsawa et al. 2010). Another complicating factor is that these interrelated dimensions are managed by different organizations. By better understanding the CEK, the underlying factors of water management practices can be improved, especially since the CEK describes the human-nature dynamics of a holistic system. The results can be used as a reference for other researches embarking on similar inquiries and the applications of integrated resources management approaches.

3 Methodology

This research used the Opak sub-basin, Mt. Merapi, case study as a research strategy. This volcano is a regular "cougher," with an eruption cycle of once every four years. However, the 2010 eruption was the result of a 100-year cycle (Jousset et al. 2012). Thus, the 2010 eruption produced more lahar due to the 160 million m^3 of volcanic materials (Newhall et al. 2000). The location is also the cultural capital of the Javanese, with the Yogyakarta Kingdom as a Special Region under the Republic of Indonesia. This condition would mean a higher possibility of survival of both tangible and intangible cultural heritage (CEK).

The units of analysis used were communities and organizations. To obtain balanced inputs, respondents included locals, decision makers, and experts that understand the CEK in this volcanic river basin. The three sub-cases as representative of communities were selected based on the post-2010 eruption and worst lahar damage report produced by the Serayu Opak River Basin Organization (BBWS SO 2014) and Disaster Management Agency of Yogyakarta Special Region (BPBD DIY). The selection criteria for the organizations represented in this research were based on the administrative boundary, activities of the organizations, and involvement in managing the volcanic river basin, which originated from national, regional, and municipal governments.

The research used qualitative approaches, inspired by ethnographic intentions (Emerson et al. 2011; Falzon 2012; Crang and Cook 2007). The primary data were gathered through 2 months of fieldwork in summer 2016, with observations of daily activities and 47 in-depth interviews using the snowballing sampling method (Flick 2008) for key persons (hamlet chiefs, community leaders, and wise persons) of the three villages, experts on water and culture, and representatives of organizations. Each interview lasted about one to two hours in Bahasa (national language) and Javanese (the first author is a native speaker of both languages). The observations (photographs, videos, and field notes) were taken for at least six hours per day in each village focusing on daily activities during a two-month period. All of the interviews were recorded, either in audio or video forms. After the rounds of interviews in each village, at the end of the observation period, a focus group was gathered, involving earlier respondents. The interviews used open-ended questions focusing on the operationalization of the CEK main elements: historical experiences, mechanism of transfer, and five-senses wisdom. The questions asked in the interviews with the focus group were as follows:

1. What kind of water-related extreme conditions occurs in this volcanic river basin and when? Hence, the historical experience indicator is used.
2. What is the form of the communities' institutionalized structures and what are the dynamics? The answers help to group type of communities and how they organize CEK.
3. What forms of CEK in water–lahar–volcano management are used and how? The forms are also known as the mechanisms of transfer.
4. What kind of local five-sense wisdom related to water–lahar–volcano management is used and how? It is answered by recognizing five-senses usage in managing water-lahar-and volcano.
5. What are the relations among historical experiences, the mechanism of transfer, and five-senses wisdom? The answer should wrap this paper, as it stands to relate the relations between the indicators of CEK.

In addition, as secondary data, this study used ancient literary from the Mataram Kingdom (*Babads* and *Serats*) and available works by several scientific disciplines (volcanologists, sociologists, anthropologists, and archeologists), policy briefs, reports, and planning documents. These secondary data were used to enhance and support the empirical findings and as an entry point to identify the CEK indicators.

After the interviews, we performed a coding process using Atlas.ti software (Friese 2019; Muhr 1991) to the dataset derived from the transcribed interviews. The software enables transparency and flexibility in the coding process, and also theory building from the dataset. The dataset was analyzed using pre-determined codes from the initial conceptual framework's variables and indicators, which later developed into axial codes. Later on, we developed patterns of interrelations between CEK and in the river basin management. These interrelations were found when two or three codes were used in the same quote derived from the interview transcriptions. This analysis aimed at answering the questions above, by means of interpreting the quotes from the interview transcriptions based on interrelations derived from connections between the codes.

4 Case Descriptions

The volcanic regions of the world consistently face hazards related to the volcano eruption cycle patterns. Indonesia has 132 active volcanoes, with the stratovolcano Mt. Merapi known as one of the most active in the world (Thouret et al. 2000). It is located 30 km north of Yogyakarta, belonging to a cross-island chain of four stratovolcanoes comprising Mt. Ungaran, Mt. Telemoyo, Mt. Merbabu, and Mt. Merapi. It has killed more than 3,000–5,000 people since 1672 (Camus et al. 2000) and many more people may have been killed within the recorded 10,000-year history of Mt. Merapi (Newhall et al. 2000). The volcano is known for its semi-continuous outpours of lava and active building of the summit dome, with its periodic dome collapse and destruction (Gertisser et al. 2012). Mt. Merapi plays an essential role in the living conditions of the Opak sub-basin.

The whole volcanic river basin is represented in two governance levels: regional level and village level (municipal depiction). The regional level includes the whole Opak sub-basin, with respondents of experts and from organizations and a total of 17 respondents. The experts were selected based on their involvement in the water management and the culture, for example, an ex-head of an agency, a prince, a water expert, and a humanist. Respondents from organizations were selected based on their capacity as the person in charge and more than five years of tenure position. The organizations were clustered into (1) national government in regional level: BBWS SO (Serayu Opak RBO), BPDAS SOP (Serayu Opak Progo Watersheds Management Authority), and BPPTKG (Center of Volcanology); (2) regional government: PU-P ESDM DIY (Public Works, Energy and Natural Resources Agency) and BPBD DIY (Disaster Management Agency); and (3) municipal government: DSDA (Water Resources Agencies) and village level governments. For the village level, the sub-cases chosen were based on the severity of lahar impacts of post-2010 eruption based on data by the BBWS SO: (1) upstream: Argomulyo; (2) midstream: Gowongan; and (3) downstream: Kebonagung. At this level, the individual interview rounds were conducted with "wise" persons (those with in-depth depth knowledge on CEK) with a total of 30 respondents. There were also focus group discussions held after the interviews as described above. Figure 2 provides the location of the study case and its sub-cases in the river basin territory of Progo Opak Serang, within the provincial boundary of the Yogyakarta Special Region (black line), and within the Opak sub-basin (red line).

5 Results

The main elements were divided into historical experiences and mechanism of transfer, but also the relation within the mechanism of transfers, which represent additional findings, namely the legitimization and symbolization process.

The first process, legitimization, is related to the knowledge generation, in which the CEK is always based on historical experiences (extreme disasters) of the natural dynamics of conservation and temporal experiences (eruption cycles and seasonal patterns). These experiences are then selected and refined through a legitimization process by the institutionalized structure and dynamics of the community before they are transferred into the next generation. One respondent, a humanist from Gadjah Mada University, provided an example wherein the Royals or the Kraton people (the Sultan and his relatives), who reign in the Government structures provide the necessary legitimacy for a certain CEK being conserved and used. This condition explains that things such as *pranata mangsa* are a product of their era and require legitimization of the highest societal order, for example, the Sultan, to exist. Based on this explanation, the researcher also found out that:

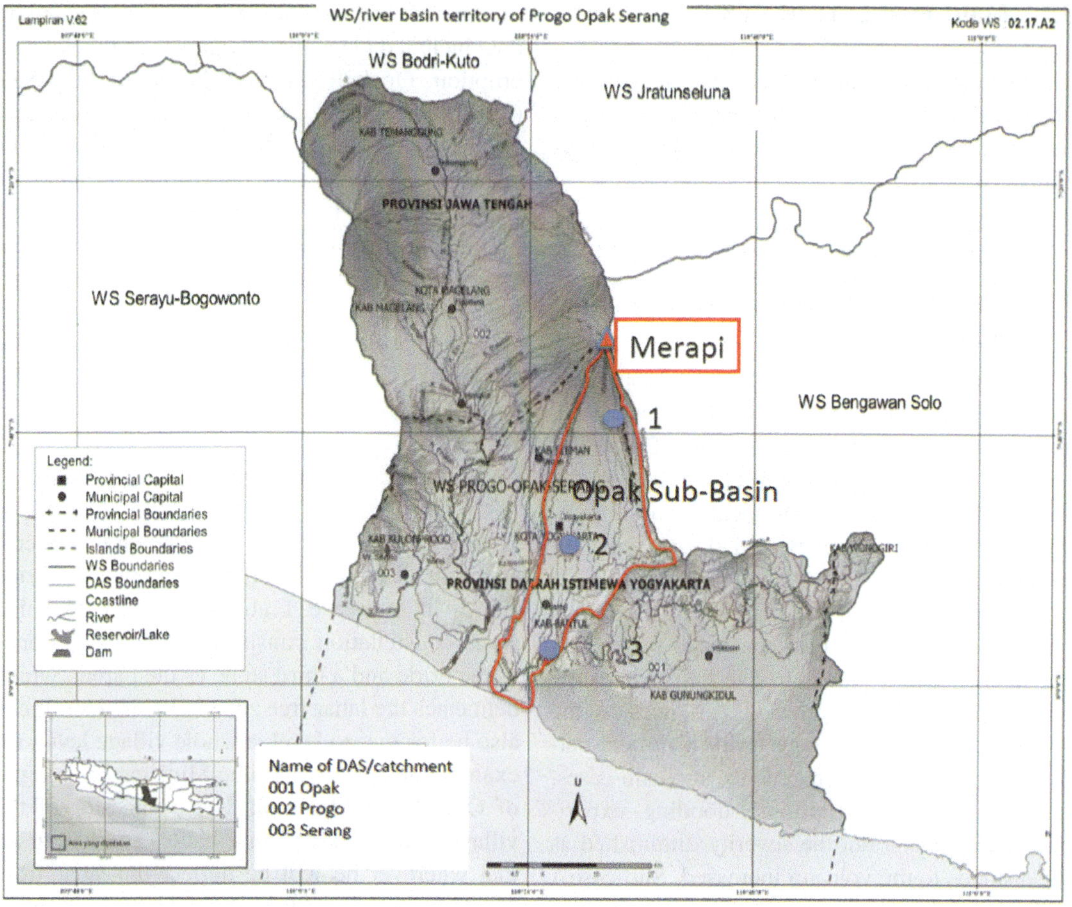

Fig. 2 Case study locations (Keppres No. 12/2012, 2012) adapted for the context of this study

> … the *pranata mangsa* is still used in the neighboring river basin (Serang and Progo) of Kulonprogo Regency as the Regent's Regulation No. 29/2016 (Dirgantara, BPDAS SOP).

In this regard, it functions as the cropping pattern for the farmers, and the regent, with the legitimacy for the municipal level, determines it annually.

The second process is symbolization, which is a standard feature of how cultural values, teachings, or philosophies manifest in daily life. The symbols can be in the form of artifacts, which remind people about the existence of these philosophies, or in the form of activities. The philosophies and cultural values are grouped into one as they always stem from the worldview. However, the symbolization of these philosophies and values take the forms of myth, artifacts, resource management practices, and five-senses wisdom. An example by a Kraton representative, Prince Wiro, explained the symbolization of the philosophy of *Manunggaling kawula Gusti* or human-nature relationship in the Tugu Monument on the unity between citizens and their leader (the Sultan). The monument also signifies the MTKKS (Mt. Merapi-Tugu (the landmark)-Kraton (the Sultan's Palace)-Krapyak (the stage) -South Java Sea) philosophical axis. This symbolization is a non-linear process. As seen from the patterns from each sub-case, this process is continually connecting with different forms of CEK.

5.1 Main Elements of CEK

The first element of any CEK is observing nature or known as *ilmu titen* in Javanese, it is such a way as to see patterns of natural dynamics, as explained below:

> It was called *ilmu titen* (phenology) in the past. In the old days, our ancestors observed (nature). Oh, on this date, this happened. And then they develop a standard, like so. (Martadi, Kebonagung Focus Group Discussion).

This research used a bottom-up approach, starting from CEK in the village level, based on each local community's respondents. The regional level CEK was explained later, based on the community respondents commenting on region-wide knowledge, experts, and organization respondents.

The comparison of the findings of the differences and similarities between these two levels are summarized in Table 1. The first part of the table explains the historical experiences as the root of CEK. At the village level, historical natural dynamics were shared by the sub-cases, including volcano eruptions, flooding experiences, and lahars, but the severity diminished as the distance to the volcano increased. Sub-case 1 still experienced pyroclastic flows, lava, and even fiery hot volcanic rocks in their village during the 2010 eruption, in addition to the continuous lahar flow in Gendol River during every rainy season from 2010 to 2017. Meanwhile, in Sub-case 2, the severity of damage most dramatically occurred in 2010–2011, when the lahar topped the Code River dikes and entered the Riverside community, submerging 1–2 m of the kampung's (informal settlement) original elevation. In Sub-case 3, the severity of damage did not interfere with the daily life, except for the lahar sedimentation in the Opak River, which silted the riverbed and thus made the river shallower. It also interfered with the villagers' income from tourism as the river could no longer be used for the Dragon Boat Festival, and the paddy fields were flooded each rainy season, especially in 2010–2015. For the regional level, the results showed that the historical experiences mentioned by the respondents and secondary data included (1) natural dynamics of the volcano eruption, Opak River, and the South Java Sea; (2) the response, which included temporal experiences of Mt. Merapi eruptions and the Javanese seasonal calendar (*pranata mangsa*), which explains the changes between the rainy and dry seasons based on living creatures behaviours, and the conservation of the volcano and the sea.

The historical experiences were later formulated into five forms of the mechanism of transfer as follows:

(1) Internalization through myths and rituals

At times, these myths are only known to a micro-coverage at hamlet level, for example, the bride stones in Sub-case 1 about the existence of a stone constellation consisting of two big stones as the bride and a third stone as the parent, which delineates the lahar free zone in the hamlet. It can also be for mezzo level, a whole village level, for example, the *Patih or* Prime Minister Jayaningrat of Old Hindu Mataram, as the ancestor of the village. He was killed in a battle, and he cursed that wherever he will be buried, the lahar from Mt. Merapi will submerge the area. His tomb is delineated as the lahar flood zone of the Opak River. This myth is celebrated in *Tambak Kali Gendol* festivities, which is held responsible for 2010 Mt. Merapi eruption taking so many lives from this village, as they saw it as a symbol of the village offering human sacrifices to the volcano, by commemorating the "curse of Jayaningrat", and the festivities have not been held there since.

In Sub-case 2, this regional-wide myth about Mt. Merapi as the ancestor and South Java Sea Queen as ruler of the Spirit Kingdom) is also known, and also the *lampor* (the sound of unseen or mystical troops marching down the river, which their loud drums), as the sign of the Queen's deities' troops going upstream to Mt. Merapi as the spirits' zone. Sub-case 2 also celebrated an annual mezzo level ritual related to the myth of the River Code as the *lampor* channel

Table 1 Summary of CEK at the regional and village levels

Locations	Yogyakarta Special Region	(1) Argomulyo (upstream)	(2) Gowongan (midstream)	(3) Kebonagung (downstream)
Historical experiences				
Natural dynamics				
Volcanic river basin	Volcano eruption, Opak river, the South Java Sea	Eruption, pyroclastic flow, lahar, flood	Flood, lahar	Flood, lahar sedimentation
Responses				
Temporal experiences	Eruptions cycle, seasonal patterns	Eruption and seasonal patterns	Flood and lahar cycles	Seasonal patterns
Conservation	The volcano and the sea	The volcano and the rivers	–	The river
Mechanism of transfer				
Worldview, philosophies, and values				
Philosophies or worldview	*Hamemayu hayuning bawono* "Beautifying the world's beauty"			
	The unity of people with the Sultan Remembering origin Mt. Merapi is my ancestor	Living with Mt. Merapi and as the Spirit Kingdom	Gendol River and Mr. Merapi as spiritual places	Being thankful to the Creator (*nrimo*)
Values	*Gotong-royong* or collective action, especially during disaster			
	–	Taboo for lahar panicking	–	Taboo of disturbing the Opak River
Internalization				
Myth	Ratu Kidul—*lampor* (mystical troops line-up making drumming sounds from the river) and the MTKKS axis	Patih Jayaningrat (ancestor of the villagers)	Ratu Kidul—*lampor* (stone mortar)	The *lumpang* (sacred stone mortar)
Rituals practices	*Labuhan* (at Mt. Merapi and the South Java Sea)	*Tambak Kali Gendol* (thanksgiving for the Gendol River) and *selametan*	*Labuhan, Merti Kali Code* (Code River thanksgiving and art performances)	*Wiwitan* (farming thanksgiving), *Nini Thowong* (sacred doll) harvesting cultural performances
Artifacts				
Macro (regional level)	MTKKS Philosophical Axis and Opak sub-basin (volcanic river basin) as a philosophical plane			
	The floodplains in the river channel or the *wedi kengser* (floodplain) as Sultan Ground			
	Ancient water network and Mataram channel			
Mezzo (municipal level)	–	The Gendol River as the deities' channel	Code River as *lampor* and spirits' channel	Opak River as the location of spirits
Micro (village level)	–	Lahar safe zone: Patih Jayaningrat tomb and the bride stones	Spatial planning: Tugu (landmark) as a part of the MKSS imaginary axis	The *lumpang* (sacred stone mortar) by the Tegal Weir
Resource management practices				

(continued)

Table 1 (continued)

Locations	Yogyakarta Special Region	(1) Argomulyo (upstream)	(2) Gowongan (midstream)	(3) Kebonagung (downstream)
Wangsit (premonition)	700-year eruption cycle, *geger bulus* (Yogyakarta's location on a small hill prevent it from flooding), 2010 lahar flows through rivers by the Sultan	On lahar flood occurrence by Sultan HB IX	–	–
Pranata mangsa (seasonal calendar)	Biophysical indicators	Main seasons (rainy and dry)	No longer used	4/12 *mangsa* (mini-seasons) identified
Direct uses	Special tree types for water conservations	Utilization of the river's base flow	Dredging of water wells after lahar flow	Opak River irrigates farmlands and fallow land in the dry season
	Sand mining for economic benefit	Sand mining for industry	Sand mining for construction and landfill	Limited traditional sand mining industry
Five-senses wisdom				
Visual	Seasonal calendar indicators	Eagle flight direction indicates lahar direction	Cloud upstream—flood time, fish indicate water quality	Worms and soil color indicate fertility, seasonal calendar, Mt. Merapi's ash as fertilizer
Sound	*Lampor* (drumming of the mystical troops) as flood early warning system, seasonal calendar	Lahar sound differs from flood	Sound of *lampor* (drumming of the mystical troops)	Cat mating, insects, birds for seasonal calendar
Touch	–	Temperature of groundwater	–	Temperature drop (dry season), soil humidity
Taste	–	Groundwater taste as water quality indicator	Taste of well water does not change post-eruption	
Smell	–	Smell of burning (pyroclastic flow), smell of iron (water quality)	River smell as water quality indicator	–

Note

Hamemayu hayuning bawono: the philosophy of the Sultanate in managing the human-nature relationship by conserving its beauty (Koentjaraningrat 1985; Olthof 2008)

Labuhan: the annual thanksgiving performed by the Sultanate to the Mt. Merapi and the South Java Sea believed to vend of bad lucks, disasters, and bad omens (Schlehe 2010, 1996)

Lampor: the sound of drumming performed by mystical troops belonging to the Queen of South Java Sea (Schlehe 2010)

Geger bulus: the hilly location of Yogyakarta described as the back of a turtle's shell (Olthof 2008; Ras et al. 1992)

MTKKS: imaginary and philosophical axis connecting the Mt. Merapi, Tugu (landmark), Kraton (Sultan Palace), Krapyak (stage), and South Java Sea, pertaining also the geological action-reaction of the volcanic eruption and seabed earthquakes (Troll et al. 2015)

Pranata mangsa: Javanese seasonal calendar, which divided the rainy and dry season into mini-seasons described with phenology in natural and animal's behaviors as indicators for the changing of the season (Retnowati et al. 2014; Daldjoeni 1984) called the *Merti Kali Code* as a way to boost the river restoration movement, which highlighted the importance of a clean river and is celebrated with cultural performances.

The *lumpang* (megalithic mortar stone) found in Opak riverbed of Sub-case 3 is related to the story of the death of the project leader and a shaman during the Tegal Weir construction in the 1990s and the myth of the Opak River as a sacred location. There are also myth-artifacts at the macro level for the whole regional level as explained in the sub-cases, for example, the *lampor,* the Ratu Kidul myth, related to the artifacts of the MTKKS philosophical axis, functioning as a sacred location by avoiding construction within the axis. The *labuhan,* an annual Thanksgiving, is held at the Islamic New Year, which also signifies the Javanese New Year at the Mt. Merapi and the Parang Kusumo Beach by the South Java Sea. During *labuhan*, the participants meditate and, aside from its spiritual virtues, the occasion is used to observe whether the surrounding environment has changed.

(2) Artifacts

Within the river basin, the artifact types also vary, but they are always accompanied by myth, rituals, or other forms of CEK. Each sub-case has its specific artifacts related to the volcanic river basin management. For Sub-case 1, these are the *Patih* Jayaningrat Tomb, the bride stones, Mt. Merapi, and the Gendol River. For Sub-case 2, the artifacts are Tugu Landmark and Code River. In Sub-case 3, they are the *lumpang* (megalithic mortar stone) and the Opak River.

For water artifacts at the village level, the following had a major impact: (1) the Mataram channel is the most important infrastructure as it provides irrigation water and inter-basin transfer from Progo Basin to the Opak sub-basin, and (2) the Opak River sub-basin with Mt. Merapi as its origin, the Opak River as the water channel from upstream to downstream, and the South Java Sea as the terminal. The other artifacts are symbolized by the MTKKS philosophical axis signifying the relationship of two forces of nature: the volcano and the sea.

Meanwhile, for the regional level, an important artifact was the city of Yogyakarta, considered as the *geger bulus* or flood free zone, which corresponds to the philosophical axis and contains the whole Opak sub-basin as a philosophical plane. Mt. Merapi, as the source of eruption, holds a key virtue in determining the management mode, whether in normal or disaster management. Another important artifact, which is scattered all over the region, including in each sub-case, is the *wedi kengser* or the floodplain. This artifact belongs to the Sultan, but the management of the floodplain belongs to the nearest adjacent plot owner, called the *cangkok.*

(3) The types of philosophies and values

The types of philosophies and values in each community differ in forms and strength. This study focuses on the type relating to the location and the natural dynamics experienced by the community. The *hamemayu hayuning bawono* (beautifying the world's beauty) philosophy is accepted at the regional level overall, but is translated into different approaches for each sub-case, and is focused on preserving the balance of the environment. Sub-case 1 took this philosophy as (1) honoring Mt. Merapi as God's force on earth: seen from a distance humans are so small compared to the volcano; (2) living in harmony with disaster, when it erupts humans should stay clear; and (3) the worldview where Mt. Merapi functions as the Spirit Kingdom. Sub-case 2 does not take this philosophy into practice, whereas Sub-case 3 interprets it as *nrimo*, being thankful to the Creator for whatever conditions that the villagers have.

For the whole region, there are several worldviews and philosophies to be considered, aside from the *hamemayu* philosophy. However, the Yogyakartans see their world in three layers of worlds (Endraswara 2013) or *bawono*: the human-God (*bawono cilik*), the environment (*bawono gede*), and the after-life (*bawono langgeng*). The context of this research allows focusing on *bawono cilik* and *bawono gede*. This philosophy guides the attitude towards other humans and the environment to be just and honoring. Other worldviews can be split into two parts: (a) *Manunggaling kawula Gusti* (unity of King with the subject), in which the people follow the Sultan's order and the Sultan listens to the people; and (b) *Sangkan paraning dumadhi* (remembering the origin and purpose of life), which is to govern one's life with wisdom with respect to the use of nature and also the fatalistic philosophy about Mt. Merapi as "my ancestor," which can be explained as "Here is where I was born, and here I will die." There are also additional values, such as *gotong-royong* or collective action, which is known throughout the regional level and in each sub-case to be used in everyday life and during a disaster. Another value is the taboo, which as shown in Sub-case 1, is the taboo against panicking during a lahar. This is related to the premonition of Sultan IX; when a lahar occurs, one should stay calm by the riverside. In Sub-case 3, the taboo relates to not disturbing the Opak River, as it is believed to seek a victim, or *tumbal*, whenever a balance of nature is disturbed there. This is supported through the death of the Tegal Weir project leader during the weir construction, and some children swimming in the Opak River being found dead.

(4) Resource management practices

At the regional level, the practices consist of *wangsit* (premonition), *pranata mangsa* (Javanese seasonal calendar), and direct resource utilization. The premonition is part of the practice, as it is considered as a traditional version of analysis conducted by the Sultan or "wise" person of the Sultanate. Some important findings on the premonition consist of the 700-year major eruption cycle (Troll et al. 2015), as the start of a new era, where 2010 is not considered within this time range. This fact means in the future, circa 2100, Mt. Merapi will have another major eruption as the momentum of radical change in the Republic of Indonesia. As explained in the artifacts, the *geger bulus* is a product of *wangsit* by Sultan I, when determining the new capital city of the Muslim Mataram. In 2010, before the lahar flowed, Sultan X had a premonition of where the lahar will flow. This premonition made him go to the six rivers flowing from Mt. Merapi dressed in his traditional clothing, stabbing those rivers with his spear as a way to remind his people to stay away from the riverside. In Sub-case 1, the story of the premonition is told about Sultan IX visiting the village and knowing when lahar flood is coming by meditating in the middle of Gendol River, which relates to the taboo of panicking during a lahar. However, the other sub-cases do not have any recorded experience of this practice.

The *pranata mangsa* (Javanese seasonal calendar) as practice on land and water resources is reported at the regional level. The practice uses biophysical indicators to determine which seasons are in place. However, Sub-case 2 no longer used this, Sub-case 1 used it for recognizing the main seasons (rainy season (*rendheng*) and dry season (*ketiga*)), whereas, in Sub-case 3, four of the twelve *mangsa* (*labuh udhan*, *rendheng*, *ketiga*, and *rendheng-pangarep kasanga*) are still acknowledged along with their five-senses wisdom (as explained in the next section). The fallow land activity is related to the *pranata mangsa* during the dry season, during which the farmers allow the land to rest.

The direct resource utilization of the lahar sediment is the sand mining activities, and temporary farming on *wedi kengser* or floodplain occurs throughout the sub-cases. In Sub-cases 1 and 3, the floodplain is used for temporary farming, but in Sub-case 2, the floodplain behind the dikes is used as an illegal settlement, and areas inside the river body are also used for temporary farming. Sand mining activities are more apparent in Sub-case 1 and used for large-scale industrial purposes. Whereas in Sub-case 2, sand mining is used to add elevation to the riverside settlement and in private construction. In Sub-case 3, traditional sand mining activities are conducted using the lahar sediment found in the riverbed. Knowledge of taking advantage of the lahar sediment has been known since ancient times, using the quote of a respondent in Sub-case 1:

> There will be a day when you will be able to auction the Gendol. (Subardjo, Argomulyo Village official).

This saying is being supported by practice, as Mt. Merapi's rivers are considered "black gold" and of the best quality for use as construction materials and are transported all over Indonesia.

For water, the practices are related to the direct use of groundwater for drinking water in Sub-case 1, and utilization of Gendol River's base flow during eruption and dredging of wells by the riverside to lower the groundwater table once more after the lahar flow in Sub-case 2. For Sub-case 3, this relates to the Opak River as a source for irrigation, fisheries, and the Batik production process, but groundwater is used for drinking water. Most respondents of the sub-cases rely on groundwater (wells) to provide drinking water since they believe that their groundwater is much cleaner and safer than the piped water. The regional level also accepts this as a fact as most of their residents use wells for their drinking water and to provide them with good water quality. Moreover, it is accepted that one should plant special trees for water conservation, such as Banyan tree *(Ficus benjamina)*, *Gayam* tree (*Inocarpus fagiferus*), *Munggur* or rain tree (*Samanea saman*), and *Ketapang* or tropical almond tree (*Terminalia catappa*).

(5) The use of five-senses wisdom

The five-senses wisdom is most sensitive in the upstream village (Sub-case 1), in terms of both volcanic activities and water resources, as this village is located closest to the mountain and they experienced the most damage in 2010. Sub-case 1 has a complete array of five-senses wisdom. An example for visual wisdom is eagle flight direction as the lahar flow indicator based on Mr. Surono's advice (Head of Volcano Center):

> When it (the eagle) flew over to the south, the lahar will flow through Gendol, when to the northeast then it's another river… (Sutrisno, Ex-Chief of Village of Argomulyo).

More related to daily life use is the visual of *uceng* (small fishes) as an indicator of water quality. However, the sound is also used for disaster mitigation, for example, the sound of the lahar ("*glenk-glenk*") differs from flood ("*kemrosak*"). The sound of *lampor*, which is still recognized in the regional level, Sub-case 1 and 2, although it can only be heard for those who are sensitive in the "sixth sense", and is used as a traditional early warning system for the coming of flood or lahar. Moreover, the *pranata mangsa*'s biophysical indicators are used to determine the seasons. In Sub-cases 1 and 3, and at the regional level, the examples are the visuals of a flowering Mango tree (*Mangifera indica*) (*labuh udhan* - beginning of the rainy season) the splitting of *Kapok* tree's *(Ceiba pentandra)* flowers (*labuh semplah* -beginning of the dry season), the sound of baby birds hatching (*rendheng* (rainy season)), cats mating, and temperature dropping (*ketiga*-dry season). Touch is used to approximate the temperature of well waters, which are too hot to be used for up to one-year post-eruption. The smell of burning air in Sub-case 1 is also used for determining the coming of a pyroclastic flow. The smell and stale taste of water following the 2010 eruption is also an indicator of iron dilution in the groundwater in Sub-case 1, whereas, in Sub-cases 2 and 3, respondents indicated no groundwater taste change after the eruption.

5.2 Formulation of CEK

The regional level and all of the sub-cases in the village level show similar patterns of groups of actors (even though different in each community) holding the legitimization processes. The formulation pattern shows the role of institutionalized structures and dynamics in the process of legitimization, rather than as part of the CEK forms (Table 2). The formulations are divided into two levels: at the regional level and at the village level. The role of this institutionalized society is to identify a certain person within the group to legitimize the product of the CEK.

Table 2 explained the existing institutionalized structure and dynamics located in both regional and subcase or village level. At the regional level, the institutionalized patron-client structure uses the philosophy of *Manunggaling kawula Gusti* (the unity of the Sultan or Kraton

Table 2 Institutionalized structure and dynamics supporting the formulation of CEK in the regional and for each subcase level

	Regional level	(1) Argomulyo (upstream)	(2) Gowongan (midstream)	(3) Kebonagung (downstream)
Institutionalized structure	Kraton and the people (patron–client)	Hamlet chief and the people's forum (social communication channel or SKSB)	*Paguyuban* RW leader (informal civic society gathering under the sub-district government) and the riverside community	Tourism village initiatives (*desa wisata*), agriculture museum initiatives, farmers group (*poktan*)
Institutionalized dynamics	Kingdom dynamics: formal and ordered	Farming and disaster activities: rural life	Urban activities	Farming and tourism activities
		Follow orders from the Sultan		

and his people). The dynamics at the regional level are also supported by the current structure of the special status of the kingdom within the authority of the regional government. This status permits the Sultan as the Governor and the *Paku Alam* (a relative of the Sultan) as Vice Governor for life. Although this position is renewed every five years, the same structure will serve the region or kingdom. For these people, the patrons are the Sultan and the Kraton people (relatives of the Sultan, for example, Princes or Princesses), making the decisions for them based on cultural knowledge and modern science as suggested by professionals in the governmental agencies. Therefore, the dynamics for each of the sub-cases follow orders from the Sultan, as he was chosen to rule the kingdom.

At the village level, the structures and dynamics differ based on the activities and initiatives of the community. Although different, the structures have similarities to the regional level in terms of the patron-client response, where leaders of the community become the patron, while the rest of the community is the client. The leader serves as the spearhead for related dynamics, be it water-related disaster management, irrigation management, or water-related tourism activities. For the community of Sub-case 1, the institutionalized structure and dynamics consist of the Sultan, his *Juru Kunci* (spiritual gatekeeper) of Mt. Merapi, and the people. The structure comprises the hamlet chiefs, the social communication channel forum, and farmer groups. Although a water user association (WUA) exists, it does not have a strong influence. The dynamics are stronger and more integrated in the case of the 2010 eruption, where the WUA, SKSB forum, village government are directly working together to save as many villagers lives as possible.

Meanwhile, Sub-case 2 is located at a riverside Kampung (urban informal housing), namely Kampung Jogoyodan. The *paguyuban rukun warga* (RW) is the institutionalized structure. The RW or *rukun warga* is the division under the village level, but the *paguyuban* RW is a self-organized initiative, which consists of several RW's within the Kampung. It mediated the RWs with the village government. Due to its riverside characteristic, located directly by the Code River (a tributary of Opak River), the *paguyuban* RW initialized the arrangement of a riverside infrastructure enhancement program to minimize flood risks or lahars and help distribute the international relief fund.

In Sub-case 3, the institutionalized structure and dynamics are the tourism village communities or *desa wisata* and farmer's group (*poktan*) for normal agricultural activities. These two groups ensure the conservation of agriculture and cultural values, including tourism, cultural, agricultural practices. Another important feature is the existence of the Agricultural Museum, where the dynamics of the communities created the pull–push factors between the *desa wisata* and the agriculture museum organizers. The WUA's

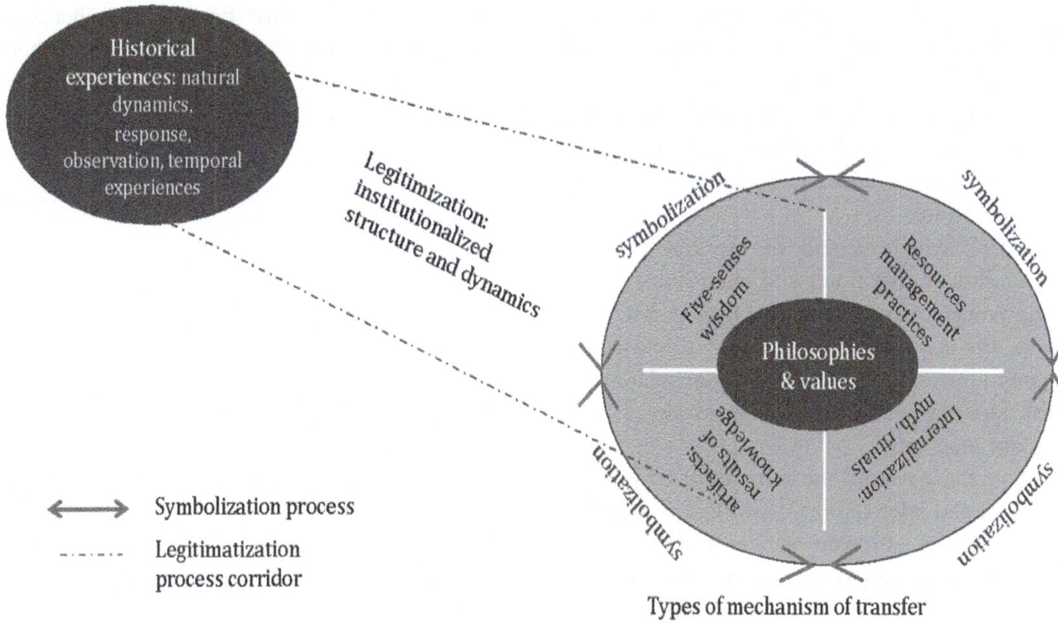

Fig. 3 Formulation of CEK based on this research

activities were more important before there was a weir, but it no longer has an active role. The *poktan* is gradually taking over the role of the WUA, as the members are more or less the same.

Although the CEK main elements were categorized using the initial conceptual framework (Fig. 1), the patterns of interrelations asked for a rearrangement of the framework (Fig. 3). The understanding of the natural dynamics is the core of historical experiences, which resulted in a response of the community, either in the form of conservation or temporal experiences. The CEK formulation is closely linked to institutionalized structure and dynamics and together they form the legitimization process. The mechanism of transfer is then left with five possible forms of CEK. The philosophies and values are set as the core of these forms, while the four alternative forms function as the "outer layer" or the "embodiment" of the CEK. The CEK in the case of Opak Sub-Basin in Fig. 3 is therefore much more dynamic than the proposed initial framework.

6 Conclusions

The CEK is in line with the nexus approach in looking at nature as one holistic system, which, in this study connects water to other resources. This study aimed to identify CEK's main elements regarding the water–lahar–volcano system, consisting of historical experience, formulation processes, and the forms of the transfer mechanism. The main findings of this research are as follows:

The CEK formulation began from the recorded historical experiences (either orally transferred or written). However, this required legitimization by certain groups to be transferred to later generations. This process is essential in transcending into philosophies, which are later developed into forms of symbolizing knowledge: the internalization of myth and rituals, the cultural artifacts, the resources management practices, and the five-senses wisdom.

The CEK is still used, some partially, such as the seasonal calendar of *pranata mangsa,* the utilization of floodplains or *wedi kengser,* and some holistically, such as Thanksgiving rituals of *labuhan* with the philosophical axis-plane as a whole river basin and the myth of Mt. Merapi and Ratu Kidul.

The CEK, which is still in use has potential contributions to the water and lahar management in the volcanic river basin of Opak in Mt. Merapi, as it is involved in the transfer of knowledge or communication strategy for water-related disasters strategy and water–lahar resources management practices.

Based on these results, these traces of CEK apply the integrated approach as interrelations between water, volcano, and lahar resources management. The CEK is an important knowledge to be conserved, as it is more accessible to the communities than the scientific approach. Examples from earlier studies (Berkes et al. 2000; Inglis 1993; Agrawal 2014; Howes and Chambers 1980; Brush 1993) have shown that conservation practice with crop pattern based on local knowledge, grazing location rotation for rainwater infiltration, and the traditional seasonal calendar are more easily used by the locals.

This study reflects on the rich knowledge the communities have to offer in understanding the condition of a volcanic river basin, also in terms of managing its water, disaster potentials, and sensitivities in 'reading' nature. This knowledge is surviving due to the existing formulation and legitimization processes brought by the Sultan and its people, which also present the notion of contested knowledge, which may have existed before, but did not survive in the current society. Our experience in performing this study shows the importance to look deeper in the cultural background of a place to find the meaning and logic of the CEK to be used in combination with scientific knowledge. Future research can take advantage of the methodology and the CEK formulation framework to assess their object of research, even develop it to further forms. An application of this research by decision makers may result in the management of this volcanic river basin to be more participative between different stakeholders and more integrated, especially for managing the interrelations of lahar, water, and eruption.

References

Agrawal A (1995) Dismantling the divide between indigenous and scientific knowledge. Dev Change 26:413–413

Agrawal A (2014) Indigenous and scientific knowledge: some critical comments. Antropologi Indonesia

BBWS SO (2014) Rencana Pengelolaan Sumber Daya Air Wilayah Sungai Progo Opak Serang. Yogyakarta

Berkes F, Colding J, Folke C (2000) Rediscovery of traditional ecological knowledge as adaptive management. Ecol Appl 10(5):1251–1262

Bignami C, Bosi V, Costantini L, Cristiani C et al (2012) Handbook for Volcanic Risk Management: an outcome from MIAVITA project. MIAVITA, Orleans, France. https://reliefweb.int/sites/reliefweb.int/files/resources/HandbookVolcRiskMgt%20hr_0.pdf. Accessed Apr 2015

Biswas AK (2004) Integrated water resources management: a reassessment: a water forum contribution. Water Int 29(2):248–256

Brush SB (1993) Indigenous knowledge of biological resources and intellectual property rights: the role of anthropology. Am Anthropol 95(3):653–671

Burton J (1995) A framework for integrated river basin management. Water Sci Technol 32(5):139–144

Camus G, Gourgaud A, Mossand-Berthommier P, Vincent P (2000) Merapi (Central Java, Indonesia): an outline of the structural and magmatological evolution, with a special emphasis on the major pyroclastic events. J Volcanol Geotherm Res 100(1):139–163

Cashman KV, Cronin SJ (2008) Welcoming a monster to the world: myths, oral tradition, and modern societal response to volcanic disasters. J Volcanol Geotherm Res 176(3):407–418

Chester DK (2005) Volcanoes, society, and culture. In: Volcanoes and the environment. Cambridge University Press, Cambridge, pp 404–439

Crang M, Cook I (2007) Doing ethnographies. Sage

Daldjoeni N (1984) Pranatamangsa, the Javanese agricultural calendar—its biclimatological and sociocultutal function in developing rural life. Environmentalist 4 (7):15–18. Accessed 2 Aug 2015

De Bélizal E, Lavigne F, Hadmoko DS, Degeai J et al (2013) Rain-triggered lahars following the 2010 eruption of Merapi volcano, Indonesia: a major risk. J Volcanol Geotherm Res 261:330–347

Degroot V (2009) Candi, space and landscape: a study on the distribution, orientation and spatial organization of Central Javanese temple remains. Sidestone Press

Delcamp A, Roberti G, de Vries BW (2016) Water in volcanoes: evolution, storage and rapid release during landslides. Bull Volcanol 78(12):87

Donovan K, Suryanto A, Utami P (2012) Mapping cultural vulnerability in volcanic regions: the practical application of social volcanology at Mt Merapi, Indonesia. Environ Hazards 11(4):303–323

Dove M (2007) Perception of local knowledge and adaptation on Mount Merapi, Central Java. In: Traditional ecological knowledge and crisis management in island Southeast Asia. Berghahn Books, New York, pp 238–262

Emerson RM, Fretz RI, Shaw LL (2011) Writing ethnographic fieldnotes. University of Chicago Press

Endraswara S (2013) Memayu hayuning bawana: laku menuju keselamatan dan kebahagiaan hidup orang Jawa

Falzon M (2012) Multi-sited ethnography: theory, praxis and locality in contemporary research. Ashgate Publishing, Ltd

Flick U (2008) Designing qualitative research. Sage Publications

Folke C (2006) Resilience: The emergence of a perspective for social–ecological systems analyses. Glob Environ Change 16(3):253–267

Friese S (2019) Qualitative data analysis with ATLAS.ti. SAGE Publications Limited

Geertz C (2000) Local knowledge: further essays in interpretive anthropology. Basic Books

Gertisser R, Charbonnier SJ, Keller J, Quidelleur X (2012) The geological evolution of Merapi volcano, Central Java, Indonesia. Bull Volcanol 74(5):1213–1233

Government of Indonesia, Inter-Agency Task Force on Water Sector Policy Reform (1999) Water Sector Adjustment Loan (WATSAL), Sectoral Environmental Assessment. E-267. Government of Indonesia, Jakarta

Graefe O (2011) River basins as new environmental regions? The depoliticization of water management. Procedia Soc Behav Sci 14:24–27

Gupta J, Pahl-Wostl C (2013) Global water governance in the context of global and multilevel governance: its need, form, and challenges. Ecol Soc 18(4)

Hooghe L, Marks G (2001) Types of multi-level governance

Howes M, Chambers R (1980) Indigenous technical knowledge: analysis, implications, and issues. In: Borkensha D, Warren D, Werner O (eds) Indigenous knowledge systems and development, pp 329–340

Iles CE, Hegerl GC (2015) Systematic change in global patterns of streamflow following volcanic eruptions. Nat Geosci 8(11):838–842

Inglis J (1993) Traditional ecological knowledge: concepts and cases. IDRC

Jaspers F (2003) Institutional arrangements for integrated river basin management. Water Policy 5(1):77–90

Jousset P, Pallister J, Boichu M, Buongiorno MF et al (2012) The 2010 explosive eruption of Java's Merapi volcano—a '100-year' event. J Volcanol Geotherm Res 241:121–135

Keppres (Presidential Decree) No. 12/2012 on River Basin Territory. Jakarta

Koentjaraningrat (1985) Javanese culture. Oxford University Press, London

Kurian M, Ardakanian R (2015) Governing the nexus. Springer, Berlin, Heidelberg

Lal R (2015) The nexus approach to managing water, soil, and waste under changing climate and growing demands on natural resources. In: Kurian M, Ardakanian R (eds) Governing the Nexus. Springer International Publishing, pp 39–60

Leung M, Santos J, Haimes Y (2003) Risk modeling, assessment, and management of lahar flow threat. Risk Anal 23(6):1323–1335

Manga M (1997) A model for discharge in spring-dominated streams and implications for the transmissivity and recharge of quaternary volcanic in the Oregon Cascades. Water Resour Res 33(8):1813–1822

Molle F (2009) River-basin planning and management: the social life of a concept. Geoforum 40(3):484–494

Muhr T (1991) ATLAS/ti—a prototype for the support of text interpretation. Qual Sociol 14(4):349–371

Newhall CG, Bronto S, Alloway B, Banks NG et al (2000) 10,000 years of explosive eruptions of Merapi Volcano, Central Java: archaeological and modern implications. J Volcanol Geotherm Res 9–50. https://www.sciencedirect.com/science/article/pii/S0377027300001323#. Accessed 1 July 2015

Nikolakis W, Grafton Q, Nygaard A (2015) Indigenous communities and climate change: a Recognition, Empowerment, and Devolution (RED) framework in the Murray-Darling Basin, Australia. J Water Clim Change jwc2015058

Ohsawa S, Saito T, Yoshikawa S, Mawatari H et al (2010) Color change of lake water at the active crater lake of Aso volcano, Yudamari, Japan: is it in response to change in water quality induced by volcanic activity? Limnology 11(3):207–215

Olthof WL (2008) Babad Tanah Jawi, mulai dari Nabi Adam sampai tahun 1647. Translated to Bahasa by Soemarsono HR. Narasi, Yogyakarta

Pahl-Wostl C, Mostert E, Tàbara D (2008) The growing importance of social learning in water resources management and sustainability science. Ecol Soc13(1):2008

Paton D, Smith L, Daly M, Johnston D (2008) Risk perception and volcanic hazard mitigation: individual and social perspectives. J Volcanol Geotherm Res 172 (3):179–188

Pierson TC, Scott KM (1985) Downstream dilution of a lahar: the transition from debris flow to hyper-concentrated streamflow. Water Resour Res 21(10):1511–1524

Ras J, Pantja Sunjata I, Supriyanto I, Ras J et al (1992) Babad Kraton menurut naskah tulisan tangan. The British Library, London Add 12320

Rasul G, Sharma B (2016) The nexus approach to water–energy–food security: an option for adaptation to climate change. Clim Policy 16(6):682–702

Retnowati A, Anantasari E, Marfai MA, Dittmann A (2014) Environmental ethics in local knowledge responding to climate change: an understanding of seasonal traditional calendar PranotoMongso and its phenology in Karst area of GunungKidul, Yogyakarta, Indonesia. Procedia Environ Sci 20:785–794

Rodolfo KS (1991) Rain-lahar generation and sediment delivery systems at Mayon Volcano, Philippines. In: Sedimentation in volcanic setting. Society for Sedimentary Geology (SEPM), SEPM Special Publication No. 45. https://archives.datapages.com/data/sepm_sp/SP45/Rain_Lahar_Generation_and_Sediment.htm

Ross A, Pickering K (2002) The politics of reintegrating Australian Aboriginal and American Indian indigenous knowledge into resource management: the dynamics of resource appropriation and cultural revival. Hum Ecol 30(2):187–214

Savenije H, van der Zaag P (2008) Integrated water resources management: concepts and issues. Phys Chem Earth Parts A/B/C 33(5):290–297

Schlehe J (1996) Reinterpretations of mystical traditions. Explanations of a volcanic eruption in Java. In: Anthropos, pp 391–409

Schlehe J (2010) Anthropology of religion: disasters and the representations of tradition and modernity. Religion 40:112–120

Snellen W, Schrevel A (eds) (2004) IWRM: for sustainable use of water 50 years of international experience with the concept of integrated water management. In: FAO/Netherlands conference on water for food and ecosystems

Steward JH (1972) Theory of culture change, the methodology of multilinear evolution. University of Illinois Press, Urbana Chicago London. Accessed 01 Aug 2015 (cultural ecology, culture core)

Thouret J, Lavigne F, Kelfoun K, Bronto S (2000) Toward a revised hazard assessment at Merapi volcano, Central Java. J Volcanol Geotherm Res 100 (1):479–502

Troll VR, Deegan FM, Jolis EM, Budd DA et al (2015) Ancient oral tradition describes volcano-earthquake interaction at Merapi volcano, Indonesia. Geogr Ann: Ser A Phys Geogr 97(1):137–166

United Nations (2015) Transforming our world: The 2030 Agenda for sustainable development A/RES/70/1. UN, New York, USA. https://sustainabledevelopment.un.org/content/documents/21252030%20Agenda%20for%20Sustainable%20Development%20web.pdf. Accessed Apr 2019

von Hofwegen P, Jaspers F (1999) Analytical framework for integrated water resources management: IHE monograph 2. Accessed 1 Sept 2015

Zalewski M (2013) Ecohydrology: process-oriented thinking towards sustainable river basins. Ecohydrol Hydrobiol 13(2):97–103

Nexus-Oriented Approach for Sharing Water Resources: Development of Eco-Industrial Parks in the Catchment of Zayandeh Rud River, Iran

Janis von Koerber, Wolf Raber, and Petra Schneider

Abstract

In order to ensure sustainable development in Iran, the government supports the development of industrial settlements with water reuse. One approach towards this goal is the Eco-Industrial Parks (EIP) concept, based on resource sharing to reduce waste and to optimize water efficiency, enabled through cooperation between industries. In the German-Iranian research activity "IWRM Zayandeh Rud" located in Central Iran, application of the EIP concept was tested and evaluated in a practical case study. The scope of the research included the development of a methodology for the quantification of industrial water volumes and qualities, and a nexus-oriented water sharing approach between water users quantified through a flux analysis with STAN (subSTance flow ANalysis). The investigation examines different approaches to substituting freshwater with reused wastewater from another industry, thus reducing the overall consumption of freshwater. To find the optimum approach, three types of linking scenarios have been considered: bilateral principle, nucleus principle, and cascade principle. The investigation results show that in addition to geographic proximity between wastewater generation and water demand, the main requirements for the connection of water fluxes are the capacity and willingness for collaboration amongst industries with different water qualities, resulting in symbiotic benefits for the connected industries.

Keywords

Water efficiency · Substance flow analysis · Industrial symbiosis · Wastewater use

1 Introduction

1.1 Industrial Symbiosis: Resource Sharing Approaches in a Circular Economy

Industrial ecology, as a subject of ecological engineering, tries to reduce the environmental burden of product life cycles, along the value chain from raw material extraction to waste management (Lifset and Graedel 2001), or even nowadays to a circular economy (McDonough and Braungart 2002; Braungart et al. 2007). Resource efficiency is a core issue, focusing on the most complete possible closure of material flows. A practical tool to facilitate this is material

J. von Koerber (✉) · P. Schneider
University of Applied Sciences Magdeburg-Stendal, Breitscheidstr. 2, 39114 Magdeburg, Germany
e-mail: janiskoerber@msn.com

J. von Koerber · W. Raber
inter3—Institute for Resource Management, Otto-Suhr-Allee 59, 10585 Berlin, Germany

S. Hülsmann and M. Jampani (eds.), *A Nexus Approach for Sustainable Development*,
https://doi.org/10.1007/978-3-030-57530-4_13

flow management, which is a goal-oriented, responsible, holistic and efficient approach for material-based systems. The goal of material flow management is to use materials taken from nature as intensively as possible in order to save resources and avoid waste. The aim is to decouple economic growth from the impact on human health and the environment associated with waste generation. Creating a more circular economy that preserves the value of products, substances and resources within the economy for as long as possible and produces as little waste as possible is a major contributor to achieving the UN Sustainable Development Goals (SDGs; United Nations 2015), in particular goal 12, responsible consumption and production, through low carbon, resource efficient and competitive economy.

One implementation approach in industrial ecology is industrial symbiosis, which involves exploiting the interactions between business, industry and the environment to improve resource efficiency. The aim is to maximize the sharing of water, energy and raw materials to reduce the amount of waste or wastewater, generally inspired by the example of a natural ecosystem in which everything is recycled. The concept of industrial symbiosis originated in the municipality of Kalundborg in Denmark (Hewes and Lyons 2008; Massard et al. 2014; Kalundborg Symbiosis 2014; Chertow and Park 2016), where the foundation stone for the first so-called Eco-Industrial Park was laid in the 1970s (Lambert and Boons 2002). Industrial symbiosis enables companies to develop multilateral solutions for material and energy flows, thereby reducing the overall environmental footprint left by economic activity in a region (Hein et al. 2015; Chertow and Park 2016).

For many years, companies and institutions around the world have sought to promote sustainability through industrial symbiosis. Due to increased interest, industrial symbiosis was considered from different perspectives in different countries. In the past, the variety of approaches led to a lack of clear characterization in the literature. Currently, the most comprehensive characterization is that of Boons et al. (2011), which builds on the definition of Chertow (2007), and takes into account both competitive advantage through the exchange of physical matter, and other social and geographical aspects. According to Chertow (2007) "*engaging traditionally separate industries in a collective approach to competitive advantage including physical exchange of materials, energy, water, and by-products. The keys to industrial symbiosis are collaboration and the synergistic possibilities offered by geographic proximity*". The characterization of Boons provides a conceptual framework for dynamic industrial cells that can grow and evolve according to new market requirements. The framework is divided into three conceptual sections. The sections consist of different sets of conditions, mechanisms of transmission and the results achieved. The conditions for the development of industrial symbiosis include specific triggers as well as local and commercial specificities, number, sector and size of the companies and characteristics of the actors involved, such as companies, administrations, planning authorities and associations.

Industrial symbiosis can take a variety of forms and is also a form of "sharing economy" (Rifkin 2011, 2014). The renunciation of ownership in favour of the acquisition of user rights characterizes the economics of sharing. "Sharing instead of owning" or "Benefiting instead of owning" are the keywords of the sharing economy (Theurl 2015). The various implementation forms for industrial symbiosis depend on the prevailing technical and geographic framework conditions. Typical forms of industrial symbiosis include (Boons et al. 2011; von Hauff 2012):

- Sharing of residues, for instance, accumulated residue/wastewater from company A can be used as raw material/resources/process water in company B;
- Sharing of emissions, for instance in the form of heat exchange between companies;
- Infrastructure sharing, for example where companies share common water treatment, emergency power supply, rail connection, etc.;

- Sharing networks, for instance where networks are shared through the formation of distribution communities, shopping communities, etc.;

In practice, the principles of a circular economy form the basis of action in industrial environments to create win–win situations between industrial partners. Businesses might have different motivations to create industrial symbiosis. The main reason and decisive factor is the economic advantage that the respective company receives from the symbiosis. Motivations for industrial symbiosis can be manyfold, including (von Hauff 2012; von Koerber 2016):

- economic advantage (monetary);
- strengthening environmental sustainability;
- image improvement;
- reduction of dependence on external resources; and
- strengthening corporate cohesion;

1.2 Industrial Ecology as Background for the Symbiosis of Business Activities

According to Isenmann (2008), it is important not only to focus on the efficient use of raw materials such as water sharing and reduction of emissions and waste but also to accept the view that nature is a viable ecological system. This process of changing perspectives from nature as a business object to a management model shows the actual path that industrial ecology pursues. Building on the approach of Lifset and Graedel (2001) three conceptual approaches have been developed:

Zero Emission Park (ZEP): The goal of Zero Emission Parks is to reduce environmental pollutants and emissions in the production of goods, to zero. The requirement to reduce emissions and waste to zero is to be understood as the ideal case. The approach here is a systematic strengthening of circular perspectives in order to remove all avoidable emissions and waste arising in the production of goods usable for other companies. The Zero Emission Approach is also used as a management standard (von Hauff 2012).

Eco-Industrial Park (EIP): The Eco-Industrial Park concept is a further development of the Zero Emission Approach, where companies in a commercial or industrial area cooperate in various fields such as environmental management, resource management, energy management, water and material management. Through the company's cooperative approach, synergies are used to reduce and overcome obstacles to industrial protection of the environment. The Eco-Industrial Park approach aims, above all, to protect the environment within an industrial park and in particular substance and energy flows. The success of this concept is strongly dependent on individual projects or individual measures, which are characterized by the three main strategies of avoiding, reducing and recycling (von Hauff 2012).

Sustainable Industrial Zones (SIZ): Sustainable Industrial Zones are further developed from Eco-Industrial Parks. Ecological and economic approaches are expanded in some areas such as rainwater management or the use of renewable energy. In addition to the ecological and economic aspects, social aspects are also taken into account (von Hauff 2012).

Transmission mechanisms are divided into two levels: social and regional industrial systems. At the societal level, the following relevant factors, in particular, have emerged: coercion, imitation, self-initiative control models, demonstration projects and learning and professionalization strategies (von Hauff 2012). At the level of regional industrial systems, institutional capacity building is in the foreground. The results can be determined by changes to social systems and surrounding ecosystems and quantified by indicators. In addition, the framework for industrial symbiosis in the mechanisms shows that the institutional capacity must be supported or promoted, since without them it is difficult to realize the potential of industrial symbiosis (von Hauff 2012). A physical link between the individual actors, whose economic activities are to be linked, is the basis for the actual implementation. Ultimately, a system must be able to mobilize resources that are not currently included in the system. Thus, a major focus that plays a role in the efficiency of industrial symbiosis is the nature of the linkage. Depending on the type of raw material or residual material for which the exchange is to be practiced within the industrial symbiosis, connections according to the bilateral, core or cascade principle are suitable for this purpose (von Koerber 2016).

Creating a more cyclical economy that preserves the value of products, substances and resources within the economy for as long as possible and produces as little waste as feasible, is a major contribution to countries' efforts to achieve a sustainable, low carbon, resource efficient and competitive economy. One of the key ethical starting points for sustainable development is the principle of social responsibility, which is reflected in the demands for inter- and intra-generational justice (Hemphill 2013). Another core element is the circulation principle, in which, analogously to closed cycles in ecosystems, the ideal picture of economic activity is developed under the objective that quantities of material and energy used in the production process do not leave the economic process (McDonough and Braungart 2002). Finally, the third element is the cooperative principle, which calls for coordination between the actors (Grice 1975; Davies 2007). The idea of closed material cycles, which is a key feature of the circulation principle according to McDonough and Braungart (2002), can usually only be achieved through the cooperation of different parties involved (in the respective value chains). While in many companies the in-house possibilities for improvement that are economically meaningful under the given conditions are almost exhausted, measures that are cooperatively implemented by several companies open up new design horizons, especially since internal use is partly impossible for technical reasons. Furthermore, development since the International Climate and Energy Event in Rio de Janeiro (Brazil) 2015 (Rio15) has emphasised the necessity of modified forms of communication and cooperation, and novel institutions for negotiating and realizing sustainable development processes, which further highlight the future significance of this cooperation principle. These principles also form the basis for industrial symbiosis.

1.3 Industrial Symbiosis as Tool for Integrated Resources Management

Integrated Water Resources Management (IWRM) is a spatial planning approach combining water management and water protection at the level of the catchment area (Grigg 2008; Global Water Partnership GWP 2000). The definition of IWRM was given by the Technical Committee (TEC, former Technical Advisory Committee, TAC), of the Global Water Partnership (GWP). It states that IWRM is "A process which promotes the coordinated development and management of water, land and related resources, in order to maximize the resultant economic and social welfare in an equitable manner without compromising the sustainability of vital ecosystems." (GWP 2000). The aim is to coordinate the influences on the water body in an interdisciplinary approach in such a way as to ensure the best possible status for humans and nature, independent of political boundaries. IWRM principles include: planning for all water sources; addressing water quantity, quality and ecosystem needs; incorporating principles of efficiency, equity and public participation; and having a multidisciplinary approach and sharing of information (GWP 2004; UNESCO 2009; Bielsa and Cazcarro 2014). IWRM is based on a management approach for balancing water demand and availability under a spatial planning approach (Grigg 2008). As the gap between water availability and water demand continues to widen, so does the competition of individual water users for scarce resources (Schneider and Avellan 2019). The overall scope of IWRM is to ensure water security on the catchment scale (Schneider and Avellan 2019), but it doesn't consider sufficiently inter-sectoral water demands. In order to develop a more coherent water resource management approach, the water-energy-food nexus was introduced (Benson et al. 2015; Muller 2015; Liu et al. 2017; Cai et al. 2018; Schneider et al. 2019a, b). The operationalization of the water-energy-food nexus concept is based on the cooperative principle through the development of adapted communication schemes like the nexus dialogue (de Strasser et al. 2016; Liu et al. 2017). In this regard, Industrial symbiosis based on water sharing can be considered a nexus approach since it represents a form of sharing economy that requires customized communication and governance schemes to unlock the resource efficiency potentials (Benson et al. 2015; Schneider et al. 2019a, b).

Globally, agriculture is the largest water user, particularly in developing countries, and industrial users from several sectors are also in competition for available water resources. This situation requires an integrated and inter-sectoral management approach, based on the analysis of existing resources and their renewability. Besides agricultural water demand, the water supply to industry, businesses and households are also at risk. To date, industrial enterprises in developing countries are dependent on freshwater resources available within the catchment area, which are used for a wide range of purposes, and as a result of continued economic growth, water demand also increases. The water-energy-food nexus was developed to solve the resulting challenges (Benson et al. 2015; Liu et al. 2017; Cai et al. 2018).

1.4 Water Resources Challenges in the Study Area

The study area focuses on the Zayandeh Rud catchment in Iran, which serves as a case study for the competing usage claims between agriculture, households and industry. German-Iranian cooperation in the area aims to develop sustainable water management for the catchment area of Zayandeh Rud River in central Iran. The river is of vital importance to the semi-arid region, but rapidly decreasing water availability and quality of surface and groundwater resources present immense challenges to those responsible and those affected. Last but not least, the Zayandeh Rud River is the most important source of drinking water in the region. The disappearance of the life-giving river is a tragedy for the entire region, not only from an economic point of view but above all from a social, health and environmental point of view. Within the scope of the German-Iranian research project "Integrated Water Resources Management Zayandeh Rud" funded by the German Federal Ministry of Education and Research (BMBF), was the task to develop a strategy for sustainable water usage along the river whilst taking various agricultural, industrial, human and environmental demands into consideration (Mohajeri et al. 2011; Ghanavizchian and Mohajeri 2013; Piadeh et. al. 2014; Mohajeri et al. 2016; Kirschke et al. 2016).

Like other countries in the Euro-Asian high mountain belt, Iran is a country with scarce water resources. The total area of Iran is 1,745,000 km^2, out of which the Zayandeh Rud catchment covers 26,000 km^2 stretching across two provinces (Mohajeri et al. 2016). Zayandeh Rud is the largest river of the Iranian Plateau in Central Iran (Fig. 1). The country has more than 70 million inhabitants and suffers from a serious water shortage. The renewable water reserves available per inhabitant per capita declined from 13,000 m^3/a 1921 to 1,400 m^3/a in 2014 (source: Iran's Energy Ministry, cited by Iran Daily 2015). The climate in Iran can be largely described as arid to semiarid. The average rainfall varies between 1,280 mm/a on the coast of the Caspian Sea and 100 mm/a in the deserts. The average annual precipitation in Iran is about 250 mm/a. Intensive farming, population growth and increasing urbanization have led to increasing water scarcity in recent decades. According

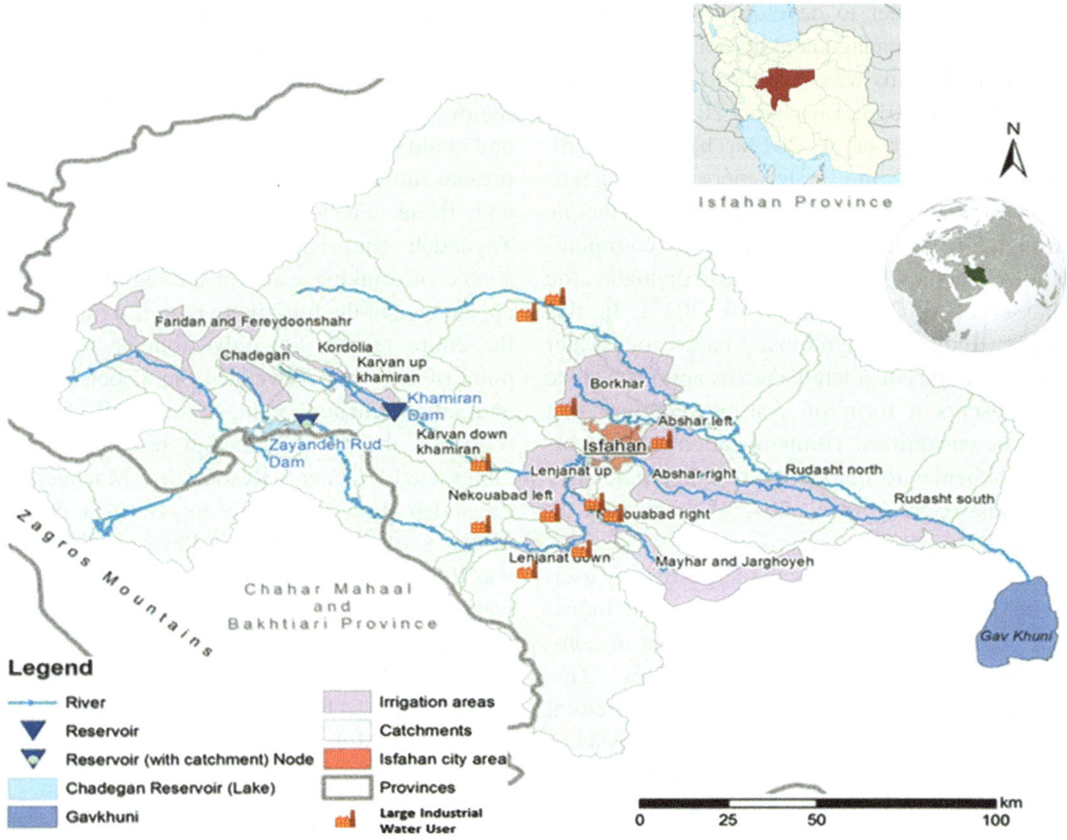

Fig. 1 Location of the Zayandeh Rud catchment in Isfahan Province, Iran (*source* inter 3, details at https://www.iwrm-zayandehrud.com/ziele/?lang=en)

to UN-Water, 96% of the population has access to clean drinking water and 100% are connected to the sewage system as of 2014 (UN-Water 2015).

In addition to agriculture, industry is an important water user in the Zayandeh Rud catchment area. More than 350,000 people work in large steel and cement plants, oil refineries and around 13,000 smaller companies. In the medium and long-term, industry is expected to expand due to regional socio-economic development, population growth and rising living standards (Raber and Mohajeri 2017). For the industrial sector, approaches are needed to reduce water consumption, reuse water within businesses and develop new water sources.

1.5 Scope of the Study

The present paper focuses on the optimization of the industrial wastewater stream using the resource efficiency potential of industrial symbiosis as an instrument of the Industrial Ecology (IE). This task requires a nexus-oriented approach for the creation of an implementable concept for sharing the available water resources. Nexus refers to a connection or series of connections linking two or more subjects or things. The Nexus approach to environmental resource management examines the inter-relatedness and interdependencies of environmental resources and their transitions and fluxes across spatial scales and between sectors (Hülsmann and

Fig. 2 Impression of model settlement Mourcheh Khort and its industries (von Koerber and Raber, Site Visit Mourcheh Khort, 31.07.2016)

Ardakanian 2014). Instead of just looking at individual components, the functioning, productivity and management of a complex system is taken into consideration. The scope of the present study included a pilot investigation of material flow management by linking the wastewater flows of different industries in a meaningful way in order to save freshwater resources and to reduce wastewater emissions. For the industrial sector, approaches are needed to reduce water consumption, reuse water within businesses and develop new water sources. The scope of the present paper was to identify wastewater reuse potentials for industrial processes and the implementation of industrial symbiosis according to the 'Eco-Industrial Park' concept. The subject of the investigations are in particular, to link the mechanisms between companies at a representative pilot site, which is the model settlement of Mourcheh Khort in the Zayandeh Rud catchment.

2 Materials and Methods

2.1 Description of the Study Site

The model settlement of Mourcheh Khort is an existing industrial park in the Zayandeh Rud catchment and located about 50 km north of the city of Isfahan (Mohajeri et al. 2016, 2018; Mohajeri and Horlemann 2017a, b). A total of approximately 500 companies are contracted, out of which about 300 already work in the food, metal, mineral, textile, plastics, paper and chemical industries, located across 582 hectares, with around 17,000 employees (von Koerber 2016). The estimated annual water requirement is approximately 4 million cubic meters. About 160 companies are connected to the local sewage treatment plant.

The water supply in the model settlement is largely decentralized. The wastewater generated from the industries is either self-treated and used for internal water reuse or green space irrigation, or discharged into the sewage system. Nearly the entire model settlement is connected to the sewer system with a mechanical biological treatment facility (von Koerber 2016). Nevertheless, some of the industries do not use the central sewage treatment plant, as disposal is cheaper by other means. Most industries supply themselves with water by means of their own groundwater well, tapping a shallow aquifer. Furthermore, industries may access and buy water from a settlement-network supplying groundwater from a deep central well. The deeper groundwater aquifer has good water quality, in contrast to the upper aquifer (particularly in regard to salinity). Industry representatives indicate that the groundwater levels of the upper aquifer have declined by 20 m over the past five years, and the water quality has deteriorated (von Koerber 2016). The annual overexploitation of groundwater resources in the Zayandeh Rud catchment recently reached about 170 million m^3 (Raber and Mohajeri 2017). Since not all industries have a deep well, a few industries are now using wastewater processed via an existing wastewater treatment plant, since it has better water quality than that of the upper aquifer (particularly in terms of salinity and softness) (von Koerber 2016). In the model settlement, the Industrial Settlement Organization (ISO) is responsible for central water supply as well as for central wastewater treatment. The still young provincial industrial settlement organizations have little experience in the coordinated, resource-efficient, and environmentally friendly design and development of growing industrial areas.

2.2 Scientific Approach

The aim of the current research was to develop concepts for the reuse of industrial wastewater and internal recycling, in combination with the reduction of water consumption. The results of the study form the fundament for a water and wastewater management development plan based on the EIP concept. The basic idea is to create networks for the meaningful common (re-) use of resources within an industrial area. The central building blocks of the scientific approach are:

- development and analysis of selected company profiles with a focus on water demand and wastewater production for the characterization and categorization of industries,
- evaluation of water demand and wastewater production in terms of quality and quantity based on the sector specific water balance (water balance for the specific products),
- analysis of material flows (water and wastewater) using the software STAN, (subSTance flow ANalysis),
- development of network models and technical modules for wastewater treatment and reuse,
- development of a methodology for the reuse of industrial wastewater as process water of different quality classes, followed by a prioritization into primary, secondary and tertiary users,
- assessment of the input–output-scheme to connect the water fluxes between the industries along the flux with the increasing pollution,
- calculation for water and wastewater saving potentials of different EIP scenarios,
- preparation of recommendations for the future development of the industrial area.

At the beginning of the investigation, hardly any data were available about water consumers and wastewater producers in Mourcheh Khort. None of the selected partner industries knew their exact water consumption since water meters are not common in the catchment area. In order to obtain a sufficient database for the characterization of the industries and derivation of feasible scenarios, a questionnaire survey and measurement campaign was carried out by the ISO. Prior to this, the industries to be inspected were selected in cooperation with the ISO. A methodology for the quantification of industrial water use in terms of wastewater disposal and freshwater demand for different production processes was developed as part of the investigation. After data collection, a flux analysis of process water was performed with STAN (subSTance flow ANalysis), a freeware tool for Material Flow Analysis (MFA) developed by the Technical University of Vienna (Cencic and Rechberger 2008).

In the first step, industries in the Mourcheh Khort Industrial Park had to be identified to establish whose water consumption was relevant. An overview of industries and their wastewater discharge rates into the sewerage system was provided by the ISO. The collected data overview includes 171 industries and represents the respective process of effluents and sanitary wastewater. Through this overview, the largest industrial wastewater producers were identified. Figure 3 gives an overview of the volumes of the largest industrial wastewater producers, out of which the red bars show the industries selected for the investigation. In addition to the industries selected for their large wastewater generation from the survey, some other industries were selected for two other criteria: Typical industry for the region, e.g. the gaz production (Iranian nougat) as well as industries which are likely to have high water consumption but no arising wastewater, e.g. stone processing. A total of 21 textile, metal, food, plastics, glass and stone industries were selected for sampling and an interview campaign.

The mixed method campaign for determining the relevant data in terms of quantity and quality of industrial wastewater was carried out in cooperation with the ISO between May and August 2016. By sampling and analysing industrial wastewater the relevant quality parameters where detected (Chemical Oxygen Demand (COD), Total Dissolved Solids (TDS), turbidity, hardness, Total Suspended Solids (TSS), Electrical Conductivity (EC), temperature). In a flanking interview backed by a questionnaire with industry representatives, the relevant production processes and specific water demands/quantities have been estimated, since there are no water meters. From the collected data water balances and substance flow sheets for the respective industries were prepared and validated through comparison with the information in the Best Available Techniques (BAT) reference documents (BREFs) of the European Commission. These have been developed by the European Union for the implementation of the Industrial Emissions Directive (IED 2010/75/EU). However, for small and medium-sized enterprises (SMEs), there is a lack of comprehensive data. According to the European Commission, SMEs are enterprises that meet the definition of a staff headcount of <50 for small enterprises (SE) and <250 for medium-sized enterprises (ME) and either the turnover ($\leq$€10 million SE/ $\leq$€10 million ME) or balance sheet total ($\leq$€10 million SE/ $\leq$€43 million ME).

The data collected for each type of industry have been synthesized in brief factsheets/characteristic papers named "industrial profile". Next to general data on raw materials, process steps, and products of the type of industry, the factsheet holds data on quantity and quality of water use and wastewater production. The concept of factsheets for different industries is of great interest to Iranian planners and decision makers, since industries may be benchmarked in terms of their water footprint and water supply and disposal schemes may be more effectively designed. In this research, the primary goal of the factsheets was to categorize industries according to their water quality demands and potentials for wastewater reuse. Therefore, three classes have been created. In the primary class, industries with high water quality requirements (e.g. food industry) or industries which produce low polluted wastewater easy to reuse (e.g. certain plastic industries) are assigned. In the tertiary class are industries with very low requirements in water quality and little wastewater production (e.g. stone processing) as well as industry with

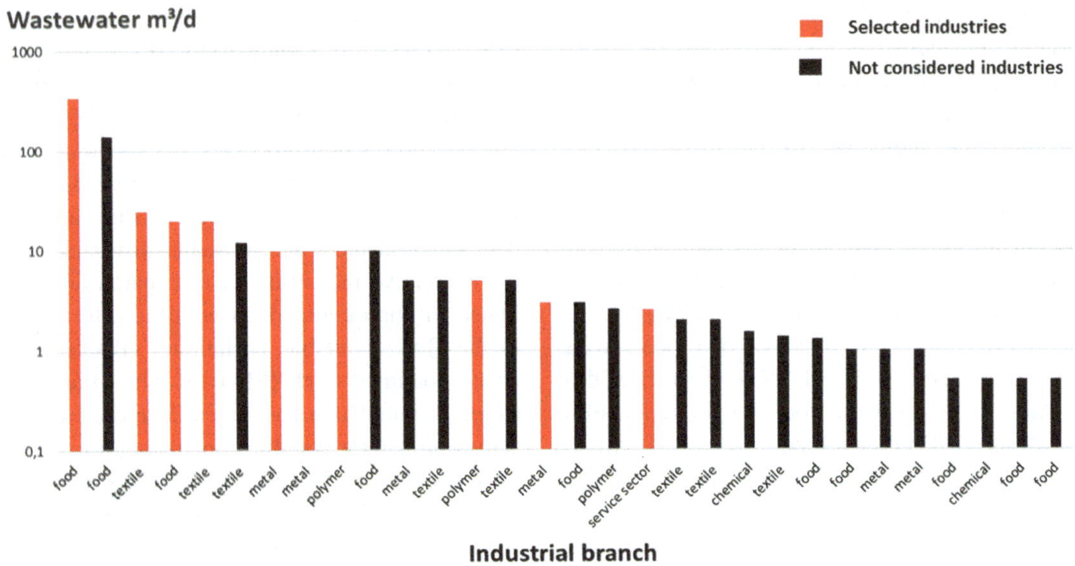

Fig. 3 Wastewater volumes of the largest industrial wastewater producers

highly polluted wastewater (e.g. dyeing facilities). The secondary class of industry served as an intermediate category.

In order to create the profiles of the individual industries, the verified data were included in the industry profiles. After the validation of the profiles, the input–output-scheme (Fig. 4) was assessed in order to find out the feasible connection of the water fluxes amongst industries along the flux chain with the increasing pollution, deriving primary, secondary and tertiary users. Figure 4 shows an input–output-scheme, how the flux chain of industries could be linked after a categorization in primary, secondary and tertiary users. After prioritization, linking principles for the respective industries have been developed as they are of fundamental importance for sharing the wastewater resources. The following options have been considered: bilateral principle, nucleus principle, and cascade principle (von Koerber 2016), as illustrated in Fig. 5. Using the company classification, its resulting wastewater, and the requirements for the process water of the respective industries, the linking options between the companies were assessed through an input–output-scheme (Fig. 4)

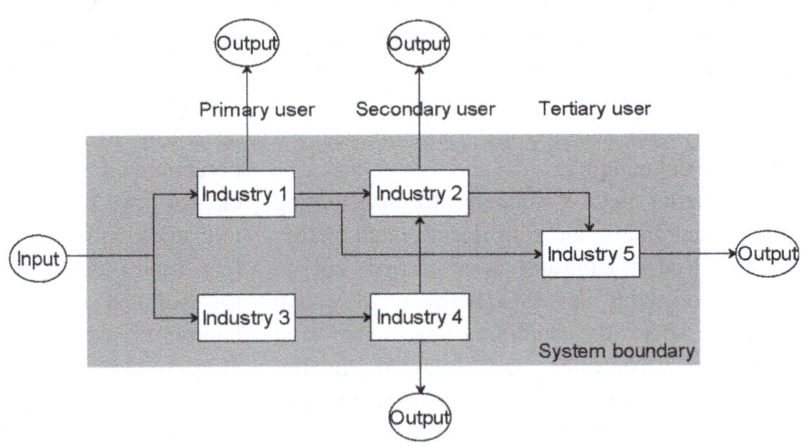

Fig. 4 General overview of the approach to the input–output-scheme and the linking mechanisms (von Koerber 2016)

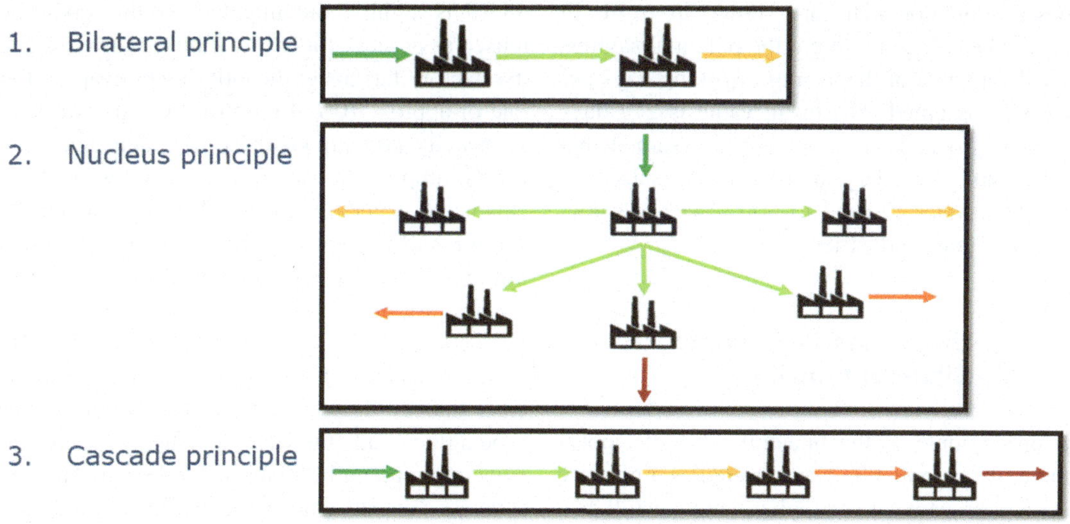

Fig. 5 Illustration of bilateral principle, nucleus principle, and cascade principle (von Koerber 2016)

according to different linking options in terms of their linking feasibility (Fig. 5).

In the illustration of the developed principles (Fig. 5) the coloured arrows indicate wastewater qualities, from very good quality (dark green) to very high wastewater pollution (dark red). The bilateral principle connects two industries in a very simple way. The wastewater from one industry is used as process water in another industry. The nucleus principle is an advancement of the bilateral principle, where the wastewater of one industry is used in several industries instead of one. The nucleus industry should have a high rate of wastewater compared to the other industries, so they can supply process water to several industries. The cascade principle is a sequence of the bilateral principle.

3 Results

The categorization, for the industries examined in the model element, revealed the following: The main primary user group is the food industry, which has a very high demand in terms of the process water quality, and for that reason, the application of (treated) wastewater in the process is not feasible. Depending on the type of food industry, the wastewater produced has a high organic load. Furthermore, two yarn manufacturers and one plastic pipe manufacturer have also been classified as primary users due to their low wastewater loadings. All measured wastewater parameters are low for the three industries. Feasible parameters are a low organic load and a low salt content of the water. Secondary users were mainly metalworking industries, such as producers of domestic components or tube manufacturers. The metal industry above all increases the salt content and the temperature of the water. Almost all tertiary users, such as dye factories and print shops, are polluting the water strongly. One exception is stone processing, where relatively heavily contaminated sewage can be processed, and no wastewater is produced since the water evaporates during the process.

On the basis of the profiles and the categorization of the model settlement industries, theoretical examples of the different linking scenarios are presented and their water conservation potentials are shown in m^3/d. In order to be able to represent the potential savings in m^3/a, a value of the average working days per year was assumed. The mean value is 277 days per year. To make it simple, it is assumed that tertiary and secondary industries may substitute their freshwater demand to 100% with wastewater from other industries.

3.1 Linking Scenario Example 1—Bilateral Principle

As an example of the bilateral principle, polyamide fiber production was selected as a primary user and process water supplier, and dyestuff as a consumer of the sewage water, which uses the wastewater of polyamide fiber production as process water (Fig. 6). In addition to water conservation, the linkage for the dye factory is advantageous in regard to the wastewater from polyamide fiber production, which has a lower hardness than the freshwater, being beneficial for the dyeing process. Besides the low hardness, the tinting process also requires a low water load through TSS. Both the hardness and the TSS meet the minimum requirements of dyeing. To simplify the representation of water flows in Fig. 6, the input was scaled to 100 m^3/d. All other values were scaled with the same factor.

Polyamide fiber production has a relatively large evaporation rate of 51.5%. The remaining 48.5% is slightly contaminated and thus could be reused as process water in various industries. The dye factory has a significantly lower evaporation rate of approx. 10%. In return, the wastewater is so heavily contaminated by colorants that it is nearly impossible to use the water without expensive water treatment. The freshwater consumption of linkage scenario example 1 is shown in Fig. 6. Thereby, a distinction is made between the status quo and the possible freshwater demand with the implementation of the linkage scenario example 1. From this graph, it becomes clear that by the connection of polyamide fiber production and the dyeing shop, freshwater consumption can be reduced by 48.5 m^3/d. This saving represents approximately 33% of the total water demand of the two industries. The daily saving of approx. 48.5 m^3/d results in a saving potential of approx. 13,000 m^3/a. Nine dyeing factories are currently located in Mourcheh Khort (ISO, Site Visit Mourcheh Khort, 2016) having total water consumption of approx. 201 m^3/d. Thus, for the dyeing plants, a water conservation potential of 56,000 m^3/a is obtained. The use of sewage from polyamide fiber production in the dyeing industry reduces the amount of total wastewater by about 53%. The higher percentages of wastewater compared to freshwater can be attributed to the high evaporation rate of freshwater. The volume of saved freshwater and wastewater is comparable, amounting to approximately 56,000 m^3/a.

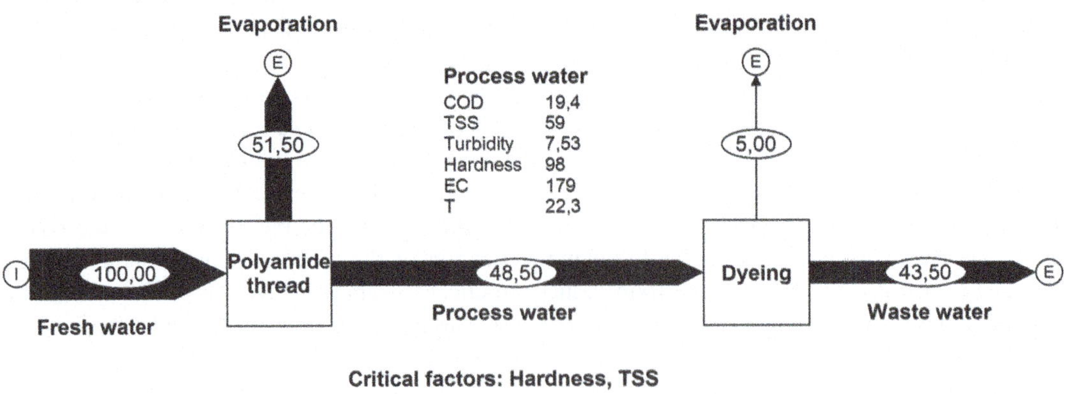

Fig. 6 Linking scenario example 1—Bilateral principle (presentation with STAN)

3.2 Linking Scenario Example 2—Nucleus Principle

In the example of the nucleus principle, the largest producer of wastewater in the industries under investigation, milk powder production, was selected as the core. The peculiarity of milk powder production is the amount of wastewater it produces, which is significantly greater than freshwater consumption. Most of the wastewater is extracted from the evaporated water of the milk. Wastewater from the dairy industry is mainly contaminated with organic matter (COD) and, therefore, must be treated before reuse, e.g. in a biological water treatment plant. By means of water treatment, it can be reused as process water in all other industries (Fig. 7).

For the linking scenario example 2, freshwater savings of 92% and sewage savings of 67% are feasible. A total of approx. 290 m^3/d (80,000 m^3/a) freshwater and wastewater can be saved. In order to make the example transferable to other cases, the water fraction introduced by the milk must be added to the freshwater input. This calculation still results in freshwater savings of around 48%.

3.3 Linking Scenario Example 3—Cascade Principle

In the example of the cascade principle, several industries were connected to one another. First, polyamide fiber production was selected as the first primary user, as in linking scenario example 1, which transfers its wastewater to another primary user: poly-pipe production. The connection of two primary users was possible because the water requirements of polymer tube production are small, particularly as a low temperature is necessary for the process. The production of polymer pipes impairs wastewater minimally as

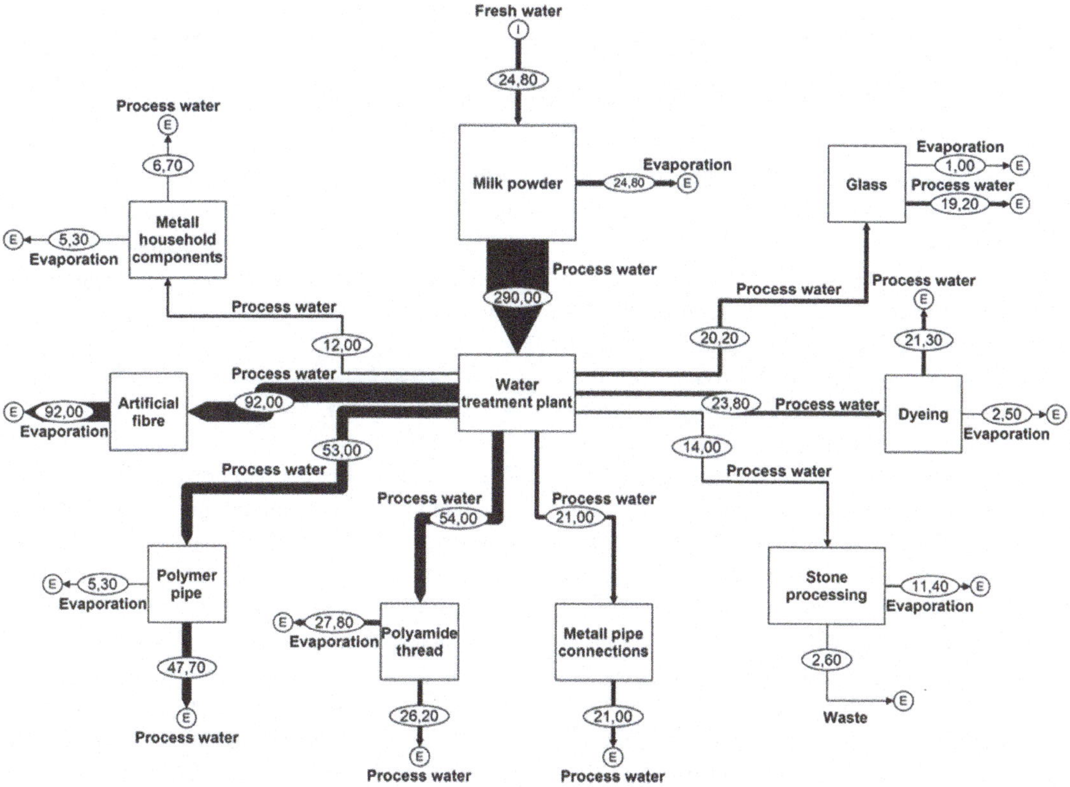

Fig. 7 Linking scenario example 2—Nucleus principle (presentation with STAN)

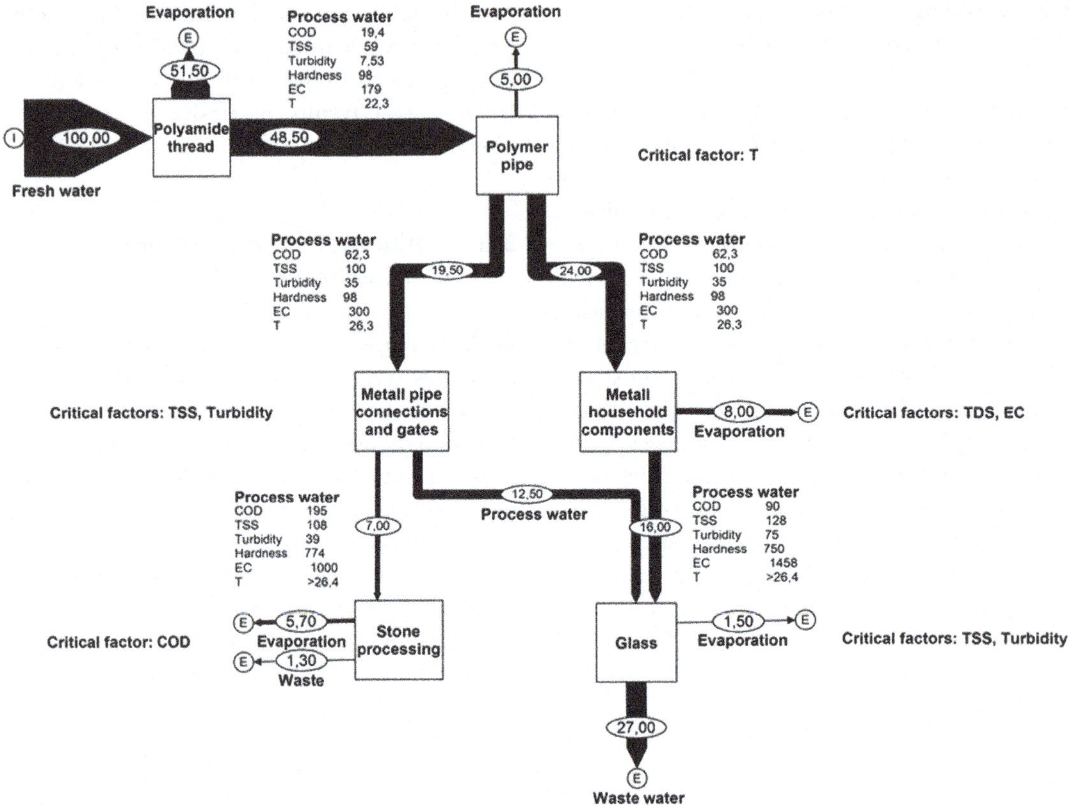

Fig. 8 Linking scenario example 3—Cascade principle (presentation with STAN)

no further pollutants are added during the production process. As a result, the wastewater can be reused as process water in two different metal industries. Both industries have been classified as secondary users. The critical factors of both industries are TSS and turbidity, the sum of the solutes (TDS) and the conductivity. All these factors are respected by the water already used by both primary users, so that the wastewater can be reused as process water in both industries. As a tertiary user, stone processing and glass production were selected. The critical factor for stone processing is COD. TSS and opacity are decisive for the production of glasses. The use of sewage from both metal processing plants is possible for both stone processing and glass production.

As in linking scenario example 1, the input was scaled to 100 m^3/d. The cascade-like arrangement of the industries and the associated use of the water can save a total of approximately 56% of the freshwater demand and approximately 83% of wastewater. In the depicted linking scenario example, approximately 128 m^3/d (35,000 m^3/a) of freshwater and wastewater could be saved.

4 Discussion

The scenarios for the EIP design in Zayandeh Rud catchment considered high inter-industrial water reuse without additional water supply for the connected water fluxes. From the results of the scenarios, the total saving potentials for the industrial settlements in the Zayandeh Rud area can be extrapolated. The current water consumption of all industrial settlements was determined in the first phase of the project and currently amounts to 27 million m^3/a. The current industrialization is expected to lead to a large growth in industrial settlements. Rough estimations based on development planning ambitions of the Iranian government suggests that water consumption will be more than doubled to 62 million m^3/a by 2025 (Raber and Mohajeri 2017). With the above mentioned hypothesis of the possibility of 100% wastewater reuse and further feasibility and full application of model scenario 1 for new industrial parks developments in the Zayandeh Rud up to 9 million m^3/a, by application of model scenario 2: 13 million m^3/a and through linking example scenario 3: 15 million m^3/a water may be saved in comparison to industrial development without any reuse concepts in industrial parks.

The water saving potentials are different in each linking scenario. At the same time, the water conservation possibilities change not only through the different types of linkages but also depend on the sector specific water balances of the products. The water saving potential can be increased by lower evaporation rates. For large evaporation rates, it is important to consider whether the water can be avoided or even recovered e.g. by condensation. Theoretically, the savings potential can also be increased by linking different scenarios.

After the development of EIP scenarios, a workshop was held with planners, operators and decision makers of industrial parks in Isfahan, Iran. In the following points, critical factors and challenges for the practical implementation of industrial symbiosis were identified and are presented here with possible solutions:

- Economic incentives,
- Change of the business philosophy,
- Conflicts in the exchange process,
- Discontinuation or failure of a linked industry,
- Distance between industries,
- Data basis for planning process.

Economic incentives: Based on the results of interviews and workshops, most of the Iranian companies in the implementation of EIPs are concerned that there are no economic incentives for the industries. Currently, the costs of energy are very low and the water is almost free of charge. Separate groundwater wells need only one license, which costs a one-time fee. There are currently no additional charges for water withdrawal. It is assumed that the investment costs for linking the industries are too large to be overcome without additional economic incentives. The possibility to change water prices is in the hands of regional policy makers. Another approach is to create economic incentives in industrial parks and to let industries pay a fixed amount per month and provide a rebate depending on the degree of environmental friendliness. In the best case, where production is highly environmentally friendly, the industry has the entire amount refunded. The level of environmental friendliness is to be assessed by means of benchmarking, taking into account both specific water consumption, the origin of the water used and the reuse of wastewater. With either of these approaches to creating an economic incentive, it is assumed that the investment costs will no longer be seen as an inhibiting factor and can be easily amortized. In addition to creating a water price, it will be necessary to significantly reduce the number of groundwater wells. There are two options for better controlling the actual water withdrawal: (a) to change the decentralized water supply system into a central water supply system, or (b) metering systems for wells. This increases the economic incentive for companies to save water.

Upgrading the business philosophy: The change of business philosophy to "nature as a symbiosis model" (Dasgupta 2008) is viewed as critical since it is assumed that the industries are not interested in the environment. All in all, there is a way of focusing on the company and the potential maximization of profits. Sustainability does not yet play a role in most industries. In order to implement the idea of nature as a symbiosis model, the company assumes that without political guidelines, only minor changes will take place. To speed up the process of rethinking the idea that the industrial park management has a focus on "natural cycles as model" and in this way create awareness for the benefits of the symbiosis.

Conflicts in the exchange process: Critical conflicts are seen in the exchange process. The concern is that the responsibility for problems in the system is pushed from one participant to another. The quality of the wastewater provided is considered as a potential conflict. In order to keep conflicts in the exchange process as low as possible from the outset, the responsibility must be clearly defined. Park management should also be used to verify compliance with the agreed qualities and to resolve conflicts, for instance through benefit sharing approaches (Dombrovski 2010; UNESCO 2010). In this way, conflicts could be avoided and the security for the participants in the exchange process could be increased. In addition, a good basis for the cooperation of the parties should be created. For example, if deterioration in the quality of wastewater is suspected, then the exchange partner should be informed directly and both can look for a solution together.

Discontinuation or failure of a linked industry: For industries, it is critical to deal with exchange processes, since the functional safety is no longer given in the event of the discontinuation or absence of industry. The structure of the link is not variable enough to react to the discontinuation and failure of industry. Planning uncertainties are already occurring in the planning process since nobody knows which industry will finally settle in the industrial park. An emergency supply must be provided for the failure of an industry in the exchange process. In order to compensate for a complete loss of an industry, a replacement industry is required that meets the requirements. In order to be able to identify potential replacement industries, a database must be available that provides links to the specific requirement profiles of the industries.

Distance between industries: The distance between the linked industries is considered as a critical factor, and the cost of linking usually increases with distance. Feasibility is questioned at longer distances. The idea of overcoming the critical factor of distance is detailed planning of the arrangement of the industries. Industries should be arranged in such a way that linked industries have shorter distances. For this detailed arrangement, a planning basis must be developed in the form of a database, which shows possible links. Without this planning basis, creating the shortest possible connection is difficult.

Image: A lack of awareness of sustainable economic activity will lead to possible damage to the image of industries that use wastewater as process water. For example, if a dyeing factory uses wastewater as a process water, then it may not be tolerated since most people think of sanitation water as wastewater. For this reason, it is assumed that many industries categorically exclude the reuse of wastewater from another industry. The damage to the image must be reversed by awareness formation towards image enhancement. The awareness of sustainable economic activity could be increased with public relations.

Data basis for planning process: The current database for the EIP planning process is considered as a challenge since the few existing data do not provide information on minimum water quality requirements. Furthermore, low validation potential is considered critical because of the small number of data records. The database has to be significantly expanded by extensive measurement campaigns and the minimum requirements must be determined by experiments. The newly collected data will be used to develop links between industries to simplify the planning process. The increase is also expected to result in a validation of the data.

5 Conclusions and Outlook

In order to control the water problem in the catchment area of the Zayandeh Rud, the available water volume must be managed carefully and overexploitation must be avoided. The current water deficit needs to be reduced in the long term. International examples of watershed development show that the deficits can be successfully reduced through sustainable management (Gray and Srinidhi 2013). Resource sensitive industrial developments by EIPs can be a part of this sustainable management. The theoretical freshwater saving potentials calculated for the various linking scenarios are very high at

33 and 56%. The analogue sewage reduction potential is even larger. However, these values are only valid for ideal conditions and must be validated in the medium and long-term. In order to reduce the total water deficit, it will be necessary to reduce water consumption in all sectors as industrial water consumption currently accounts for approximately 2% of total water withdrawal in the catchment.

The practical implementation of an EIP requires a relentless problem analysis, viable solutions and the courage to overcome a potential system challenge. The complexity of ecosystems and industrial systems is currently too large to allow a full understanding of all factors or to allow management which incorporates planning down to every last detail. The implementation of EIPs in existing industrial parks is possible but is difficult to implement due to the already existing structures and arrangements and therefore, comparatively unlikely. The potential for linkage according to the bilateral principle is significantly larger in this case because of the lower conflict potential and the lower complexity compared to linkage according to the nucleus or cascade principles.

The EIP concept should be the design philosophy for new industrial park developments, which is based on benefit-sharing strategies (Soliev et. al. 2015; Jalilov et. al. 2015; Belinskij 2015; Hensengerth 2015). The further development of the benefit-sharing concept on the watershed scale may be conceptualized as "benefit shed" (Avellan et. al. 2017; Schneider and Avellan 2019), through the assessment of resource flows within a Nexus context. A benefit shed is apparent when at least two physical resource systems interact and thus form a Nexus system, where stakeholders involved aim to reveal mutual benefits through increased efficiency (Avellan et al. 2017; Schneider and Avellan 2019; Schneider et al. 2019a, b), which is particularly apparent in an industrial symbiosis for wastewater sharing. Benefit sharing approaches as well as the benefit shed analysis process present promising options for ensuring water security. In the future, further activities could be developed to evolve research from an IWRM to a Water-Waste-Energy-Food Nexus. In order to reduce water scarcity, the further scope might consider re-linking all saved water volumes of the respective sectors considered under a Water-Energy-Food Nexus approach to ensure water security in the long-term (Schneider et al. 2019a).

In further research, the potential to expand the EIP concept and industrial symbiosis for all resource flows between companies (e.g. energy, material, heat) should be elaborated. Water as a universal processing and transportation medium is a good starting point for elaborating potentials and prerequisites for installing EIPs. For holistic sustainable development, other resources are of great importance and should be targeted. Based on the well-known types of industrial symbiosis such as EIP, Eco-innovation Parks, Regional Recycling or Utilization Network, Zero-Emission Parks, Cradle to Cradle Business Innovation and Improvement Zones, or Sustainable Industrial and Business Areas, the concept of industrial symbiosis also offers great potential for the new developments of Industry 4.0. In Industry 4.0, production meshes with state-of-the-art information and communication technology, driven by the rapidly increasing digitization of the economy and society.

Acknowledgements The investigation was funded by the German Federal Ministry of Education and Research (BMBF) within its funding measure "Integrated Water Resources Management" as part of the framework programme "Research for Sustainability". Reference code: 02WM1353A-G.

References

Avellan T, Roidt M, Emmer A, von Koerber J, Schneider P, Raber W (2017) Making the water-soil-waste nexus work: framing the boundaries of resource flows; Sustainability 9:1881. 10.3390/su9101881

Belinskij A (2015) Water-energy-food nexus within the framework of international water law. Water 7:5396–5415. https://doi.org/10.3390/w7105396

Benson D, Gain AK, Rouillard JJ (2015) Water governance in a comparative perspective: from IWRM to a "Nexus" approach? Water Altern-An Interdisc J Water Polit Dev 8:756–773

Bielsa J, Cazcarro I (2014) Implementing integrated water resources management in the Ebro River Basin: from theory to facts. Sustainability 7:441–464. https://doi.org/10.3390/su7010441

Boons F, Spekkink W, Mouzakitis Y (2011) The dynamics of industrial symbiosis: a proposal for a conceptual framework based upon a comprehensive literature review. J Cleaner Prod S:905–911

Braungart M, McDonough W, Bollinger A (2007) Cradle-to-cradle design: creating healthy emissions a strategy for eco-effective product and system design. J Cleaner Prod 15(13–14):1337–1348

Cai X, Wallington K, Shafiee-Jood M, Marston L (2018) Understanding and managing the food-energy-water nexus—opportunities for water resources research. Adv Water Resour 111:259–273. https://doi.org/10.1016/j.advwatres.2017.11.014

Cencic O, Rechberger H (2008) Material flow analysis with software STAN. J Environ Eng Manage 18(1):5

Chertow M (2007) "Uncovering" industrial symbiosis. J Ind Ecol S:11–30

Chertow M, Park J (2016) Scholarship and practice in industrial symbiosis: 1989–2014. In: Clift R, Druckman A (eds) Taking stock of industrial ecology, Springer, Cham, pp 87–116. https://doi.org/10.1007/978-3-319-20571-7_5. ISBN 978–3–319–20570–0

Dasgupta P (2008) The place of nature in economic development. The Beijer Institute of Ecological Economics, Beijer Discussion Paper Series 216, Royal Swedish Academy of Sciences, Sweden, p 14

Davies BL (2007) Grice's cooperative principle: meaning and rationality. J Pragmatics 39(12):2308–2331. ISSN 0378-2166. https://doi.org/10.1016/j.pragma.2007.09.002

De Strasser L, Lipponen A, Howells M, Stec S, Bréthaut C (2016) A methodology to assess the water energy food ecosystems nexus in transboundary river basins. Water 8:59. https://doi.org/10.3390/w8020059

Dombrovki I (2010) Benefit-sharing in transboundary water management through intra-water sector issue linkage? In: Lundqvist J (ed) On the water front. Stockholm International Water Institute (SIWI), pp 25–31

Ghanavizchian S, Mohajeri S (2013) Coping with culture—conducting participation projects in Iran. In: Schön S, Mohajeri S, Dierkes M (Hg.): Machen Kläranlagen glücklich? Ein Panorama grenzüberschreitender Infrastrukturforschung, Berlin, inter 3, S, pp 108–115

Gray E, Srinidhi A (2013) Watershed Development in India: Economic Valuation and Adaptation Considerations. Working paper. World Resources Institute, Washington, DC. http://www.wri.org/publication/watershed-development-india-economic-valuation-adaptation-considerations

Grice P (1975) Logic and conversation. In: Cole P, Morgan J (eds) Syntax and semantics. 3: Speech acts. Academic Press, New York, pp 41–58

Grigg NS (2008) Integrated water resources management: balancing views and improving practice. Water Int 33 (3):279–292

Global Water Partnership GWP (2000) Integrated water resources management, TAC Background Papers No. 4, Stockholm, Sweden, ISBN: 91-630-9229-8

Global Water Partnership GWP (2004) Catalyzing change: a handbook for developing Integrated Water Resources Management (IWRM) and water efficiency strategies. http://www.un.org/esa/sustdev/csd/csd13/documents/bground_5.pdf. Accessed 10 Jan 2017

Hemphill T (2013) The ISO 26000 guidance on social responsibility international standard: what are the business governance implications? Corporate Governance: The Int J Bus Soc 13(3):305–317. https://doi.org/10.1108/CG-08-2011-0062

Hein AM, Jankovic M, Farel R, Yannou B (2015) A conceptual framework for eco-industrial parks. In: Proceedings of the ASME 2015 international design engineering technical conferences & computers and information in engineering conference IDETC/CIE 2015

Hensengerth H (2015) Where is the power? Transnational networks, authority and the dispute over the Xayaburi dam on the lower mekong mainstream. Water Int 40 (5–6):911–928

Hewes A, Lyons DI (2008) The humanistic side of eco-industrial parks: champions and the role of trust. Reg Stud 42(10):1329–1342. https://doi.org/10.1080/00343400701654079

Hülsmann S, Ardakanian R (eds) (2014) Advancing a nexus approach to the sustainable management of water, soil and waste (white book). UNU-FLORES, Dresden

Iran Daily (2015) http://www.iran-daily.com/News/121143.html. Accessed 17 March 2019

Isenmann R (2008) Settling the boundaries and highlighting the scientific profile of Industrial Ecology. Information Technologies in Environmental Engineering, Special Issue January, 1. Jg. Heft 1, pp 32–39

Jalilov JM, Varis O (2015) Keskinen M (2015) Sharing benefits in transboundary rivers: an experimental case study of Central Asian water-energy-agriculture nexus. Water 7:4778–4805. https://doi.org/10.3390/w7094778

Kalundborg Symbiosis (2014) Kalundborg symbiosis 40th anniversary: Grafisk forum A/S. Denmark

Kirschke S, Horlemann L, Brenda M, Deffner J, Jokisch A, Mohajeri S, Onigkeit J (2016) Benefits and barriers of participation: experiences of applied research projects in integrated water resources management. In: Borchardt D, Bogardi JJ, Ibisch RB (eds) Integrated water resources management: concept, research and implementation. Springer International Publishing Switzerland, ISBN: 978–3–319–25069–4 (Print) 978–3–319–25071–7 (Online), pp 303–331

Lambert AJD, Boons FA (2002) Eco-industrial parks: stimulating sustainable development in mixed industrial parks. Technovation 22(8):471–484

Lifset R, Graedel TE (2001) Industrial ecology: goals and definitions. In: Ayres RU, Ayres L (eds) Handbook for industrial ecology. Edward Elgar, Brookfield. http://www.elgaronline.com/view/1840645067.00009.xml

Liu J, Yang H, Cudennec C, Gain AK, Hoff H, Lawford R, Qi J, de Strasser L, Yillia PT, Zheng C (2017) Challenges in operationalizing the water–energy–food nexus. Hydrol Sci J 62:1714–1720. https://doi.org/10.1080/02626667.2017.1353695

Massard G, Jacquat O, Zürcher D (2014) International survey on eco-innovation parks. Learning from experiences on the spatial dimension of eco-innovation, Federal Office for the Environment and the ERANET ECO-INNOVERA, Bern. Environmental Studies, No. 1402

McDonough W, Braungart M (2002) Cradle to cradle: remaking the way we make things. North Point Press, New York

Mohajeri S, Nuñez von Voigt T (2011) Integrated water resources management in Iran. Isfahan, Iran: successfully shaping international technology and knowledge transfer. Water Wastewater Technology WWT International. Trade J Water Wastewater Manage. Special issue 2011. HUSS-MEDIEN GmbH, Berlin, pp 29–32

Mohajeri S, Horlemann L, Sklorz S, Kaltofen M, Ghanavizchian S, Nuñez von Voigt T (2016) Integrated water resource management in Isfahan: The Zayandeh Rud Catchment. In: Borchardt D, Bogardi JJ, Ibisch RB (eds) Integrated water resources management: concept, research and implementation. Springer International Publishing Switzerland, ISBN: 978–3–319–25069–4 (Print) 978–3–319–25071–7 (Online), pp 603–627

Mohajeri S, Horlemann L (2017a) Projekt “IWRM Zayandeh Rud” in Isfahan/Iran. Fachzeitschrift Für Leitungsbau, Umwelttechnik Und Tiefbau Nr. 6 (17):20–22

Mohajeri S, Horlemann L (eds) (2017b) Reviving the Dying giant—integrated water resources management in the Zayandeh Rud Catchment. Springer Nature, Cham/Schweiz, Iran

Mohajeri S, Horlemann L, Torfeh MA, Besalatpour AA (2018) Bringing an IWRM concept into practice—a case study from the Zayandeh Rud River Basin/Iran. In: KA International Special Edition 2018/19, GFA Hennef/Germany

Muller M (2015) The “Nexus” as a step back towards a more coherent water resource management paradigm. Water Altern-An Interdisc J Water Polit Dev 8:675–694

Piadeh F, Moghaddam MRA, Mardan S (2014) Present situation of wastewater treatment in the Iranian industrial estates: recycle and reuse as a solution for achieving goals of eco-industrial parks. Resour, Conserv Recycl 172–178. https://doi.org/10.1016/j.resconrec.2014.06.004

Raber W, Mohajeri S (2017) Current and future industrial water use in the Zayandeh Rud Catchment. In: Mohajeri S, Horlemann L (eds) Reviving the dying giant—integrated water resource management in the Zayandeh Rud Catchment, Iran; Springer International Publishing, Cham, Switzerland, pp 95–115

Rifkin J (2011) The third industrial revolution: how lateral power is transforming energy, the economy, and the world. Martin’s Press, St

Rifkin J (2014) The zero marginal cost society: the internet of things, the collaborative commons, and the eclipse of capitalism. Martin’s Press, St

Schneider P, Avellan T (2019) Water security and sustainability. In: Leal Filho W (eds) Encyclopedia of sustainability in higher education. Springer, Cham. ISBN 978–3–319–63951–2. https://doi.org/10.1007/978-3-319-63951-2

Schneider P, Avellan T, Le Hung A (2019a) Water-energy-food nexus and sustainability. In: Leal FW (ed) Encyclopedia of sustainability in higher education, Springer, Cham. ISBN 978–3–319–63951–2. https://doi.org/10.1007/978-3-319-63951-2

Schneider P, Folkens L, Meyer A, Fauk T (2019b) Sustainability and dimensions of a nexus approach in a sharing economy. Sustainability 11(3):909. https://doi.org/10.3390/su11030909

Soliev I, Wegerich K, Kazbekov J (2015) The costs of benefit sharing: historical and institutional analysis of shared water development in the Ferghana Valley, the Syr Darya Basin. Water 7:2728-2752. https://doi.org/10.3390/w7062728

Theurl T (2015) Ökonomie des Teilens: Governance konsequent zu Ende gedacht. Wirtschaftsdienst 95:87–91

United Nations (2015) Transforming our world: the 2030 Agenda for Sustainable Development, United Nations—Sustainable Development knowledge platform. Accessed 25 Sep 2015

United Nations Educational, Scientific and Cultural Organization UNESCO (2009) IWRM guidelines at River Basin Level—Part 2–1: The guidelines for IWRM coordination. http://unesdoc.unesco.org/images/0018/001864/186418e.pdf. Accessed 10 Jan 2017

United Nations Educational, Scientific and Cultural Organization (UNESCO) (2010) Sharing water, sharing benefits: working towards effective transboundary water resources management, a graduate/professional skills-building workbook. Compiled and edited by Prof. Aaron T. Wolf

UN-Water (2015) World Water Development Report. https://www.unwater.org/publications/world-water-development-report-2015/. Accessed 15 March 2019

von Hauff M (2012) Industrial ecology management: Nachhaltige Entwicklung Durch Unternehmensverbünde, Gabler Verlag. ISBN-13: 978–3834923615

von Koerber J (2016) Concept for the resource-efficient use of water using the example of the development of Eco Industrial Parks in Iran (in German), Master thesis at the University of Applied Sciences Magdeburg-Stendal

City-to-City Learning Within City Networks to Cater City Needs to Climate Adaptation—Results of a Preliminary Study

Wolfgang Haupt, Chris Zevenbergen, and Sebastiaan van Herk

Abstract

Over the past years, a considerable number of cities joined together in networks that address climate more and more increasingly also adaptation. The presented research focuses on one key aspect of those networks: The sharing of knowledge and experience among the member cities, referred to as city-to-city learning. Due to climate change cities worldwide are facing enormous challenges that require an acceleration of the learning process of how to respond to them. Transnational city networks can help to provide a platform for learning from each other and connect cities to work together. However, the questions are how useful and effective these knowledge sharing processes are in practice, and how existing networks can facilitate and improve them. This was examined through interviews with key stakeholders representing cities that joined transnational networks. Main results were that cities are interested in learning from their peer cities that face similar challenges, and to follow the examples and adopt the strategies of pioneering cities. The contribution of networks were particularly seen in (i) facilitating knowledge sharing, (ii) promoting the adaptation and resilience work within the own city council, (iii) providing (easier) access to funding and, most importantly, (iv) enabling the establishment of informal city-to-city relationships based on mutual trust. Furthermore, some city representatives asked for a stronger involvement of the private sector in order to finance the municipalities' climate change action, whereby the network could act as a facilitator.

Keywords

Adaptation · City-to-city learning · Peer learning Policy learning · Resilience · Transnational city networks

1 Introduction

Cities nowadays are facing unprecedented challenges and opportunities. Climate change is increasingly putting pressure on cities manifesting in a higher incidence of flooding, heatwaves, water scarcity and the like. Hosting the majority of the world's population and being at the forefront of many issues deriving from climate change also results in an increased awareness of

W. Haupt (✉)
Leibniz Institute for Research on Society and Space, Erkner, Germany
e-mail: wolfgang.haupt@leibniz-irs.de

C. Zevenbergen
IHE Delft Institute for Water Education, Delft, The Netherlands

S. van Herk
Bax and Company, Barcelona, Spain

S. Hülsmann and M. Jampani (eds.), *A Nexus Approach for Sustainable Development*,
https://doi.org/10.1007/978-3-030-57530-4_14

the cities' role and responsibility in coping with these challenges. Responding to the challenges, and seizing the opportunities, is placing new and complex demands on municipal decision makers. Furthermore, it requires a more integrated approach (Nexus Approach, see e.g. Kurian and Ardakanian 2015), for instance through mainstreaming policies and investments. Climate change adaptation is not a sector of its own but applies to a wide range of domains (e.g. transport and infrastructure, flood protection, institutional capacity) and is critically linked to a complex system of interlinked resources. As outlined by the UNU Institute for Integrated Management of Material Fluxes and of Resources (UNU-FLORES) applying a Nexus Approach means that "instead of just looking at individual components, the functioning, productivity, and management of a complex system is taken into consideration" (UNU-FLORES nd). Therefore, integrated resources management, adopting nexus thinking should be instrumental to cope with virtually all aspects of global change (Hettiarachchi and Ardakanian 2016). Indeed, during the 2017 Dresden Nexus Conference (DNC) case studies from Dresden, Prague, Brno and Pilsen could demonstrate the urgent need and success of nexus-oriented approaches in climate change adaptation planning (see Dresden Nexus Conference 2017).

One promising way to meet and coordinate those more and more complex needs is the engagement of key municipal policymakers in peer learning exchanges. The window of opportunity to respond to climate change is limited, and the financial means of cities are often restricted. For that reason, there is neither enough time nor are there sufficient resources for extensive (isolated) trial and error responses. The emphasis on accelerated learning stems from the insight that combined processes of peer learning and learning from experiments and full-scale pilots will be crucial. Increasingly, these knowledge and experience sharing and learning processes are referred to as city-to-city learning (C2CL) (Seymoar et al 2009; van Herk et al. 2016a; Lundby and Sjöberg 2013; Koop and van Leeuwen 2017). C2CL exchanges are increasingly being promoted and facilitated by several transnational climate city network initiatives on which this chapter focuses. Against a background of nation states struggling to deal with major global issues, such as climate change, cities are increasingly positioning themselves as alternative global actors and problem solvers (Barber 2013; Sassen 2013). The emergence of transnational networks in which cities join together globally are one of the expressions that clearly reflect this shift. Transnational city networks, such as the C40 Climate Leadership Group, Local Governments for Sustainability (ICLEI), 100 Resilient Cities (100RCN) or the Making Cities Resilient Campaign of the United Nations Office for Disaster Risk Reduction (UNISDR) and many others, offer an avenue for cities that are building resilience and to achieve the United Nations (UN) Sustainable Development Goals (SDGs). This relates in particular to the goals number eleven (Sustainable cities and communities) and thirteen (Climate action) (see UN nd). Given the interrelatedness of SDGs, it has been argued that similar to addressing challenges posed by global change (see above), an integrated approach is required to achieve SDGs (Hülsmann and Ardakanian 2018). Above all, the contribution of these C2CL networks can be seen by connecting cities to each other and to potential sources of funds and knowledge (van Herk et al. 2016b).

The steadily growing number of transnational city networks, the high number of cities joining them globally and their potential significance for tackling climate change stands in a certain contradiction to the rather little research that has been conducted on them. Generally, a more careful analysis of those networks is urgently required (Bouteligier 2013; Lee 2013; Gordon and Acuto 2015; Gordon 2016). Furthermore, it needs to be better understood why municipalities decided to join them and what they expected from the membership (Gore 2010; Niederhafner 2013).

Particular attention is paid to an under-explored but not less important feature of transnational city networks: The sharing of knowledge and experiences among member cities, which was identified as the main feature of those networks by previous research (Bulkeley et al. 2003; Andonova et al. 2009; Newig et al. 2010; Feldman 2012; Bouteligier 2013; Zeppel 2013; Fenton and Busch 2016). Given the described urgent need to adapt to climate change, the simple sharing of knowledge and experience is not enough. To keep pace and cope with the rapid changes occurring in cities nowadays requires accelerated learning through learning from each other (peer learning). Also for this reason, the understanding of those learning processes through international cooperation needs to be enhanced (Vinke-de Kruijf and Pahl Wostl 2016). Besides, Newig et al. (2010) outlined that the effects of those transnational networks 'on governance-related variables such as learning have been under-researched, leaving the question of "do networks matter?"'

Overall, this chapter aims to critically understand the relationship between transnational city networks and cities climate governance, unveiling which are the learning processes offered by or leveraged from the networks. Therefore, it analyses (i) in which forms C2CL is taking place among member cities, (ii) as to how useful it is perceived by the involved municipal policy makers and (iii) to what extent the networks are catering to the city needs, while leveraging learning processes to become more climate resilient. These questions have been addressed through 23 in-depth interviews with municipal stakeholders across 16 countries, aiming to provide a global picture.

2 Understanding City-to-City Learning

2.1 Definition, Objective and Learning Levels of City-to-City Learning

Academic definitions of C2CL are quite rare. For Koop and van Leeuwen (2015) it aims at 'improving the implementation capacities of cities by sharing information'. Here, 'cities are encouraged to participate in learning alliances to actively share knowledge and experiences on the implementation of state-of-the-art technologies' (Koop and van Leeuwen 2017, 412). Lundby and Sjöberg (2013) describe it as a form of learning where ideas, experiences and support are shared between cities. Fisher (2014) pointed out that C2CL is often performed within smaller projects that prioritise C2C exchanges on a certain issue. It seems that also in academic publications the term is often used in a not very precise manner, not uncommonly just as a bullet point or as part of a listing taking for granted its meaning. It is referred to C2CL in the fields of smart cities (Calzada 2016; Campbell 2012), low carbon urbanism (Thornbush 2016; Beermann 2014), resilience (Mackee and Giggins 2015; Haigh et al. 2014), water use (Koop and van Leeuwen 2015, 2017), intelligent transport systems (Kolosz and Grant-Muller 2016) but also more remote areas such as mega events in cities (Azzali 2016) or international security (Engelke 2013).

The term C2CL was first used by Seymoar et al. (2009) in the framework of a study that examined the knowledge sharing processes between cities that joined the PLUS sustainability network. Main results were 'that local administrators learn mainly from their peer group' and that 'Networks emerge as a key strategy for speeding the exploitation of knowledge and learning, both formally and informally, in and between local governments' (Seymoar et al. 2009). In the meantime, the term C2CL was substantially shaped by IHE Delft (Institute for Hydrological Education) for UNISDR in the context of disaster resilience. They see the objective of C2CL in building.

> greater institutional and human capacity required to accelerate progress towards disaster resilience. A distinction from other capacity building services, such as technical training, is that cities will learn together. The promise for cities is to become learning organizations by engaging in city learning networks (Van Herk et al. 2016a).

UNISDR regards accelerated learning among municipal decision makers through exchanges (peer learning) as a key component of a cities' response to the opportunities and challenges caused by a rapidly changing urban environment (ibd.). Van Herk et al. (2016a) distinguished three levels of learning. Firstly, *reactive learning* meaning 'the discovery of existing knowledge, from searching for lessons learned by other cities'. Secondly, *pro-active learning* is 'based on inspiration and joint creation of new knowledge. This will take place during exchange activities (e.g., sharing of information, good practices and ideas amongst peers) and the execution of joint projects'. Lastly, there is 'learning from the learning process itself to secure the capacity to learn' defined as *meta-learning*. Here, two decisive components are continuous monitoring and evaluation (Van Herk et al. 2016a). Van Herk et al. (2016a) listed a number of reasons why cities should engage in C2CL processes, namely that it: (i) 'accelerates the transfer of knowledge and experience from one or more cities, to one or numerous others', (ii) 'fosters the joint creation of new knowledge in cities, in turn empowering local governments: Creating a collective voice', (iii) 'reinforces local networks by stimulating active participation of policy makers, practitioners, the private sector, NGOs and citizens, paving a way for development of beneficial approaches and multi- stakeholder partnerships', (iv) 'enhances the collective memory of participating cities', (v) 'provides opportunities to leverage and aggregate resources for joint activities (e.g. research programs)' and (vi) 'increases the capacity of cities to learn overall'.

2.2 City-to-City Learning Within a Broader Context

The knowledge and experience exchanges and learning among government officials have been investigated intensively, which has resulted in a considerable number of different literature streams. Generally, C2CL can be understood as a specific form of peer learning and fits into the literature on policy learning, lesson drawing but also shares few similarities with theoretical frameworks such as policy transfer and policy mobility. Bennett and Howlett (1992) noted that there is a 'definitional ambiguity' manifesting in different authors using the same terms for learning. For instance, when discussing the relationship between policy learning and policy change, while having very different notions of it. Without attempting to go into detail, this paragraph aims to introduce the most relevant synonymous literature and differentiate it from C2CL as understood in this research.

As mentioned before, the logic behind C2CL mainly stems from peer learning. According to Andrews and Manning (2016, 4) peer learning should be understood as 'a potentially powerful way of sharing knowledge about doing public sector reform' and describe it as follows:

> This learning involves individuals exchanging knowledge and experience with each other, and diffusing this learning back to their organisations to ensure an impact—at scale—on reform initiatives. While peer learning entails complex organisational logistics, it avoids the risk of focusing on process rather than product. It recognises that ultimately learning takes place between individuals and it facilitates interpersonal interchanges that are well- matched and that are based on trust and commitment.

Andrews and Manning (2016) also highlighted the importance of the 'political' in peer learning since public sector management cannot be seen isolated from political influences. They emphasise that 'Peer learning replaces abstract notions of "vision" and "political will" with an emphasis on practical problem-solving' (Andrews and Manning 2016).

In a recent literature review on policy learning in the social sciences, the term was defined as 'the updating of beliefs based on lived or witnessed experiences, analysis or social interaction' (Dunlop and Radaelli 2013, 599). Technically speaking policy learning takes place 'when policy-makers adjust their cognitive understanding of policy development and modify policy in the light of knowledge gained from past policy experience' (Stone 2004). A subcategory of policy learning is government learning which describes 'the process by which governments increase their intelligence and sophistication and in this manner enhance the effectiveness of their actions (Etheredge 1981, cited in Bennett and Howlett 1992). This essentially outlines how C2CL should be understood.

Just like government learning also lesson drawing comes very close to how C2CL is understood in this chapter. Rose (1991) defined it as 'a particular type of learning in which policy-makers learn from both, the positive and negative experiences of others: 'confronted with common problems, policy-makers in cities, regional governments, and nations can learn from how their counterparts elsewhere respond. More than that, it raises the possibility that policy-makers can draw lessons that will help them deal better with their own problems' (Rose 1991). Also, lesson drawing should be understood as a framework 'which favors an analytical approach in which learning from evidence-based actions, rather than just copying "best practices," is a preferred form of policy transfer and considered key to successful policy change' (Radaelli 2004 and Rose 2002 cited in Montero 2017).

Another often applied theoretical framework when examining the sharing of information and knowledge is policy transfer. Policy transfer documents and explains the (international) movement of policies (Lovell 2016) and aims to understand the different factors that restrict or facilitate this transfer process (Dolowitz and Marsh 1996, 2000). It relates to a notion of some national policymakers who follow the assumption that importing 'innovatory policy developed elsewhere in the belief that it will be similarly successful in a different context' (Stone 1999). An approach that mainly emerged as a critical response to policy transfer is policy mobility. It highlights the mobile and changeable character of policies, assuming that 'policies rarely travel as complete "packages," they move in bits and pieces—as selective discourses, inchoate ideas, and synthesised models—and they therefore "arrive" not as replicas but as policies already-in-transformation' (Peck and Theodore 2010, 170). In sum, policy mobility acknowledges how knowledge is mobilised and aims at understanding how ideas become hegemonic and global (McCann 2011). What significantly distinguishes C2CL from both, policy transfer and policy mobility is the rather minor role of the actual policy. Instead C2CL, as understood in this chapter, focuses on the learning processes among individuals of different cities and how they can be established and optimised.

To sum it up, the authors understand C2CL as a process of mutual learning among municipal policymakers and their peers in other cities that face comparable challenges and work in similar environments. Those learning exchanges are of particular importance because of rapidly occurring changes, with one of the main causes being climate change, and the consequential necessity to adapt fast. The emphasis is on a two-way traffic flow of information. This means that the involved individuals should mainly learn and improve together. Also, it is important to keep in mind that C2CL never means city A is learning from city B but person A from city A is learning from and with person B from city B. Ultimately, the social and personal skills of the involved stakeholders will determine the outcome and development of learning exchanges among cities.

3 Exploring Transnational City Networks that Address Climate Change Adaptation

3.1 Networks in the Academic Literature

Cities are increasingly playing a key role within the global debates and policies about climate governance (Bulkeley 2010; Rosenzweig et al. 2010). The start was made in the early nineties when a few pioneering municipalities framed the first transnational climate change networks. Their aim was to share data, experiences and roadmaps about their policies (Bulkeley and Schroeder 2012). Subsequently, in the last decade, there has been a consistent growth of cities' networks (ibd.). The key move was about local governments taking the lead ahead of their national governments, aspiring to put their names on the global map (Pattberg and Widerberg 2015). Increasingly considered as global climate governors (Gordon, 2016) cities started facing an emerging tension between contributing to meaningful global climate governance and addressing the specific and practical local challenges (Gordon and Acuto 2015).

The shift from government to network governance provides local institutions access to flows of opportunities, while 'creating the illusion of empowering all even if in practice, accountability, legitimacy, legality and equity are compromised as the most powerful actors influence the whole governance process' (Gupta et al. 2015, 217). Beyond the need of proving with specific metrics and indicators the impact on the ground of those network governance, the trend is that municipalities have globally become more connected to each other, since together having more influence than they would separately (Andonova et al. 2009). Indeed, 'the analysis of various networks shows that environmental challenges are being addressed through network management efforts because individual local governments lack the capacity or resources to address some issues without the cooperation of neighbouring municipalities while national governments deliver incentives' (Giest and Howlett 2013). For instance, European networks were able to establish a variety of knowledge transfer and best practice learning mechanisms over the years by enabling collaboration (Bradford 2004). Also, since experimenting with different possible solutions is a feature of urban responses to climate change across the world, sharing information, practices and results of those pioneering activities could be seen as a driver for networking (Bulkeley and Castán Broto 2012). Policy learning, from other cities and professionals, has been indeed reported as one of the core 'soft and intangible results of city networks' (Pattberg and Widerberg 2015).

In 2005 Selin and Vandeveer predicted that 'Information about regional climate change and policy experiences will move beyond the region via the many (often overlapping) networks of activists environmental and public health NGOs, professional organizations, and networks of state and local officials and administrators (Selin and Vandeever 2005). The large number of best practice examples circulated by different networks (e.g., the Benchmark of Excellence for Covenant of Mayors signatories) suggests that this has become reality (Haupt 2018). However, putting the emphasis on learning it has to be noted that the actual usefulness and applicability of best practices, particularly when shared among 'unequal' partners, is highly questioned in the academic literature (e.g. Wolman and Page 2002; Stead 2012). For instance, a survey of local government officials in the UK conducted by Wolman and Page in 2002 indicated that only a small minority of them believed that best practices play a large or significant role in their decision-making.

Urban climate change governance still mainly focuses on the mitigation agenda. This misbalance was pointed out by Bulkeley and Schroeder (2012), and it appears that it is still the case. De Gregorio Hurtado et al. (2015) analysed the urban climate change policies in Spain and Italy —the two countries with the by far highest participation rate in the Covenant of Mayors initiative—and found out that the vast majority of local action plans still have to be considered as pure mitigation plans. However, in the meantime, some networks have paid more attention to adaptation, for instance, the Covenant of Mayors, the new Global Covenant of Mayors or ICLEI. Besides, new networks focusing on adaptation only such as 100RC or Asian Cities Climate Change Resilience Network (ACCCRN), have emerged. This development is well consistent with the general debate, where the climate summit in Paris 2015 triggered a shift from mitigation to adaptation by promoting the idea to consider them both as equal components of climate change action (UNFCCC 2015).

3.2 Different Types of Networks

This section provides an overview of the city networks regarded in this chapter and aims to highlight the main differences among them. It aims to illustrate that there is not 'the city network'. Quite the contrary, city networks addressing climate change vary significantly, for instance in terms of size, organizational structure or geographical and thematic scope (Haupt and Coppola 2019). Table 1 gives an overview of the networks addressed in the study and provides some basic information (network size, countries represented in the network and the year of the networks establishment).

There were three main selection criteria for networks to be analysed: It should be a transnational network, it should (also) tackle resilience and/or climate adaptation and it should have a certain focus on learning and knowledge sharing among its members. The listed networks were inspired by a compilation of transnational climate networks developed by Giest and Howlett (2013). However, the networks listed that did not address climate adaptation or resilience, (e.g. 'only' climate change mitigation) were excluded. Additionally, new networks that emerged later on and fulfilled the listed requirements were added. However, no claim is made that this is a complete listing. The information and data on the different networks presented here derive from openly accessible sources provided by the networks (websites and reports).

Looking at the genesis of the different transnational city networks, we distinguish the following four different types of networks: Networks (i) initiated by cities, (ii) launched by a supranational organization, (iii) established by a foundation and (iv) that resulted from the merge of already existing networks. ICLEI, C40, EUROCITIES and Energy Cities are examples for a number of cities joining together in order to establish a network. While ICLEI, C40 and Energy Cities had a focus on sustainability and climate change related issues from the beginning on, EUROCITIES was founded to advocate for a leading role of local governments (C40 2018; EUROCITIES 2017; Energy Cities nd). The membership in EUROCITIES and C40 is restricted to cities with a certain population, whereas Energy Cities and ICLEI are open to municipalities of all sizes (ibd.). Examples for networks that were established by a supranational organ are the resilience and climate change adaptation initiatives Making Cities Resilient Campaign (UNISDR) and Mayors Adapt (European Commission), Cleaner Energies Saving Mediterranean Cities (n.d), UNISDR (ndb). In

Table 1 Geographic scope, thematic focus and financing of the city networks

Transnational municipal network	Founding year	Spatial reach	Number of members
EUROCITIES	1986	Europe	194
Citynet	1987	Asia	88
Climate Alliance	1990	Europe	1719
Energy Cities	1990	Europe	2730
*Local Governments for Sustainability **(ICLEI)***	1990	Global	798
*United Nations International Strategy for Disaster Reduction **(UNISDR)***	1999	Global	3862
URBACT	2003	Europe	550
*C40 Leadership Group **(C40)***	2005	Global	91
*Asian Cities Climate Change Resilience Network **(ACCCRN)***	2008	Asia	12
*Covenant of Mayors **(CoM)***	2008	Europe	7755
*Cities Development Initiative For Asia **(CDIA)***	2007	Asia	140
*100 Resilient Cities **(100RC)***	2013	Global	100
Compact of Mayors	2014	Global	7516

Source Citynet (nd), Climate Alliance (nd), Compact of Mayors (nd), UCLG (nda), UNSIDR (ndb), Covenant of Mayors (nd), Energy Cities (nd), ICLEI (nd), URBACT (nd), CDIA (2014), EUROCITIES (2017), ACCCRN (2018), 100RC (2018), C40 (2018)

the meantime, Mayors Adapt was fully integrated into the Covenant of Mayors, another European Commission initiative that originally solely focussed on climate change mitigation (Cleaner Energies Saving Mediterranean Cities nd). Both 100RC and ACCCRN were launched by the Rockefeller Foundation to promote the reinforcement of urban resilience (100RC 2018; ACCCRN 2018).

The Compact of Mayors represents a special case since it is an initiative that tries to unite a number of already existing networks, including ICLEI, C40, CityNet UCLG and the Covenant of Mayors (Compact of Mayors nd). The idea is not to replace existing networks but to establish a standardised way of reporting and monitoring of a cities climate action measures, for instance through robust and transparent data collection standards (ibd.). Most city networks try to make their member cities efforts and success representable and measurable through different monitoring and evaluation systems. However, this is not an easy task and this has also been identified as one big research gap on city networks and urban climate governance (see Bulkeley 2010; Bulkeley and Schroeder 2012; Giest and Howlett 2013; James and Verrest 2015).

As Table 2 illustrates, there are various further criteria for a more in-depth classification of the networks, like their geographic scope (regional, global), thematic focus (resilience, adaptation, beyond that), and the financing of the network (membership fee, no membership fee).

A general pattern that could be observed was that city networks with a membership fee do not require strict commitments from their members. The situation is different in (some) networks without a membership fee. For instance, Mayors Adapt or the Compact of Mayors do not ask for paying a membership fee but then again expect the joining municipalities to set up an action plan that details their strategy to reach the adaptation and mitigation goals (Covenant of Mayors, Compact of Mayors nd). If a municipality was unable to fulfil its obligations or missed the action plan submission deadline, Mayors Adapt or the Covenant of Mayors removed them from the list of participants and instead list them as 'on hold' (see Covenant of Mayors nd). In the 100RC network there is a more classic 'give and take' relationship between the network and its members: The cities have to assign a person in charge, the Chief Resilience Officer (CRO) financed by 100RC. Then they have to develop a

Table 2 Geographic scope, thematic focus and financing of the city networks

Network	Geographic Scope	Focus on resilience or adaptation	Broader focus	Membership fee
100RC	Global	X		
C40	Global		X	
UCLG	Global		X	X
UNISDR	Global	X		X
Compact of Mayors	Global		X	
ICLEI	Global		X	X
Energy Cities	Europe		X	X
EUROCITIES	Europe		X	X
Mayors Adapt	Europe	X		
ACCCRN	Asia	X		X
CityNet	Asia		X	X

Source 100RC (2018), ACCRN (2018), EUROCITIES (2017), ICLEI (nd), C40 (2018), CityNet (nd), Compact of Mayors (nd), Covenant of Mayors (nd), Energy Cities (nd), UCLG (nda), UNSIDR (ndb)

suitable resilience strategy and for which they receive financial, logistical and expert guidance and support in return (100RC 2018).

4 Methodology

The results presented in this chapter are based on key stakeholder interviews that were carried out between April 2016 and 2017. The target group was representatives of cities that joined transnational city networks (e.g. CROs, environmental coordinators, consultants or engineers or climate change managers). Overall, 23 interviews with municipal policymakers were conducted, on-site, via phone, Skype or email. The interview material contains data from all continents and from 16 different countries, with interviewees from the USA and the UK (both three) being the most numerous ones (see Fig. 1). Table 3 summarises the number of corresponding network member cities represented by the interviewed stakeholders.

The interview partners were identified through desk-research, existing contacts or recommendations of the interviewees. The core questions asked to the city representatives were about: (i) the reasons to participate in transnational city networks and the perceived added value of a membership, (ii) their perceptions about C2CL (in which forms is it taking place? How important is it and how is it supported by the networks their city joined and (iii) the main barriers but also imaginable improvements for networked C2CL processes.

Since deriving from qualitative interviews, the evidence presented in this chapter should not be considered as irrefutable facts, but rather as a mapping of the perceptions of key municipal stakeholders about the network functions and C2CL features. It has to be taken into consideration that the interview partners only shared the information they were willing and allowed to share. However, for the stated research aims the most suitable and qualified interview partners were chosen: The municipal policymakers that are responsible for urban resilience or adaptation and for managing the network activities and responsibilities.

5 Findings

Why are cities joining transnational city networks and what is the perceived added value of a membership?

Asked about their municipalities' motivation to participate in transnational networks, the most common responses were around the mission of

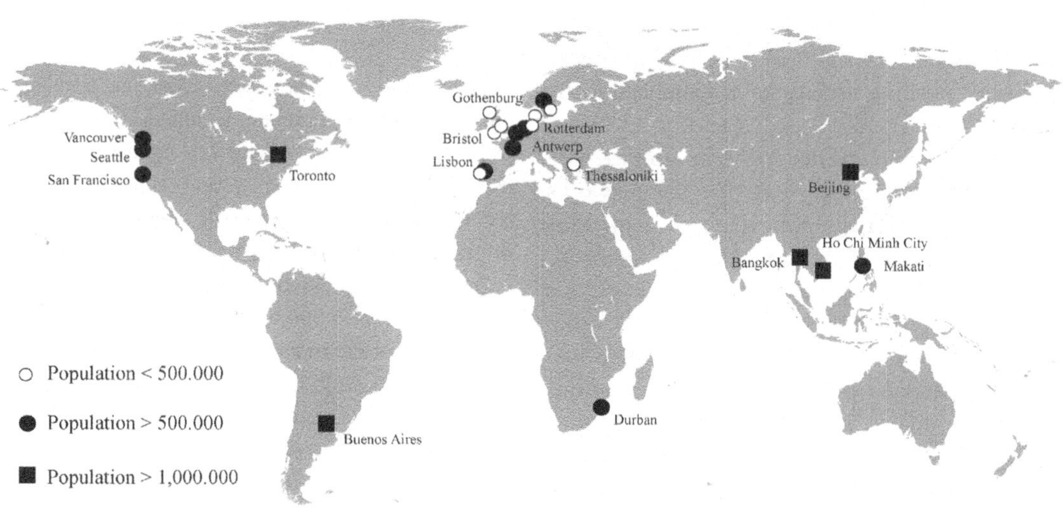

Fig. 1 Cities represented by the interviewed municipal policy makers, *source* Own figure

Table 3 Number of corresponding network members among the interviewed cities

ICLEI	C40	Compact of Mayors	100RC	Mayors Adapt	Energy Cities	UNISDR	UCLG	CityNet	EUROCITIES
16	12	12	11	7	6	6	4	3	2

Source Own table

putting their cities' work into a broader (international) context, in order to get (easier) access to funding and also more recognition at the national and international level. For cities that aim to position themselves as frontrunners in resilience or climate adaptation, joining a related network was often seen as a necessary and strategic move. In this case, the network served as a platform to showcase their progress or success. For cities with a certain importance, for instance, the (regional) capitals or the regions or countries biggest city, a network membership was also seen as a necessary formality: These cities often felt 'obliged' to lead and 'give a good example' to others, as the Program Leader Sustainable City of Antwerp explained. Almost all interview partners emphasised the positive role of networks in creating a platform for knowledge sharing among cities, and many pointed out that learning was an important reason to join them. Besides, getting (easier) access to funding and the chance to easily get in touch with colleagues from abroad, which allowed regarding similar issues from a perspective different from the own national one, were identified as further positive aspects of a membership by some.

A positive outcome that has been stressed by a number of interviewees was that the networks connected municipal policymakers and thus helped to create valuable personal contacts. Being a member of a transnational network is a formal activity, whereas the contacts that result due to a city's membership can also evolve into more informal and personal partnerships. In this regard, the major importance of mutual trust among stakeholders and face-to-face interactions were identified as the determining factor for the establishment of such partnerships on a personal level.

According to many interviewees, an important asset of being part of a transnational network is that it helped a lot to raise the profile of their work and the topic of climate change adaptation or resilience within their own city council. This was even more impactful, when decision makers joined network meetings, like in the case of Vejle where members of the city council, jointly with the CRO, attended a 100RC event in New Orleans. The Climate Change Project Manager of Cascais (Greater Lisbon) emphasised that putting his own work in a broader context through the participation in transnational networks helped to increase the understanding and recognition of the issue of climate change in other relevant municipalities departments. Through this, departments that have access to some funding, for instance the civil protection for flooding hazard measures, can be convinced to include climate change response into the justification of their work proposals.

What kind of knowledge is transferred and what knowledge are cities interested in?

The kind of knowledge that is transferred between network members included information on projects or initiatives, advice and updates on how to use shared methodologies or technologies or to receive technical assistance. It could also be more implicit and informal, but hence not less important knowledge, such as advice and on how to prepare applications for a network or a funding program. Here, particularly members from the same country, who also understood the national and cultural context, were pointed out as valuable contacts.

Among the interviewees, sharing best practices and expertise enjoyed popularity since it allowed a more efficient use of resources. However, it was differentiated between studying best practice examples summarised on a one or a few pages long document on the one, and the close interaction with the best practice peer-city, on the other hand. The simple studying of a best practice document was described as being only of limited use for their own city (if at all). On the contrary, deep diving into a best practice, for instance through a discussion or question and answer interaction, or even through a peer visit on-site was described as much more useful. In this context, a city representative of Vancouver explained that the city sent personnel to Philadelphia in order to build on Philadelphia's expertise in engineering approaches to greening infrastructure, which was described as very useful for the development of Vancouver's climate adaptation program. One interviewee also stated that the sharing of best practices gave credibility about innovative approaches, especially when they were implemented in a city with a 'big' name,' which could be very helpful to convince decision makers of their own city. As mentioned above, many frontrunner cities joined transnational networks also in order to have a platform to present their cities. In return, many members saw one asset of a transnational network in the opportunity to get in touch with and learn from such a frontrunner city.

Which tools do networks offer to its members in order to facilitate city-to-city learning?

All regarded networks emphasise the importance of knowledge sharing among cities and provide them with different tools to trigger knowledge exchanges. Most networks encourage their members to share best practices, be it through a formal best-practice sharing platform or through connecting cities to each other. Additionally, most networks regularly organise webinars and update and inform their members through a network newsletter. Apart from that most networks also offer formal programs or projects that aim to facilitate C2CL (see Table 4). The City Sharing Initiative of UNISDR or the Mayors Adapt Twinning Programme are examples for a network connecting cities working on similar issues and facing similar challenges (see Mayors Adapt 2015; UNISDR 2016). During a multi-day study tour organised by the mentor city, Copenhagen in the framework of the Mayors Adapt City Twinning Programme the representatives of the learning city Antwerp had the opportunity to better understand Copenhagen's strategy to deal with cloudbursts. The Program Leader Sustainable City of Antwerp described that this in-depth exchange provided very valuable input for the upcoming development of the city's water plan.

Webinars were described as a valuable tool, provided that there is a clear learning objective and enough time for follow up questions. It was mentioned that conferences could be an effective instrument to meet many peers at once. In that respect, it was emphasised, that C2CL could be achieved most easily through a mix of tools, meaning that stakeholders could meet first in big conferences to get updated on the global reflection on a topic. Then in the next step, municipalities that face similar problems and share the same background could collaborate more in-depth.

What role does the private sector play in climate adaptation and how are networks facilitating its integration?

The lack of financial resources to do networking and to implement climate policies was reported as a problem in many cases. This is one reason why a number of interviewees stressed the need to stronger involve the private sector, which could also be facilitated by the networks. Indeed, some networks also try to involve the private sector.

In order 'to facilitate long-term commitments from the private sector in support of ICLEI's Mission and Strategic Plan', the network promotes the establishment of corporate partnerships open to commercial, for-profit organizations, businesses, and corporations (ICLEI and UN HABITAT nd). Current corporate partners of ICLEI are Mckinsey and Siemens (ibd.). Lately, C40's aim has become 'supporting early

Table 4 Formal city-to-city learning programs or projects offered by transnational climate networks

Network	C2CL facilitated through...
100RC	Network Exchange program
UCLG	Mediterranean City to City Migration project[a]
UNISDR	City-to-City Sharing Initiative
ICLEI	Urban-LEDS
Energy Cities	IMAGINE, DISPLAY, ENGAGE
EUROCITIES	In depth peer review visits among partner cities (CASCADE)
Mayors Adapt	Mayors Adapt City Twinning Programme
ACCCRN	Shared learning approach

Source Own table based on Covenant of Mayors (2017), 100RC (2016), Orleans Reed et al. (2013), Energy Cities (nd), EUROCITIES (nd), UCLG (ndb), UNISDR (nda)

[a]The focus of UCLG goes far beyond climate change response and covers a high number of topics relevant for the local level. The Mediterranean City to City Migration project does not tackle climate adaptation but is a good example for a network connecting members of different city administrations to work jointly on shared issues (UCLG ndb)

engagement between cities and the private sector to accelerate the deployment of climate solutions'. Therefore, in cooperation with the cleantech cluster CLEAN and the International Cleantech Network, C40 established the City Solutions Platform in order to trigger and facilitate 'early engagement between cities and the private sector to accelerate the deployment of climate solutions' (C40 2017). 100RC aims to help establish a private sector marketplace for resilience tools. They should enable providing the member cities with the 'critical, often innovative tools to help cities around the world become more resilient to the shocks and stresses that are a growing part of the twenty-first century' (Young 2014). The list of partners is long and steadily growing. Currently (April 2017), 100RC has 86 partners, of which many are from the private sector (e.g. Cisco, Jacobs, Microsoft, Siemens and Veolia) (100RC 2018). Having its focus on disaster resilience, UNISDR is well aware of the 'lack of regulation and incentives for private disaster risk reduction' and is possessed with the question of 'How disaster risk is considered and managed in capital investments, supply chains and operations will be a decisive factor on achieving the Sendai Framework targets' (UNISDR and African Aid and Development Forum 2016). To tackle this issue, UNISDR established the Private Sector Alliance for Disaster Resilient Societies ARISE that aims to encourage the creation of private public partnerships, initially consisting of over 140 private sector entities and affiliates (ibd.).

However, at this point, many interviewees did not regard the current private sectors engagement as sufficient and criticised that many networks missed the opportunity to involve private partners. C40, for instance, has been numerously pointed out as a network that only dedicates minimal attention to this gap, although it was widely recognised as a weakness. However, it was acknowledged that C40 is now starting to tackle this point. The situation was a bit different in 100RC, where the private sector involvement was perceived as quite big. One CRO confirmed that it was also an overall goal of the network to build a marketplace, not just for public administrations to share their knowledge but also for business to develop solutions. Another CRO pointed out that bringing in private and growing markets was so deeply a part of 100RC that it has become a difficult relationship, since the private sector is supposed to be mostly concerned with the bottom line. Also, the growing number of businesses involved in the network made it more and more difficult to oversee the whole process.

What are the obstacles for city-to-city learning in networks and what opportunities for improvements exist?

The most common obstacles that hindered C2CL were limited resources (time, staff, funding for networking and attending meetings in person, technical expertise) that prevented the policymakers from contributing more to the networks. Also, language barriers could be a minor but limiting obstacle. Not all stakeholders spoke English and technical documents were often only available in the national languages (and translating them constituted a significant investment of time). Another general threat for transnational networks, especially for those that demand the payment of a membership fee, could be posed by administrative or political changes that might lead to a shift in (political) priorities. This could result in a threat for networking continuity.

Additionally, the interviewed stakeholders were asked about their recommendations for how the networks they joined could improve C2CL. What some interviewees suggested was to create an easily accessible database with meaningful information about the member cities or to provide metrics for measuring the benefits of the networks. Others stated that they would appreciate it if there was collaboration or at least coordination across different networks. This would allow using the resources at disposal in a more integrated manner. It was also stressed that having the opportunity to listen to and interact with experts, which could be arranged by networks, could be very helpful. Further suggestions for improvement included: (i) more assistance on how to correctly apply for funding, (ii) the establishment of standardised training courses on climate change adaptation, (iii) the inclusion of regional governments, (iv) a stronger linkage of the network activities to the Conference of the Parties (COP) discussion, (v) measures to increase the recognition of networks and (vi) extending networks to executives and politicians or the establishment of a network coach, who could focus on maximizing the networks potential.

6 Concluding Remarks

C2CL is of all times, but only since the last years, it has been explicitly being named, studied, documented (if at all). It thus remains a rather young concept with a limited base of practice, case studies, and guidance. To make deliberate and strategic change and to accelerate the uptake of good practices, the results reveal that it is generally recognised that cities need to engage in C2CL networks and have to learn from lessons of other cities. The learning aspect was also pointed out as one important reason to participate in transnational city networks. Networks, such as UNISDR's Making Cities Resilient Campaign, Rockefeller Foundation's 100 Resilient Cities, EU Mayors Adapt, C40 or ICLEI, are currently providing the platforms to build and share evidence around adaptation in practice.

Cities engaging in networked learning are looking for practical exchanges on progress. The involved stakeholders want to learn from and share knowledge with peers from cities that face similar challenges. Being part of a transnational network and engaging in C2C exchanges within these can help to promote and position the climate adaptation and resilience work within the own city council and thus helps to receive the critical support from the political side. There is wide recognition that personal contacts based on mutual trust between the stakeholders, that represent the cities engaged in the knowledge exchanges, are crucial, and many interviewed stakeholders confirmed that city networks helped to create these. Stronger involvement of the private sector was requested by some interviewees and is part of the working agenda of some networks, however with a very different priority from network to network. How to successfully involve business in climate adaptation remains a critical and mainly unanswered question. This points to the need to examine and evaluate the outcomes of different network practices.

While this chapter has offered a rather global and generic perspective of C2CL in transnational city networks, it is crucial to gain a better understanding of the usefulness of the various specific network programs. Firstly, in relation to how they concretely support C2CL, and secondly, how and to what extent they cater the cities needs to climate adaptation and in general transformation towards sustainability, e.g., by adopting a nexus approach to resources management. This future research could for example build on the Mayors Adapt Twinning Programme or the City-to-City Sharing Initiative of UNISDR, where cities facing similar challenges are brought together for in-depth peer learning exchanges. Besides, the research could also be expanded to programs or projects that were not set up by city networks but have similar formats and objectives such as the North Sea Region (NSR) Interreg project Blue Green Infrastructure through Social Innovation (BEGIN) (Interreg nd).

The interviews showed that many cities tend to focus their attention on pioneering or frontrunner cities, aiming at learning from those models. However, no matter how progressive and successful these few frontrunners are, the majority of cities which need to urgently adapt to climate change and become more resilient are indeed the non frontrunners. This raises the critical question of how frontrunners can truly help other cities in following their example? Can the successful measures of a global frontrunner city just be downscaled or transferred to a different context? Besides, a strict division into a 'learning' and a 'teaching' city to a certain extent also contradicts with the existing definitions of C2CL that emphasizes the importance of sharing and exchanging knowledge in order to learn and improve together.

References

100RC (2016) 100RC network exchange program. https://www.100resilientcities.org/100rc-network-exchange-program/

100RC (2018) 100 resilient cities. https://www.100resilientcities.org/

ACCCRN (2018) Our history. https://www.acccrn.net/about-acccrn/history

Andonova L, Betsill M, Bulkeley H (2009) Transnational climate governance. Glob Environ Polit 9(2):52–73

Andrews M, Manning N (2016) A guide to peer-to-peer learning. How to make peer-to-peer support and learning effective in the public sector?, Effective Institutions Platform

Azzali S (2016) Mega-events and urban planning: Doha as a case study. Urban Des Int 1:3–12

Barber BR (2013) If Mayors ruled the world: dysfunctional nations, rising cities. Yale University Press, New Haven, USA

Beermann J (2014) Urban partnerships in low-carbon development: opportunities and challenges of an emerging trend in global climate politics. URBE. Rev Bras de Gest Urbana (Braz J Urban Manag) 6 (2):170–183

Bennett CJ, Howlett M (1992) The lessons of learning: reconciling theories of policy learning and policy change. Policy Sci 25(3):275–294

Bouteligier S (2013) Inequality in new global governance arrangements: the North–South divide in transnational municipal networks. Innov: Eur J Soc Sci Res 26 (3):251–267

Bradford N (2004) Place matters and multi-level governance: perspectives on a new urban policy paradigm. Policy Options 25:39–44

Bulkeley H (2010) Cities and the governing of climate change. Annu Rev Environ Resour 35:229–253

Bulkeley H, Castán Broto V (2012) Government by experiment? Global cities and the governing of climate change. Trans Inst Br Geogr 38(3):361–375

Bulkeley H, Davies A, Evans B, Gibbs D, Kern K, Theobald K (2003) Environmental governance and transnational municipal networks in Europe. J Environ Plan Policy 5(3):235–254

Bulkeley H, Schroeder H (2012) Beyond state/non-state divides: global cities and the governing of climate change. Eur J Int Relat 18(4):743–766

C40 (2018) C40 cities climate leadership group. https://www.c40.org

Calzada I (2016) (Un)Plugging smart cities with urban transformations: towards multi-stakeholder city-regional complex urbanity? URBS, Rev de Estud Urbanos y Cienc Soc J 6(2):25–45

Campbell T (2012) Beyond smart cities: how cities network, learn and innovate. London

CDIA (2014) CDIA investing in Asia's future. https://cdia.asia/

CityNet (nd) About. https://citynet-ap.org

Covenant of Mayors (2017) Twinning programme on adaptation to climate change: the main outcomes in key figures!. https://eumayors.eu/news-and-events/news/1448-twinning-programme-on-adaptation-to-climate-change-the-main-outcomes-in-key-figures.html

Covenant of Mayors (n.d) Covenant of Mayors for climate and energy. https://www.covenantofmayors.eu/

Cleaner Energies Saving Mediterranean Cities (nd) New 2030 framework and integration of adaptation. https://www.covenantofmayors.eu/about/covenant-of-mayors_en.html

Compact of Mayors (nd) About. https://www.globalcovenantofmayors.org

De Gregorio Hurtado S, Olazabal M, Salvia M, Pietrapertosa F, Olazabal E, Geneletti D, D'Alonzo V, Di Leo S, Reckien D (2015) Understanding how and why cities engage with climate policy: an analysis of local climate action in Spain and Italy. TeMA, Special (ECCA) 23–46

Dresden Nexus Conference (2017) State of the nexus approach 2017. UNU-FLORES, DNC Conference Report Dresden. https://collections.unu.edu/eserv/UNU:6340/DNC17_Report.pdf

Dolowitz D, Marsh D (1996) Who learns what from whom: a review of the policy transfer literature. Polit Stud 44(2):343–357

Dolowitz D, Marsh D (2000) Learning from abroad: the role of policy transfer in contemporary policy-making. Governance 13(1):5–23

Dunlop CA, Radaelli CM (2013) Systematizing policy learning: from monolith to dimensions. Polit Stud 61 (3):599–619

Energy Cities (nd) About us. https://energy-cities.eu/

Engelke P (2013) Foreign policy for an urban world: global governance and the rise of cities. Washington D.C, Atlantic Council—Brent Scowcroft Center on International Security

Etheredge LS (1981) Government learning: an overview. In: Long S (ed) The handbook of political behavior, vol 2. Pergamon, New York

EUROCITIES (nd) CASCADE—cities exchanging on local energy leadership. https://www.eurocities.eu/eurocities/projects/CASCADE-Cities-exchanging-on-local-energy-leadership&tpl=home

EUROCITIES (2017) Home. https://www.eurocities.eu/eurocities/about_us

Feldman DL (2012) The future of environmental networks—governance and civil society in a global context. Futures 44(9):787–796

Fenton P, Busch H (2016) Identifying the "usual suspects—assessing patterns of representation in local environmental initiatives." Chall Sustain 4(2):1–4

Fisher S (2014) Exploring nascent climate policies in Indian cities: a role for policy mobilities? Int J Urban Sustain Dev 154–173

Giest S, Howlett M (2013) Comparative climate change governance: lessons from european transnational municipal network management efforts. Environ Policy Gov 23(6):341–353

Gordon D (2016) The politics of accountability in networked urban climate governance. Glob Environ Polit 16(2):82–100

Gordon D, Acuto M (2015) If cities are the solution, what are the problems? The promise and perils of urban climate leadership. In: Toly N, Schroeder H, Johnson C (eds) The urban climate challenge: rethinking the role of cities in the global climate regime. Routledge, pp 63–81

Gore C (2010) The limits and opportunities of networks: municipalities and Canadian climate change policy. Rev Policy Res 27(1):27–46

Gupta J, Pfeffer K, Ros-Tonen M, Verrest H (2015) An inclusive development perspective on the geographies of urban governance. In: Gupta J, Pfeffer K, Ros-Tonen M, Verrest H (eds) Geographies of urban governance. Springer International Publishing, Cham, Switzerland, pp 217–228

Haigh RP, Amaratunga D, Keraminiyage K, Thayaparan M (2014) 4th international conference on building resilience incorporating the 3rd annual conference of the ANDROID disaster resilience network (Book of Abstract)

Haupt W (2018) European municipalities engaging in climate change mitigation and adaptation networks: examining the case of the Covenant of Mayors. In: Sharifi A, Yamagata Y (eds) Resilience-oriented urban planning. Springer International Publishing, Cham, pp 93–110. https://doi.org/10.1007/978-3-319-75798-8_5

Haupt W, Coppola A (2019) Climate governance in transnational municipal networks: advancing a potential agenda for analysis and typology. Int J Urban Sustainable Dev 11(2):123–140. https://doi.org/10.1080/19463138.2019.1583235

Hettiarachchi H, Ardakanian R (2016) Managing water, soil, and waste in the context of global change. In: Hettiarachchi H, Ardakanian R (eds) Environmental resource management and the nexus approach: managing water, soil, and waste in the context of global change, pp 1–7. Springer International Publishing, Cham. https://doi.org/10.1007/978-3-319-28593-1_1

Hülsmann S, Ardakanian R (2018) The nexus approach as tool for achieving SDGs: trends and needs. In: Hülsmann S, Ardakanian R (eds) Managing water, soil and waste resources to achieve sustainable development goals. Springer International Publishing, Cham, pp 1–9. https://doi.org/10.1007/978-3-319-75163-4_1

ICLEI (nd) ICLEI local governments for sustainability. https://iclei.org/

ICLEI and UN HABITAT (nd) Urban low emission development strategies. https://urbanleds.iclei.org

Interreg (nd) BEGIN—Blue Green Infrastructures through Social Innovation. https://northsearegion.eu/begin/

James P, Verrest H (2015) Beyond the network effect: towards an alternative understanding of global urban organizations. In: Gupta J, Pfeffer K, Ros-Tonen M, Verrest H (eds) Geographies of urban governance. Springer International Publishing, Cham, Switzerland, pp 65–84

Kolosz BW, Grant-Muller S (2016) Sustainability assessment approaches for intelligent transport systems: the state of the art. IET Intell Transp Syst 10(5):287–297

Koop SHA, van Leeuwen CJ (2015) Assessment of the sustainability of water resources management: a critical review of the city blueprint approach. Water Resour Manag 29(15):5649–5670

Koop SHA, van Leeuwen CJ (2017) The challenges of water, waste and climate change in cities. Environ Dev Sustain 19(2):385–418

Kurian M, Ardakanian R (eds) (2015) Governing the nexus—water, soil and waste resources under conditions of global change. Springer, Berlin, Heidelberg

Lee T (2013) Global cities and transnational climate change networks. Glob Environ Polit 13(1):108–127

Lovell H (2016) The role of international policy transfer within the multiple streams approach: the case of smart electricity metering in Australia. Public Adm 94 (3):754–768

Lundby L, Sjöberg H (2013) An evaluation of city-to-city learning in the campaign making cities resilient—matching criteria and implementation. Lund, Sweden, Department of Fire Safety Engineering and Systems Safety Lund University, p 36

Mackee J, Giggins H (2015) Proceedings of the 5th international conference on building resilience. Newcastle, Australia

Mayors Adapt (2015) Deadline extended: Mayors Adapt City Twinning Programme. https://www.eurocities.eu/eurocities/news/Deadline-extended-Mayors-Adapt-City-Twinning-Programme-WSPO-9Y9BZ4

McCann EJ (2011) Urban policy mobilities and global circuits of knowledge: toward a research agenda. Ann Assoc Am Geogr 101(1):107–130

Montero S (2017) Study tours and inter-city policy learning: Mobilizing Bogotá's transportation policies in Guadalajara. Environ Plan A 49(2):332–350

Newig J, Günther D, Pahl-Wostl C (2010) Synapses in the network: learning in governance networks in the context of environmental management. Ecol Soc 15(24)

Niederhafner S (2013) Comparing functions of transnational city networks in Europe and Asia. Asia Europe J 11(4):377–396

Orleans Reed S, Friend R, Toan VC, Thinphanga P, Sutarto R, Singh D (2013) "Shared learning" for building urban climate resilience – experiences from Asian cities. Env Urbanization 25(2):393-412. https://doi.org/10.1177/0956247813501136

Pattberg P, Widerberg O (2015) Theorising global environmental governance: key findings and future questions. J Int Stud 43(2):684–705

Peck J, Theodore N (2010) Mobilizing policy: models, methods, and mutations. Geoforum 41(2):169–174

Radaelli CM (2004) The diffusion of regulatory impact analysis—best practice or lesson-drawing? Eur J Polit Res 43(5):723–747

Rose R (1991) What is lesson-drawing? J Public Policy 11(1):3–30

Rose R (2002) When all other conditions are not equal: the context for drawing lessons. In: Jones Finer C (ed) Social policy reform in socialist market china: lessons for and from abroad. Aldershot, Ashgate

Rosenzweig C, Solecki W, Hammer S, Mehrotra S (2010) Cities lead the way in climate–change action. Nature 467:909–911

Sassen S (2013) A focus on cities takes us beyond existing governance frameworks. In: Kaldor M, Stiglitz JE (eds) The quest for security: protection without protectionism and the challenge of global governance, vol 2. Columbia University Press, New York City, pp 238–259

Selin H, Vandeveer S (2005) Canadian-U.S. environmental cooperation: climate change networks and regional action. Environ Polit 35(2):353–378

Seymoar N-K, Mullard Z, Winstanley M (2009) City-to-city learning. https://sustainablecities.net/wp-content/uploads/2015/10/city-to-city-learning.pdf

Stead D (2012) Best practices and policy transfer in spatial planning. Plan Pract Res 27(1):103–116

Stone D (1999) Learning lessons and transferring policy across time, space and disciplines. Politics 19(1):51–59

Stone D (2004) Transfer agents and global networks in the 'transnationalization' of policy. J Eur Public Policy 11(3):545–566

Thornbush MJ (2016) Urban greening for low carbon cities—introduction to the special issue. AIMS Environ Sci 3(1):133–139

UCLG (nda) About us. https://www.uclg.org

UCLG (ndb) Mediterranean city-to-city migration profiles and dialogue. https://www.uclg.org/sites/default/files/8.5.eng_paris_eb_c2c_migration_project.pdf

UN (nd) About the sustainable development goals. https://www.un.org/sustainabledevelopment/sustainable-development-goals/

UNFCCC (2015) Report of the Conference of the Parties on its twenty-first session, held in Paris from 30 November to 13 December 2015. https://unfccc.int/documents

UNISDR (2016) Terms of references: city-to-city learning exchanges call for proposals 2016–2017. https://www.preventionweb.net/files/50315_c2ctermsofreference.pdf

UNISDR (nda) City-to-city learning exchange background information. https://www.preventionweb.net/files/50315_c2cbackground.pdf

UNISDR (ndb) Making cities resilient. https://www.unisdr.org/we/campaign/cities

UNU-FLORES (nd) The nexus approach. https://flores.unu.edu/en/research/nexus

van Herk S, Gersonius B, Zevenbergen C, Rijke J (2016a) City exchange for disaster resilience. https://www.un-ihe.org/sites/default/files/city_to_city_flyer.pdf

van Herk S, Hollmann C, Zevenbergen C (2016b) How can urban resilience networks cater to city needs? https://www.urbanresilienceresearch.net/2016/06/23/how-can-urban-resilience-networks-cater-to-city-needs-2/

Vinke-de Kruijf J, Pahl Wostl C (2016) A multi-level perspective on learning about climate change adaptation through international cooperation. Environ Sci Policy 66:242–249

Wolman H, Page E (2002) Policy transfer among local governments. An information theory approach. Governance 15(4):477–501

Young M (2014) 100RC adds four private sector partners to build urban resilience. https://www.100resilientcities.org/100rc-adds-four-private-sector-partners-to-build-urban-resilience/

Zeppel H (2013) The ICLEI cities for climate protection programme: local government networks in urban climate governance. In: Cadman T (eds) Climate change and global policy regimes. Springer, pp 217–231

A Participatory Multi-Stakeholder Approach to Implementing the Agenda 2030 for Sustainable Development: Theoretical Basis and Empirical Findings

Sebastian Eichhorn, Moritz Hans, and Martin Schön-Chanishvili

Abstract

The Agenda 2030 provides a universal target system covering a broad thematic spectrum including the key areas of people, planet, prosperity, peace and partnership. There is a consensus among policy-makers and scholars that the implementation of the global goals relies on both an integrated (nexus) approach and the participation of all levels, especially the local level. Therefore, the paper focuses on the research question of how integrated management and participatory multi-stakeholder approach can be systematically realized at the local level, thus contributing to the implementation of the Agenda 2030. The authors aim to answer this question on the basis of experiences and results of a project which took place between January 2016 and April 2018 in North Rhine-Westphalia (Germany). In this two-and-a-half-year process, 15 municipalities developed individual Integrated Municipal Sustainability Strategies (IMSS) in the context of the Agenda 2030. The paper presents the methodical approach of the participation process and the results of a project-internal process-oriented monitoring and evaluation. The results show that the participation design has a direct influence on the outcome of the process. The central influencing factors identified are (1) political support from the heads of the administration, (2) the avoidance of creating parallel working structures by forming project-related working groups, (3) the independent working of the working groups, (4) the composition of steering committees (the representation of relevant stakeholders), (5) the involvement of important political decision-makers, and (6) an open negotiation process that ensures the participation of all participants/stakeholders.

Keywords

Participatory approach · Stakeholder engagement · SDG implementation · Municipal sustainability strategies

1 Introduction

For more than two decades, sustainable development (SD) has been part of the political and scientific discourse. Appropriate groundwork

S. Eichhorn (✉) · M. Hans
Research Institute for Regional and Urban Development, Brüderweg 22-24, 44135 Dortmund, Germany
e-mail: Sebastian.eichhorn@ils-forschung.de

M. Hans
e-mail: mhans@stadtdo.de

M. Hans
City of Dortmund, Betenstraße 19, 44137 Dortmund, Germany

M. Schön-Chanishvili
Germanwatch E.V, Stresemannstr. 72, 10963 Berlin, Germany
e-mail: schoen-chanishvili@germanwatch.org

S. Hülsmann and M. Jampani (eds.), *A Nexus Approach for Sustainable Development*,
https://doi.org/10.1007/978-3-030-57530-4_15

such as the "Limits to Growth" of the Club of Rome and the Brundtland Report from 1987 triggered a sincere discussion among experts and politicians worldwide about the way social and political systems must change in order to ensure the protection of the global ecosystems (Donella et al. 1972; Brundtland and Khalid 1987). By adopting the Rio Declaration and the Agenda 21, the UN member states officially declared the principles of SD to be the foundation of the further socio-economic development of humanity (United Nations 1992). In addition, the UN member states worked out an output-orientated development plan mainly for developing countries in their Millennium Declaration in 2000 (United Nations 2000). The eight Millennium Development Goals (MDGs) focused on main challenges, such as hunger, poverty and the provision of basic health services. By the end of their term in 2015, the MDGs were widely confirmed as a means for successfully managing development cooperation (Martens 2013).

In New York in 2015, the heads of state of all UN member states agreed on the Agenda 2030 as the replacement for the MDGs and as a common vision and plan for a global transformation that represents both a synthesis and a new step in the scientific discourse and political process. The plan, entitled "Transforming Our World: The Agenda 2030 for Sustainable Development", includes four important aspects. The Agenda 2030

1. Contains a differentiated target system for achieving the common vision. It includes 17 goals, 169 targets and at least one indicator for each target. It calls for a "data revolution" to make the indicators measurable worldwide;
2. Covers an extremely wide range of topics, summed up in five key areas: people, planet, prosperity, peace and partnership;
3. Represents a consolidation of the sustainability agenda ("Rio process") and development agenda (MDGs);
4. Has a universal character, declaring all countries as partners in achieving the goals and targets (shared responsibility of nation-states and all sub-levels).

Looking at the Sustainable Development Goals (SDGs), it becomes clear that in particular, two basic prerequisites are necessary for successful implementation and target achievement. On one hand, the thematic range of the target system requires the participation of different actors (United Nations 2015), mainly because public funds alone will not be sufficient for achieving this common vision of a sustainable world nor will government action alone be enough to give the needed impetus. The ownership of different stakeholders is needed. Subsequently, an integrated and participatory multi-stakeholder approach is of crucial importance for the implementation of the Agenda 2030. Therefore, the Agenda 2030 addresses multi-stakeholder cooperation in different forms. The Agenda 2030 includes common appeals ("All countries and all stakeholders… will implement this plan"), conceptually focuses on the aspect of the partnership ("with the participation of all countries, all stakeholders and all people") and—most importantly—provide targets for the effective cooperation between government structures, businesses and civil society. With regard to cities, these targets focus on sustainable urban development (SDG 11, Target 11.3: "By 2030, enhance inclusive and sustainable urbanization and capacity for participatory, integrated and sustainable human settlement planning and management"), as well as on the importance of a multi-stakeholder approach for effective North–South cooperation (SDG 17, Targets 17.16 and 17.17).

On the other hand, many targets of the Agenda 2030 have a degree of concretization, which depends on action and implementation at the local level—in towns, settlements, and smaller and larger cities. The topics concerned range from education (SDG 4, Target 4.a "Build and upgrade education facilities…") and traffic safety (SDG 3, Target 3.6 "… halve the number of global deaths and injuries from road traffic accidents.") to access to open and green areas (SDG 11, Target 11.7 "…provide universal access to safe, inclusive and accessible, green and public spaces…"). The New Urban Agenda

(NUA) agreed upon at the Habitat III cities conference in Quito, Ecuador in 2016, also highlights "sustainable urban development as a critical step for realizing sustainable development in an integrated and coordinated manner at global, regional, national, sub-national, and local levels, with the participation of all relevant actors" (United Nations 2017). In its main report from 2016, the German Advisory Council on Global Change (WBGU) suggests that cities "should play a decisive role in the 'century of the cities'—as key drivers of the transformation towards sustainability" (Kraas et al. 2016). On one hand, cities trigger technological innovation and social mobility. On the other hand, they incite climate change and excessive resource consumption, while their populations are exposed to the consequences of heat waves or heavy rain, increasing land consumption and motorized traffic, as well as poverty and challenges of social inequality (Ministerinnen und Minister der Bundesländer 2007).

Therefore, the Agenda 2030 implies a threefold responsibility along its core dimensions of implementation, which are also particularly relevant for the local level. First is the dimension of domestic policy: for example, halving the national poverty rate (SDG 1, target 1.2) also involves a challenge for countries that are highly economically developed. German municipalities have the opportunity to positively influence this goal. They can make use of local social policies, such as early childcare, or of non-formal and professional education offers to improve opportunities for the deprived. Second, local political action and collective behavior patterns can have a lasting impact on other countries. For example, German municipalities can help to improve the living and working conditions in the Global South through fair trade (SDG 12, target 12.7) or a reduction of their greenhouse gas emissions (SDG 7, target 7.2 and 7.3) or consumption of natural resources (SDG 8, target 8.4). In particular, public procurement policies of German administrations tend to make growing contributions in this field (SKEW 2015). Third, municipalities may become actively involved in other countries in the field of SD, for example, through partnerships, projects or technical cooperation (SDG 17, target 17.16 and 17.17). Many German municipalities engage in concrete bilateral projects, e.g., for common climate change mitigation policies (Website SKEW 2017). Although in many German municipalities, different and thoroughly successful approaches to implement SD can be identified, SD is rarely understood as a core business of local action and implemented in an integrated manner.

Therefore, this paper focuses on answering the research question of how integrated management and participatory multi-stakeholder approach can be systematically realized at the local level, thus contributing to the implementation of the Agenda 2030. This is done on the basis of experiences and results of the model project titled "Global Nachhaltige Kommune in Nordrhein-Westfalen" (GNK NRW, Eng. Global Sustainable Municipality in North Rhine-Westphalia), which took place between January 2016 and April 2018. In this two-and-a-half-year process, 15 municipalities in North Rhine-Westphalia (Germany) developed individual Integrated Municipal Sustainability Strategies (IMSS) in the context of the Agenda 2030. The GNK NRW project was carried out by the Sustainability Network of North Rhine-Westphalia (LAG 21 NRW) and the Service Agency Communities in One World (SKEW) from Engagement Global and was funded by the Federal Ministry for Economic Cooperation and Development (BMZ).[1]

The paper presents the methodical approach to the participation process and the results of a project-internal process-oriented monitoring and evaluation. The project results presented here are based on an approach of participatory urban planning for the LAG 21 NRW, as explained in Sects. 2 and 3. How the approach was implemented is explained in Sect. 4. Subsequently, the design and empirical findings of the accompanying monitoring and evaluation are presented in

[1]The SKEW of Engagement Global provides support to German municipalities engaging in development cooperation. SKEW works on behalf of the Federal Ministry for Economic Cooperation and Development (BMZ).

Sect. 5. Section 6 concludes and summarizes the main findings of the project.

2 Participation as a Key Element of Cooperative Planning

The planning and design of people's living environment have a direct impact on the quality of life of residents and on future developments. The participation of citizens or representatives of interest groups in planning and development is increasingly seen, promoted and encouraged as a meaningful addition to planning processes. Therefore, planning is currently understood primarily as a process of communication and consensus building (see Ritter 1998; Selle 1996; Healey 2004; Poister and Streib 2005).

Since Arnstein's "ladder of citizen participation", a large variety of studies have focused on the objectives, forms and intensities of participation (Arnstein 1969; Biggs 1989; Farrington 1998; Goetz and Gaventa 2001; Johnson et al. 2004; Lawrence 2006). Participation can thus be distinguished not only in terms of the stakeholders who promote participation but also in terms of different objectives and intensities, the project phases in which they occur, the forms or activities of participation, as well as the underlying topics. On one hand, influencing factors for successful participation are analyzed, and on the other hand, the effects on the outcome of a project or planning process are examined. Studies show that the benefits of participation directly depend on the participation form and design: (1) a professional support for the process, (2) the application of structured methods for aggregating information, (3) the provision of information, (4) a balanced power relationship between participants, and (5) a similar influence on the negotiation and decision-making process for all participants (Drazkiewicz et al. 2015; Jobert et al. 2007; Kastens and Newig 2008; de Vente et al. 2016). Applied studies prove the positive effects of participation in general (Drazkiewicz et al. 2015; Irvin and Stansbury 2004). Richards et al. (2004) assume that stakeholder participation can increase trust and confidence in decisions if the process is transparent and conflicts are considered seriously. It is also claimed that participation may increase the likelihood that decisions are perceived to be holistic and fair, accounting for a diversity of values and needs. Wallerstein (1999) analyzes the empowerment of stakeholders by co-generating knowledge with researchers. According to de Vente et al. (2016), the involvement of stakeholders is linked to the successful achievement of goals and a higher quality of decisions by improving the negotiation process. It is assumed that the results of such dialogue processes are often more broadly accepted and can decisively enrich the outcome of a planning process because of higher ownership of the stakeholders involved.

In planning, participation is regularly required by law.[2] This requirement applies to spatial planning, (transport) infrastructure planning or residential and environmental planning. In Germany, e.g., participation in spatial planning processes is regulated by paragraphs 3 and 4 of the Federal Building Code (BauGB). Here, concrete procedure steps are described to ensure that the public in general, as well as public and other authorities, can react adequately to planning. Practice, however, shows that the actual participation, especially of citizens, is often rather low or too little. Looking for reasons for this phenomenon, citizens, associations and authorities identified the following issues to be addressed (Bock 2017):

- Participation comes too late, at a point when the project has already achieved spatial and design-technological concretization.
- Restricting participation in formal procedures prevents citizens from openly and seriously discussing the advantages and disadvantages of a project or project alternatives.
- The participation of citizens is not achieved continuously overall procedural stages of planning.
- Decisions are not always communicated transparently.

[2]In this paper, the legally binding participation is called the linear planning approach.

Subsequently, linear planning approaches are not always appropriate to initiate transformation processes because they do not consider the existing expertise, interests, and resources of important external stakeholders in a serious way (see Fig. 1). Although these linear planning processes are routine and efficient in terms of proven and legally regulated procedural steps, the reasons given can lead to the resistance of affected actors, delays, higher costs and sometimes even to the complete abandonment of a project.

Thus, for municipalities to be able to perceive their role as actors of transformation, the manner of participation and decision-making must be re-adjusted. Since the success of municipal efforts depends on a broad consensus among all relevant actors, the mobilization of citizens is a central prerequisite for the successful implementation of SD and the SDGs. In its main report from 2011, the WBGU supports the assessment of common planning comprehension and criticizes the lack of actions with a long-term orientation. The WBGU asks for a new social contract, which must be accompanied by a new participatory culture (Schellnhuber et al. 2011). Target 16.7 also underlines this aspect of SD and aims at "responsive, inclusive, participatory and representative decision-making at all levels".[3]

Cooperative planning approaches[4] are intended to meet these requirements, as they are based on the finding that municipalities are confronted with multi-causal problems and faced with an active, opinion-oriented and heterogeneous city society (Selle 2000). In contrast to linear planning approaches, cooperative planning approaches are not legally regulated and can be adapted differently according to the respective occasion (see paragraph 6, No. 11 BauGB). These processes, in which the public authority generally takes the initiative, strongly depend on voluntariness and on the principle of cooperative handling responsibilities. In cooperative planning processes, different social actors with diverging interests are usually involved: (1) representatives of the administration, (2) politicians, (3) civil society actors (e.g., churches, associations), (4) science and (5) the economy. While in the formal planning process, the different actors work together only at predetermined points (see paragraphs 3 and 4 of the BauGB: early public participation and formal participation), cooperative planning processes can provide the early and continuous participation of all interest groups (see Fig. 1). In addition, these interest groups are not separated during the participation process. Cooperative planning processes stipulate that all actors discuss issues eye-to-eye and decide consensually in a common arena. As shown in Fig. 1, all actors are involved in all steps of the process. Note that the actors involved are not limited to the development of goals and measures but make arrangements for joint implementation.

3 Towards a Sustainability Strategy at the Local Level

The participation of actors is based on the cooperative planning approach and occurs throughout the entire GNK NRW process. In addition, the LAG 21 NRW approach embeds the cooperative planning process in an integrated management system. Therefore, the development and implementation of IMSS is designed as a continuous improvement process (CIP). The CIP ensures the effectiveness and success control of IMSS and aims at implementation, the achievement of objectives, constant improvement and further development regarding the formulated content. According to the LAG 21 NRW approach, the development and implementation of IMSS is divided into six main steps (see Fig. 2): (1) working structure, (2) analysis, (3) development/revision of a sustainability strategy, (4) (first) sustainability report and formal decision, (5) implementation and monitoring and (6) evaluation and update.

With regard to single fields of action, management tools have been used for several years

[3]SDG 16: Promote peaceful and inclusive societies for sustainable development, provide access to justice for all and build effective, accountable and inclusive institutions at all levels.

[4]In this paper, non-binding public participation is called the cooperative planning approach.

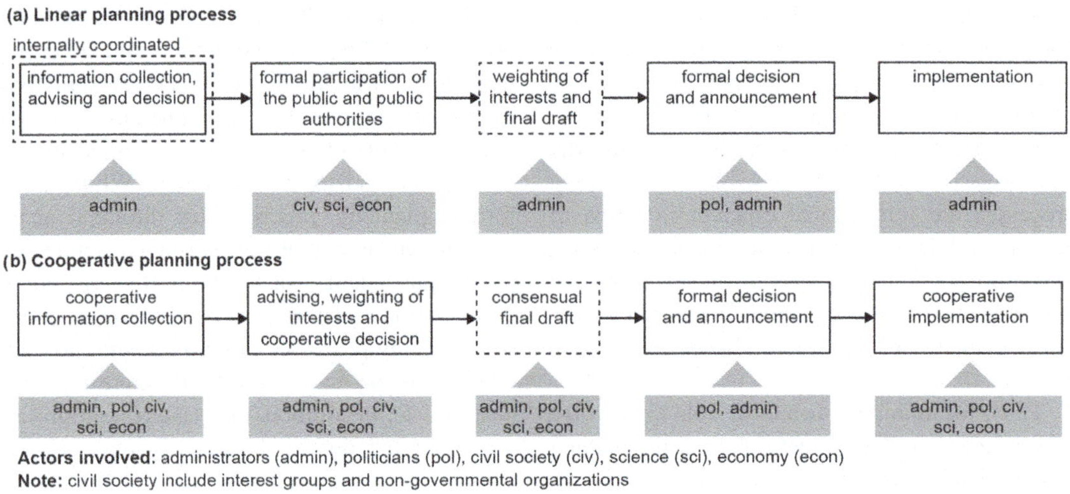

Fig. 1 Linear and cooperative planning processes; *Source* Own illustration

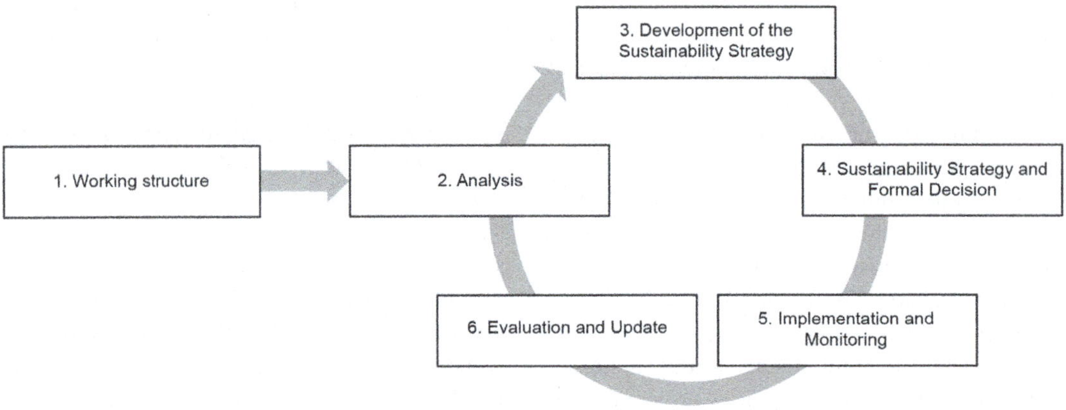

Fig. 2 Continuous improvement process (CIP); *Source* Own illustration

(e.g., Integrated Water Management (IWRM) (Al Radif 1999), Integrated Natural Resource Management (INRM) (Hagmann et al. 2002), Water-Food-Energy Nexus Approach (WEF) (Bazilian et al. 2011), and a Nexus Approach to the sustainable management of water, soil, and waste (Hettiarachchi and Ardakanian 2016; Hülsmann and Ardakanian 2018).

1. Working structure: Recent studies emphasize the interrelatedness between the goals and targets of the Agenda 2030 (Bringezu et al. 2016; Mans et al. 2016; Tosun and Leininger 2017). Therefore, Smajgl (2018) and Hoff (2018) note the importance of an integrated (nexus) approach that takes into account the positive and negative interactions (synergies and conflicts) to achieve the SDGs. The LAG 21 NRW project approach envisages the formation of the following interdisciplinary working groups to develop and implement IMSS in the context of the Agenda 2030: coordination, core group and steering committee. The different groups are responsible

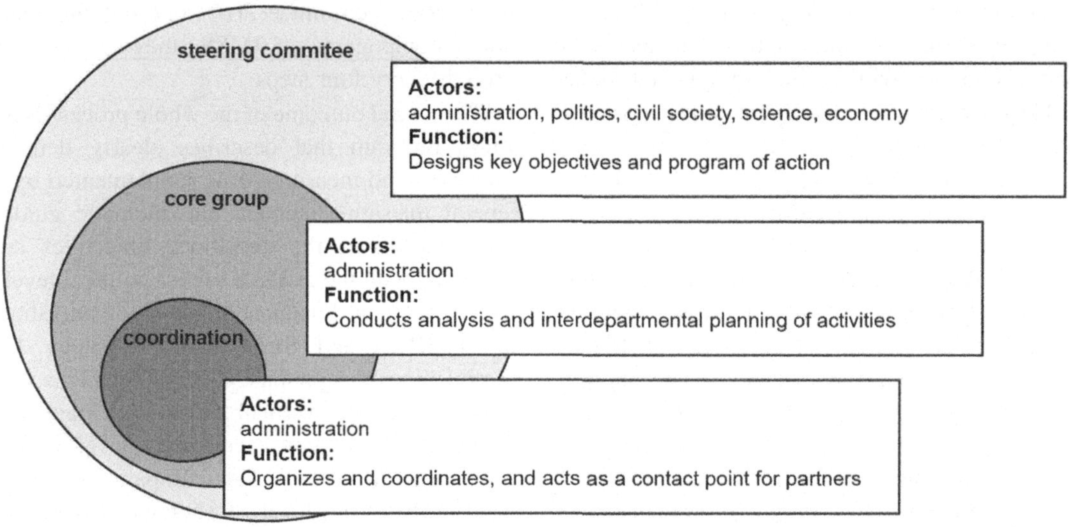

Fig. 3 Working structure; *Source* Own illustration

for the organization of the process, its development and implementation, and the CIP of the IMSS.

Coordination

The coordination consists of a coordinator and a substitute. The group supports the process of developing and implementing the IMSS and is the central point of contact for the various stakeholders involved. Moreover, it ensures results and provides information on the progress of the process for interested parties and is responsible for continuous internal and external communication (public relations). This coordination is part of the core group (see Fig. 3).

Core Group

The core group is an administrative team. In regard to the integrated nature of the SDGs and interdisciplinary working groups, the core group consists of persons from different departments (e.g., planning, health, environment, and transport). The main task of the core group is the preparation of analyses and the interdepartmental planning of activities. Regarding the project focus, additional administrators or external experts can be temporarily added to provide further expertise in the working process. The core group is part of the steering committee (see Fig. 3).

Steering Committee

The steering committee is composed of stakeholders (see Fig. 3). Stakeholders are defined as multipliers and represent different social interests. Analogous to the core group, the steering committee is also built up in an interdisciplinary manner, including potential actors such as economic enterprises, scientific institutions, nongovernmental organizations (NGOs), associations and churches. Regarding the project focus, the steering committee can also be adapted, supplemented and expanded. As an integral part, representatives of the political bodies should be involved. The steering committee is the central feature of the LAG 21 NRW approach: here, it links the integrated management system to the cooperative planning approach by involving all relevant stakeholders during the whole planning process in one negotiation process (see Fig. 1). Considering the specific (specialist) expertise gathered, the steering committee designs the objectives and measures of the IMSS. Together with the core group, the steering committee aims for the horizontal integration of different disciplines and stakeholders in order to identify emerging conflicts as early as possible and to seek consensual solutions to them. The extensive and continuous participation of relevant stakeholders serves the holistic and transparent balancing of different goals.

2. Analysis: Analysis is used to describe the current state of a municipality. On the one hand, statistical data (quantitative analysis) and, on the other hand, existing local concepts and strategies, projects, (international) city or project partnerships, as well as political decisions (qualitative analysis) are used and evaluated. The analysis and merging of the quantitative and qualitative data lead to a gap analysis. So far, non-treated or only insufficiently treated fields in the context of an SD and the SDGs are to be systematically recorded. This process can be summarized with the following question: Which fields of an SD and the SDGs are already covered well, which are covered to a lesser extent, and which are not? An analysis of strengths, weaknesses, opportunities and threats (SWOT) is generated on the basis of the gap analysis. The SWOT analysis presents a positioning analysis based on input parameters (trend developments and local activities), action options and thematic fields for the implementation of SD within the context of the Agenda 2030 at the local level.

3. Sustainability strategy: The sustainability strategy is developed (or revised if a sustainability strategy already exists) in a participatory process with different stakeholders from politics, administration, civil society actors, science and economy (see Fig. 3). The IMSS includes four components, which become more concrete from left to right (from guidelines to measures). Each component is developed with the help of a four-step participation process (see Fig. 4).

First, a draft is determined by all the actors in the steering committee. This draft is processed and condensed by the facilitator (LAG 21 NRW). This revised version is sent to the core group, which reviews the draft and, if necessary, revises it. This check particularly includes feasibility and relevance. In the final step, the revised draft is discussed, revised and adapted by all actors in the steering committee. To work out the individual components of IMSS, they are all run through every four steps.

The central outcome of the whole process is an action program that describes clearly defined objectives and measures. It is supplemented by a general mission statement and thematic guidelines. Moreover, the developed objectives are compared with the goals of higher political levels (SDGs, German Sustainable Development Strategy (DHNS) and Sustainability Strategy for North Rhine-Westphalia (NHS NRW)). This step serves to create vertical integration and the development of consistent cross-level target systems (Hoff 2018). The comparison occurs particularly at the level of strategic and operational objectives. On this basis, common objectives and possible synergies between the different political levels can be highlighted and promoted directly.

4. (First) sustainability report and formal decision: The IMSS is documented in a sustainability report. On the one hand, the report contains a detailed process description on the development of each strategy and, on the other hand, an action program. The first sustainability report is formally adopted by the formal decision-making bodies (committees and council). The formal decision promotes the political legitimacy of the IMSS and, hence, is considered as a basic principle in all fields of local development.

5. Implementation and monitoring: After the successful formal decision, the implementation of the objectives and measures established in the action program begins. The implementation phase is accompanied by indicator-based monitoring, which is used to measure the degree of success of the strategy.

6. Evaluation and update: In the sense of the CIP, the strategy is updated in regular rotation. The update is based on evaluation results, which occur against the background of current developments (errors, trends, etc.), as well as the degree of achievement of the formulated objectives. The evaluation results are recorded in

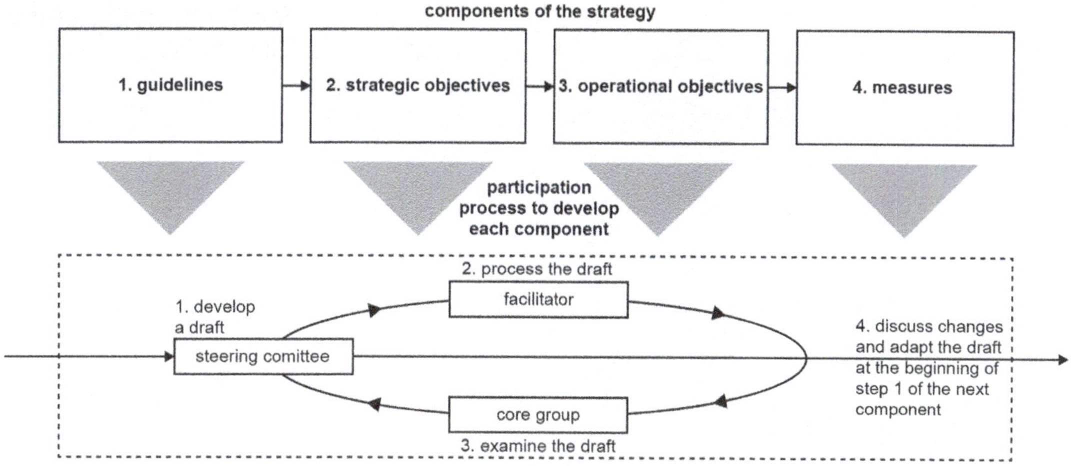

Fig. 4 Components of IMSS and four-step participation process; *Source* Own illustration

progress reports. In line with the original sustainability report, the reports are formally adopted.

4 Case Study—Towards Municipal Sustainability Strategies in North Rhine-Westphalia

The presented cooperative planning approach with its CIP serves as the basis for the Europe-wide unique model project GNK NRW, which was carried out in 15 model municipalities in NRW, Germany (see Fig. 5). Even though NRW is the most-populated federal state in Germany, with approximately 17.7 million inhabitants and a high population density of 524 people per km^2, it has very heterogeneous settlement structures. While urban structures such as the Rhine Ruhr metropolitan region are very condensed, there are several comparatively rural regions in NRW (e.g., Sauerland with 198 people per km^2). Hence, the attending municipalities were faced with different demographic, economic and structural initial situations that had to be considered by way of individual analysis and adapted process support.

Beginning in early 2016, the 15 municipalities involved, representing 2.5 million people, transferred the SDGs to the local level by elaborating on an individual IMSS with specific local objectives in the context of the Agenda 2030. Following the principles "within the municipality for the municipality" and "within the municipality for the world", the process of IMSS development had two overall objectives:

(1) Considering local challenges and an urban development that links ecological, social and economic concerns regarding the planetary boundaries; and
(2) Promoting the One World perspective and the associated responsibility towards people in other parts of the world, especially those in the Global South.

During the 28-month project duration, the 15 model municipalities developed sustainability strategies that are not focused solely on sectoral subjects such as climate and energy, economy and employment or education; they also worked out comprehensive target systems with strategic and operational objectives and defined responsible actors and necessary resources for their implementation. As a result of the two-and-a-half-year process, the municipalities now own an integrated sustainability strategy, which was developed in corporate responsibility with different stakeholders, decided and adopted by the formal decision-making bodies, and will be updated in a regular rotation. Sharing these experiences even beyond municipalities being part of the project requires taking part in

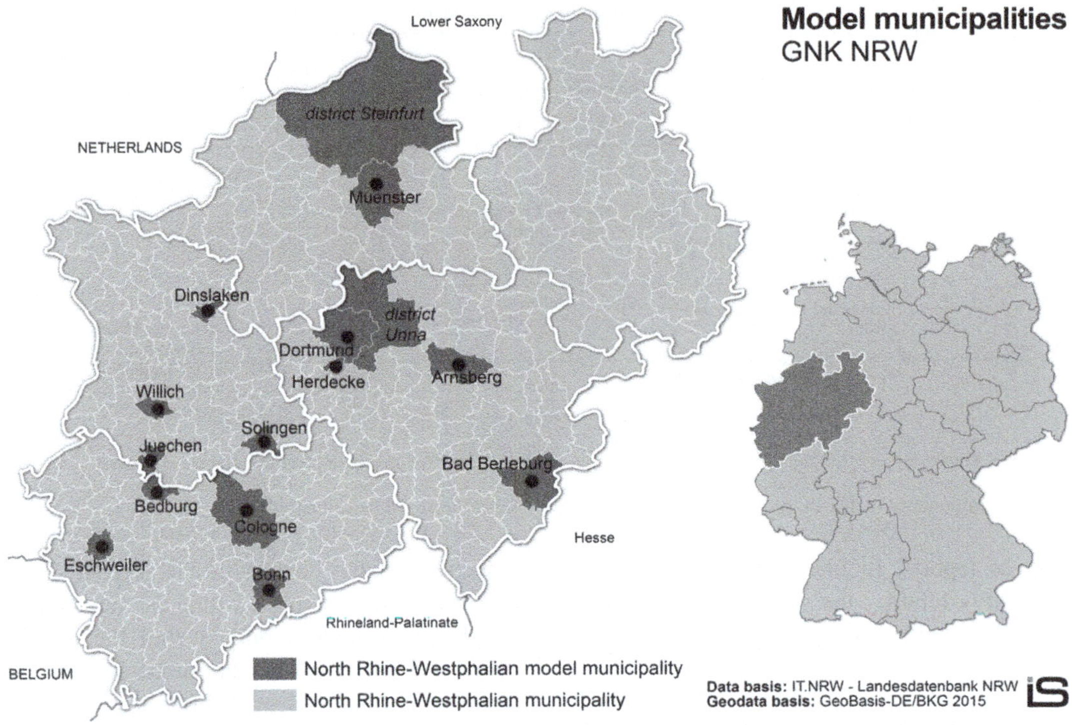

Fig. 5 Model municipalities of the GNK NRW project; *Source* Own illustration

platforms and mechanisms devoted to knowledge transfer, as implemented in city to city learning initiatives (Haupt et al. 2021).

Accompanying Monitoring and Evaluation

To guarantee effective and successful processes and enable the systematic capture of learning effects regarding the transformation of the SDGs to the local level (the model character of the project), the GNK NRW project was accompanied by continuous internal monitoring and evaluation of the main working steps. Thereby, monitoring and evaluation serve different purposes:

The monitoring provides interim results with regard to the working processes in the steering committees. Through a continual review of results that were compiled during the main steps of the presented CIP, it is possible to check whether the work of the municipal steering committees complies with the requirements of the LAG 21 NRW project approach. Thus, it enables early detection of erroneous developments so that countermeasures may be initiated during the working steps if necessary.

The evaluation serves to verify whether and how the project's overall objectives and those defined for the main working steps have been met. In addition, reasons for deviations from the exemplary approach and the omission of objectives are identified to derive extensions and adjustments to the LAG 21 NRW project approach, if it appears necessary.

To meet these requirements, the internal monitoring and evaluation were based on a multi-method approach that integrates different perspectives:

- The basis of this approach is provided by an analysis of the documented results that were compiled during the main steps of the CIP presented (see Sect. 3), for example, attendance lists for the steering committee meetings, logs to record the working processes in the committees, the selected topics of the IMSS or references between the developed strategic objectives and the SDGs. These products of the working process were examined to check whether they correspond to the LAG 21 NRW approach and previously defined objectives.

- Structured interviews with project team members have been conducted to capture a comprehensive view of the project processes in the specific municipalities. They first took place after the second meeting of the steering committees and contained questions about, e.g.,
  - Cooperation of the steering committee members
  - Quality of working results
  - Intensity of participation of different actor groups
- Standardized surveys of the municipal project coordination and their proxies were implemented with an online survey tool twice (n = 20/18). While the first survey was conducted after the second meeting of the steering committees, the second occurred after the developed target system was adopted by the committee members (last meeting of the steering committee). The following issues were queried in both surveys to enable a comparison:
  - Satisfaction with the achieved results
  - Impacts of the project in the municipalities
  - Challenges and requirements regarding their role as project coordinators and the work of the steering committee
  - Satisfaction with advisory service of the LAG 21 NRW
- A standardized survey of all steering committee members was implemented after the last meetings (n = 197). Here, the survey also contained the following content:
  - Satisfaction with the achieved results
  - Impacts of the project in the municipalities
  - Challenges and requirements regarding the work of the steering committee
  - Satisfaction with advisory services of the LAG 21 NRW

- Impressions of local working processes were guaranteed through participatory observation. For this purpose, two steering committee meetings were observed for every main working step of the committees.

The evaluation of whether the working processes of the steering committees were effective and successful was conducted with the help of success indicators that were determined in

Table 1 Example of the evaluation of the steering committees

Object of investigation	Success indicator	Database
Size and composition	– Size and composition correspond to the LAG 21 NRW approach – All relevant actors in the groups participate – There is attendance of • One World promoters • Heads of administration • All political groups	– Attendance lists of steering committee meetings – Logs – Participatory observation
Cooperation	– Steering committees work cooperatively, parallel structures are avoided	– Standardized surveys of the municipal project coordination – Structured interviews with project team members – Participatory observation
Results	– Steering committees produce results that correspond to the project objectives • In accordance with working instructions • In relation to SD • In consideration of the planetary boundaries	– Products of main working steps – Logs / written comments on results – Participatory observation

Source own illustration

advance of the committees' working processes. Table 1 provides an example of the investigated objects, corresponding success indicators and database used.

5 Findings on the Participatory Process

Developing and especially implementing integrated sustainability strategies at the local level can be successful in the long term only if they are co-developed by the local community and meet the necessary acceptance and support of the citizens, civil society, administration and politics (see Sect. 3). For this reason, the working structure of the LAG 21 NRW approach provides a wide participatory process in the form of a steering committee that is composed of local stakeholders and meets at least five times during the project period. Initial findings of the accompanying monitoring and evaluation concerning the establishment, composition and work of the steering committees are presented below.

Establishment, Size and Composition of the Steering Committees in the GNK NRW Project

Because the steering committee serves as the central body for developing the IMSS, its personnel composition and the selection of attending actors requires careful consideration. For this reason, the first step in establishing the municipal working groups (see Sect. 3) included an extensive actor mapping to identify relevant local stakeholders who should participate in the process due to their specific resources. After a joint brainstorming with the LAG 21 NRW, the project coordination and managing personnel of the municipality, all actors identified from politics, administration and civil society were categorized according to the following criteria:

- Available resources (e.g., know-how, human resources, networks, finances, and leverage)
- Thematic proximity/affinity to sustainability topics

With the help of this group assessment, potential members of both the core group and the steering committee were identified. These suggestions were finally discussed in the course of a detailed kick-off meeting with the project coordination, potential members of the core group, managing personnel of the municipality and the LAG 21 NRW. Thereupon, the potential members of the steering committee were invited to help develop the IMSS as volunteers in a cooperative planning process.

The monitoring and evaluation of the actual size and composition of the steering committees were based on attendance lists, which were signed by members every time the committees met. To analyze the committees' composition, the participants were also invited to note their profession and the institution they represent. On this basis, the participants were assigned to one of the actor groups (administration, politics, civil society, science and the economy).

As shown in Fig. 6, representatives of the administration—including the core group and coordination—formed the majority within the steering committees. However, civil society actors, for example, those from NGOs, associations or churches, were also strongly represented so that the target system of the IMSS also reflects societal interests to an appreciable extent. In addition to these dominant groups, stakeholders from the local economy and science participated in the process of IMSS development in almost every municipality. With shares of 5 percent and 4 percent, the presence of these stakeholders is considerably smaller, and particularly in the smaller model municipalities, it presents an eye-catching effect. Moreover, representatives from the major political groups participated in all municipal steering committees.

The total number of steering committee-members differed between the 15 model municipalities. While in some municipalities, only 14 to 15 people, including core group and coordination, attended the steering committee meetings, other committees consisted of 30 actors or more. A clear link between the population size of a

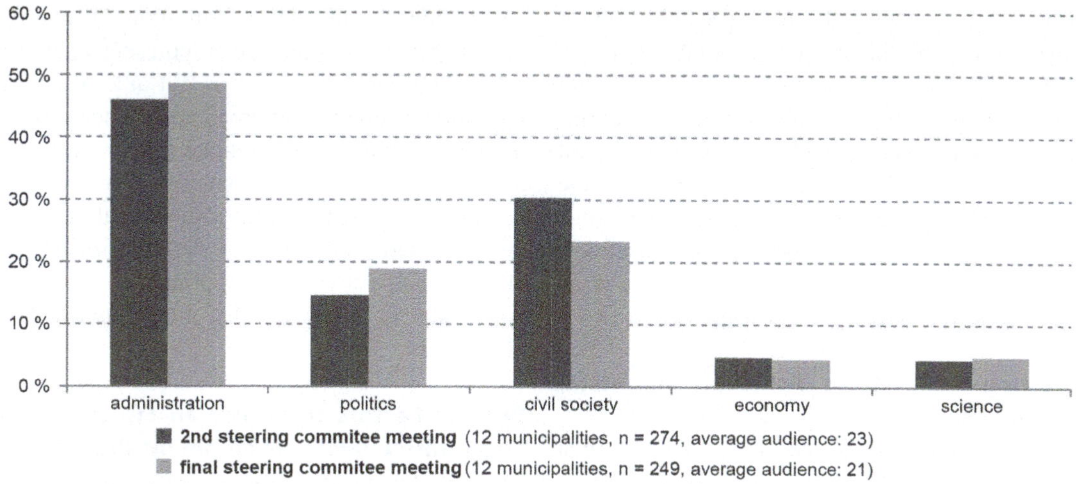

Fig. 6 Average composition of the steering committees in the model municipalities; *Source* Own illustration

municipality and the size of the steering committee cannot be identified.

During the project process, the average group size remained almost constant; from the second to the last meetings of the municipal steering committees, it decreased only slightly from 23 to 21 participants (see Fig. 6). However, slight shifts among the different actor groups could be determined. While the attendance of civil society actors decreased slightly, hardly any changes were observed in the number of actors from science and the economy. In contrast, the number of administrative staff and political representatives increased moderately. In particular, the increasing attendance of politicians is a desirable and important trend during the project process because it is most likely that the intense involvement of political actors improves the chances for a (unanimously) formal decision regarding the IMSS.

Through the presented size and composition of the steering committees (see Fig. 6), it was possible to include professional competencies—represented mostly by administrative and scientific members—as well as civic knowledge and interests into the process of developing IMSS. Thereby, the numerical superiority of administrative staff appeared not to be obstructive, as it enabled the integrative consideration of different departments (horizontal integration). The relatively small number of economic stakeholders also does not constitute a problem as long as it can be assured that at least a few of them take an active part in the strategy process so that their interests and assessments are also taken into account during discussions. The attendance of scientific actors is advantageous to integrate their specific expertise and perspectives on the latest developments, but in the steering committees, this was highly dependent on the presence of scientific institutions in geographical proximity to the model municipalities. Bringing together administration, politics and civil society to cooperatively develop IMSS is one of the project's key elements. Therefore, at least a small number of representatives from every actor group should be involved in the steering committees; otherwise, important aspects of a comprehensive municipal development may be disregarded.

The experiences from GNK NRW show that bringing together an appropriate steering committee is a very sensitive process in most municipalities because they often strive for a wide-ranging committee composition to guarantee a linkage with all local stakeholders. In this context, it is quite important to note that, if possible, the steering committee should not exceed 30 participants in order to ensure the working ability of the board—particularly against the background of the limited timeframe of the project. Especially in bigger cities, minor deviations from this reference point are likely and acceptable. To check whether the size and composition of the steering committees correspond to the LAG 21 NRW approach and if all relevant stakeholders participate during the whole IMSS development process, continuous monitoring as described above is highly useful. It allows the recognition of erroneous trends at an early stage to react in time if necessary.

Steering Committees Achieved Successful Outcomes

With the help of the steering committees that correspond to the LAG 21 NRW project approach in terms of size and composition in almost all municipalities, 12 municipalities developed IMSS, including specific strategic and operational targets, as well as measures. In addition, three municipalities developed analytical documents for reporting their sustainable developments and updates of existing sustainability strategies. The resulting IMSS covers a wide range of SDGs. The target systems developed cooperatively by the steering committees contribute not only to the 17 global goals but also directly to attaining 96 of the 169 targets of the Agenda 2030. In total, over 1,350 references between the municipal strategic objectives and the 169 targets of the Agenda 2030 have been identified. Thus, the size and composition of the steering committees have been proven to be valuable to achieve results that correspond to the overall project goals (see Sect. 4). Although some intense discussions within and between the participating actor groups appeared, they almost always remained constructive. Only very few conflicts appeared to be able to slow down the overall process considerably or even endanger it. In light of these overall results, it can be noted that the working structure of the LAG 21 NRW project approach is well suited for developing IMSS in a cooperative manner.

According to the limited amount of stakeholders who took an active part in the steering committees, the LAG 21 NRW project approach is highly suitable to work cooperatively on IMSS development and create commitment among institutions and organizations represented by the committee members, but it is not able to create widespread support and broad acceptance among the general public. To facilitate this process, the implementation of "sustainability conferences" has proven valuable. In the course of these conferences, citizens were not only informed but also invited to develop appropriate measures that contribute to achieving the elaborated IMSS objectives.

Success Factors to Ensure an Effective and Cooperative IMSS Development Process

Through the presented quantitative analysis of the steering committees' composition, the qualitative query of the project team members' experiences and the survey of municipal coordination, the following key success factors of ensuring a cooperative process of IMSS development and effective working in the steering committees were identified:

- Political support from the heads of the administration, as well as their presence on the steering committees, are essential components of effective working processes in the municipalities. First, it leads to a stronger identification of public administration members with the whole process of implementing IMSS when supported by the head of the administration. It also contributes to a formal decision on the sustainability strategy. Second, majors and department heads may sometimes settle disputes better than external moderators because of their local reputation. However, it must be guaranteed that their presence does not lead to a situation in which contrasting opinions will not be heard (also see de Vente et al. 2016). In the GNK NRW project, heads of the administration attended steering committee meetings in almost half of the municipalities.
- For the overall project process, it is helpful to avoid creating parallel structures when forming the working structures of the LAG 21 NRW project approach. In fact, the diverse

and complex administrative processes require a flexible adaption to local specifics and should allow an affiliation to existing committees. Experiences from the GNK NRW model project and many other municipal management processes that were accompanied by the LAG 21 NRW show that existing working structures are clearly more accepted than new ones.

- The coordination and core groups must be able to independently organize, explain and justify the main working steps to develop IMSS. Furthermore, it is important that the coordination and core groups act coherently in the steering committees. Diverging interests and conflicts should be discussed in other contexts, e.g., in core group meetings or at a bilateral level.
- The composition of the steering committee should be well balanced so that individual interests do not dominate the cooperative IMSS development process. Furthermore, when selecting stakeholders, it is necessary to ensure that actors from all actor groups (administration, politics, civil society, economy and science) attend steering committee meetings in order to include a wide range of local expertise, ensure local ownership of the IMSS and guarantee a strong commitment of all stakeholders from the very start.
- Against the background of the Agenda 2030, the attendance of the so-called One World Promoters, for example, from One World Net NRW,[5] is particularly relevant. While most of the involved actors represent more or less generic fields of municipal development, One World Promoters help to integrate objectives and measures that expand the view to the Global South.
- An early and intense involvement of important political decision-makers is one of the key success factors identified. The continuous involvement of political actors increases identification with the process, as their potential influence rises during the development of IMSS. Regarding the invitation to join the steering committees, it should be considered that the chosen politicians have sufficient influence within their parties to increase acceptance in the political arena of decision-making. Thus, the flow of information into the political groups facilitates the formal decision regarding the IMSS. In this context, it is also advisable to raise awareness amongst local politicians with the help of concerted lobbying.
- To foster inter-municipal learning and knowledge sharing among the model municipalities, seven networking conferences specifically addressed to the project coordination were organized. Further, project-external municipalities were invited to join the intercommunication at certain points. These opportunities were highly utilized so that networking conferences must also be considered as essential success factors of the GNK NRW project.

As the development and implementation of the IMSS is designed as a CIP (see Sect. 3), the continuation of the cooperative processes in the steering committees after having adopted the first IMSS (e.g., for monitoring and evaluation) is an essential factor for successfully implementing the Agenda 2030 in the long term. The conducted survey of project coordinators shows that a consensus target system, clear political guidance and the provision of specific human and financial resources are equally important for this purpose as ambitious but realizable targets and visible achievements.

6 Conclusions and Outlook

With the ecological, social and economic dimensions, the Agenda 2030 affects many fields of action in which different stakeholders operate. Therefore, developing IMSS in this context depends on an integrated approach and the

[5]The One World Net NRW is the umbrella association of the development associations in North Rhine-Westphalia. The main tasks of the One World Network are consulting One World Engagement, education and outreach projects, as well as networking and training on all aspects of One World work in North Rhine Westphalia.

support of relevant stakeholders. As shown, the LAG 21 NRW approach takes up this prerequisite by creating interdisciplinary working groups in which relevant stakeholders come together.

The project approach thus clearly overlaps with the Nexus approach, which also aims at an integrated perspective to bridge sectors and consider interrelated resources (Biermann et al. 2017). While the Nexus approach focuses on specific interlinkages between the SDGs and, in particular, on the sustainable use and management of environmental resources (e. g., Water-Food-Energy Nexus Approach), the sustainability strategies developed in the GNK NRW do not directly focus on SDGs connected to natural resources. Based on the actors involved, the model municipalities have set different priorities in strategy development and thus focus on different SDGs.

Since 2017, the project approach has also been implemented in other federal states in Germany, e.g., in Thuringia, Saarland and Schleswig–Holstein. It is also recognized in the update to the German Sustainable Development Strategy and was recently highlighted in Brussels during panel debates about linkages between SD processes on the local level and the policy framework at the EU level.

The results of the project-internal process-oriented monitoring and evaluation show that the process is influenced by its participatory approach (see Sect. 5): (1) Politics and administration must be involved as important decision-makers. Inclusion should not result in other actors being hindered in their participation. (2) Civil society stakeholders have tacit knowledge. The process must take advantage of this important expertise to enrich it. (3) To this end, the development and negotiation process has to be open and should not be predefined by actors from administration and politics. A basic prerequisite is that all participants are perceived as equal actors because broad acceptance can be achieved only if all stakeholders jointly discuss and negotiate the objectives and contents. (4) It is shown that complex content is discussed in integrated planning approaches. Synergies and trade-offs between target dimensions become apparent. Involving actors with different professional perspectives and goals is likely to lead to more conflicts. In contrast to sectoral planning processes, conflicting target dimensions (e.g., affordable living vs. land consumption or inner urban development vs. adaptation to climate change) can be considered and shared jointly from the outset. (5) The processes in the 15 model municipalities show that the development of IMSS is a challenge for all actors involved. Content and methodological preparation is an important prerequisite for a successful process. It must be ensured that all participants are ultimately convinced of the results. Clearly, structured methods as well as the transparent and professional derivation of results foster acceptance.

Nevertheless, there is no blueprint for successful participation. The involvement of different actors is—depending on the situation—always subject to these actors' own dynamics. Undoubtedly, key success factors have already been identified by research and best practices, but it must be recognized that compliance with these factors does not necessarily guarantee successful participation and the desired outcome. On one hand, extensive participation of actors and interest groups can lead to a loss of control of the process. On the other hand, citizens involved can also represent private interests and therefore unilaterally influence the process. Overall, extensive participation can create a more expensive and time-consuming process than decision making by a single administrator (Marzuki 2015). In the author's opinion, especially long-term and far-reaching changes, e.g., the transformation to sustainability requires the active involvement of different municipal actors. Otherwise, the broad acceptance of such processes will be highly unlikely.

References

Al Radif A (1999) Integrated water resources management (IWRM): an approach to face the challenges of the next century and to avert future crises. Desalination 124:145–153. https://doi.org/10.1016/S0011-9164(99)00099-5

Arnstein SR (1969) A ladder of citizen participation. J Am Inst Planners 35(4):216–224

Bazilian M, Rogner H, Howells M, Hermann S, Arent D, Gielen D, Steduto P, Mueller A, Komor P, Tol R, Yumkella K (2011) Considering the energy, water and food nexus: towards an integrated modelling approach. Energy Policy 39:7896–7906. https://doi.org/10.1016/j.enpol.2011.09.039

Biermann F, Kanie N, Kim RE (2017) Global governance by goal-setting: the novel approach of the UN Sustainable Development Goals. Curr Opin Environ Sustain 26:26–31

Biggs SD (1989) Resource-poor farmer participation in research: a synthesis of experiences from nine national agricultural research systems. OFCOR-Comparative Study (Países Bajos)

Bock S (2017) Erfolgsbedingungen kommunaler Bürgerbeteiligung. Universität Potsdam. https://publishup.uni-potsdam.de/opus4-ubp/frontdoor/deliver/index/docId/10476/file/kwischr10_online_s103-116.pdf

Bringezu S, Potočnik J, Schandl H, Lu Y, Ramaswami A, Swilling M, Suh S (2016) Multi-scale governance of sustainable natural resource use: challenges and opportunities for monitoring and institutional development at the national and global level. Sustainability 8(778)

Brundtland GH, Khalid M (1987) Our common future. New York

de Vente J, Reed M, Stringer L, Valente S, Newig J (2016) How does the context and design of participatory decision making processes affect their outcomes? Evidence from sustainable land management in global drylands. Ecol Soc 21(2)

Donella H, Meadows DL, Randers J, Behrens WW III (1972) The limits to growth. Universe, New York

Drazkiewicz A, Challies E, Newig J (2015) Public participation and local environmental planning: testing factors influencing decision quality and implementation in four case studies from Germany. Land Use Policy 46:211–222

Farrington J (1998) Organisational roles in farmer participatory research and extension: lessons from the last decade. Nat Res Perspect 27:1–4

Goetz A, Gaventa J (2001) Bringing citizen voice and client focus into service delivery. IDS Working Paper 138, Brighton: Institute of Development Studies

Hagmann JR, Chuma E, Murwira K, Connolly M, Ficarelli P (2002) Success factors in integrated natural resource management R&D: lessons from practice. Conserv Ecol 5(2):29. https://www.consecol.org/vol5/iss2/art29/

Haupt W, Zevenbergen C, van Herk S (2021) City-to-City learning within city networks to cater city needs to climate adaptation—results of a preliminary study. In: Hülsmann S, Jampani M (eds) A nexus approach for sustainable development. Springer Nature, Cham, Switzerland. https://doi.org/10.1007/978-3-030-57530-4_14

Hettiarachchi H, Ardakanian R (eds) (2016) Environmental resource management and the nexus approach: managing water, soil, and waste in the context of global change. Springer, Cham

Healey P (2004) The treatment of space and place in the new strategic spatial planning in Europe. Int J Urban Reg Res 28:45–67. https://doi.org/10.1111/j.0309-1317.2004.00502.x

Hoff H (2018) Integrated SDG implementation—how a cross-scale (vertical) and cross-regional nexus approach can complement cross-sectoral (horizontal) integration. In: Hülsmann S, Ardakanian R (eds) Managing water, soil and waste resources to achieve sustainable development goals. Springer, Cham

Hülsmann S, Ardakanian R (2018) The nexus approach as tool for achieving SDGs: trends and needs. In: Hülsmann S, Ardakanian R (eds) Managing water, soil and waste resources to achieve sustainable development goals. Springer, Cham

Irvin RA, Stansbury J (2004) Citizen participation in decision making: is it worth the effort? Public Adm Rev 64(1):55–65

Johnson N, Lilja N, Ashby A, Garcia JA (2004) Practice of participatory research and gender analysis in natural resource management. Nat Resour Forum 28:189–200

Jobert A, Laborgne P, Mimler S (2007) Local acceptance of wind energy: factors of success identified in French and German case studies. Energy Policy 35(5):2751–2760

Kastens B, Newig J (2008) Will participation foster the successful implementation of the water framework directive? The case of agricultural groundwater protection in northwest Germany. Local Environ 13 (1):27–41

Kraas F et al (2016) Humanity on the move: unlocking the transformative power of cities. WBGU, Berlin

Lawrence A (2006) No personal motive? Volunteers, biodiversity, and the false dichotomies of participation. Ethics, Place Environ 9:279–298

Mans N, Griggs D, Visbeck M (2016) Map the interactions between sustainable development goals: Mans Nilsson, Dave Griggs and Martin Visbeck present a simple way of rating relationships between the targets to highlight priorities for integrated policy. Nature 534 (7607):320ff

Martens J (2013) Die Post-2015-Agenda. Gemeinsame Ziele für Entwicklung und Nachhaltigkeit, Global Governance Spotlight January

Marzuki A (2015) Challenges in the public participation and the decision making process. Sociologija i prostor 53(1 (201)):21–39. https://doi.org/10.5673/sip.53.1.2

Ministerinnen und Minister der Bundesländer (2007) Leipzig Charta zur nachhaltigen europäischen Stadt. Informationen zur Raumentwicklung 4 (2010)

Poister TH, Streib G (2005) Elements of strategic planning and management in municipal government: status after two decades. Public Adm Rev 65:45–56. https://doi.org/10.1111/j.1540-6210.2005.00429.x

Richards C, Blackstock KL, Carter CE (2004) Practical approaches to participation SERG Policy Brief No. 1. Macauley Land Use Research Institute, Aberdeen

Ritter E-H (1998) Stellenwert der Planung in Staat und Gesellschaft. In: Akademie für Raumforschung und Landesplanung (eds) Methoden und Instrumente der räumlichen Planung, Hannover, 6–22
Schellnhuber HJ, Messner D, Leggewie C, Leinfelder R, Nakicenovic N, Rahmstorf S, Schlacke S, Schmid J, Schubert R (2011) World in transition: a social contract for sustainability. WBGU, Berlin
Selle K (1996) Planung und Kommunikation: Gestaltung von Planungsprozessen in Quartier. Stadt und Landschaft. Bauverlag, Wiesbaden
Selle K (2000) Was? Wer? Wie? Warum? Voraussetzungen und Möglichkeiten einer nachhaltigen Kommunikation: Arbeitsmaterialien für Studium und Praxis. Dortmunder Vertrieb für Bau- und Planungsliteratur, Dortmund
Servicestelle Kommunen in der Einen Welt (SKEW) von Engagement Global (2015) Kommunal fair handeln: Möglichkeiten für fairen Handel und faires Beschaffungswesen. Frankfurt a. M.
Smajgl A (2018) Participatory processes and integrated modelling supporting nexus implementations. In: Hülsmann S, Ardakanian R (eds) Managing water, soil and waste resources to achieve sustainable development goals. Springer, Cham
Tosun J, Leininger J (2017) Global challenges. 1, 1700036. https://doi.org/10.1002/gch2.201700036
United Nations (1992) Agenda 21. The United Nations program for action from Rio, Rio de Janeiro
United Nations (2000) United Nations Millennium Declaration. New York
United Nations (2015) Transforming our world: the 2030 Agenda for Sustainable Development. New York
United Nations (2017) New Urban Agenda. Quito. https://habitat3.org/wp-content/uploads/NUA-English.pdf
Wallerstein N (1999) Power between the evaluator and the community: research relationships within New Mexico's healthier communities. Soc Sci Med 49:39–53
Website Service Agency Communities in One World (SKEW) (2017) Municipal climate partnerships. https://skew.engagement-global.de/municipal-climate-partnerships.html. Accessed 05 May 2017

Legal Reference

Federal Building Code (Baugesetzbuch) in the version amended by the Act to Amend the Federal Building Code and to Reorder Spatial Planning Law [BauROG], issued on August 18th 1997 (BGBl. I p. 2081)

Correction to: The Energy-Water Nexus in Iran: The Political Economy of Energy Subsidies for Groundwater Pumping

Tinoush Jamali Jaghdani and
Vasyl Kvartiuk

Correction to:
Chapter "The Energy-Water Nexus in Iran: The Political Economy of Energy Subsidies for Groundwater Pumping" in: S. Hülsmann and M. Jampani (eds.), A Nexus Approach for Sustainable Development, https://doi.org/10.1007/978-3-030-57530-4_16

The original version of the book was inadvertently published with an incorrect sentence in chapter "The Energy-Water Nexus in Iran: The Political Economy of Energy Subsidies for Groundwater Pumping". The sentence has been corrected now as in given below: The correction chapter and the book have been updated with changes.

From:

"Moazedi et al. (2011) have demonstrated a non-linear relationship between groundwater levels and energy use. They conclude that higher energy subsidies should be provided as groundwater levels drop. However, this recommendation does not consider the workings of Iranian politics."

To:

"Moazedi et al. (2011) have demonstrated a non-linear relationship between groundwater levels and energy use. It can be concluded from this finding that higher energy subsidies should be provided as groundwater levels drop in order to keep the same level of production."

The updated version of this chapter can be found at
https://doi.org/10.1007/978-3-030-57530-4_8

S. Hülsmann and M. Jampani (eds.), *A Nexus Approach for Sustainable Development*,
https://doi.org/10.1007/978-3-030-57530-4_16

The manufacturer's authorised representative in the EU is Springer Nature Customer Service Centre GmbH, Europaplatz 3, 69115 Heidelberg, Germany. If you have any concerns regarding our products, please contact ProductSafety@springernature.com

Printed and bound by CPI Group (UK) Ltd, Croydon, CR0 4YY
30/09/2026
02225287-0019